PVC: Production, Properties and Uses

Also from The Institute of Materials

Plasticisers: Principles and Practice
ALAN S. WILSON

PVC: Production, Properties and Uses

GEORGE MATTHEWS

THE INSTITUTE OF MATERIALS

Book 587
First Published in 1996 by
The Institute of Materials
1 Carlton House Terrace
London SW1Y 5DB

ISBN 0 901716 59 6

Typeset in Great Britain by
Fakenham Photosetting
Printed and bound in Great Britain at
The University Press, Cambridge

Contents

Acknowledgments

The 17 chapters on PVC in the original Plastics Institute monograph (*Vinyl and Allied Polymers: Volume two*) were first conceived by the author during the early 1960s, and were largely based on his experience with the Plastics Division of Imperial Chemical Industries Limited from 1946 to 1964. Thus many colleagues, customers, and other associates contributed information which found its way into those chapters, and it would be quite impossible to name them all, even if, indeed, it was possible to remember them. Their help, whether consciously or unconsciously given, was and is none the less greatly appreciated, and it is hoped that any reader who feels his work has been incorporated into this text without specific acknowledgement or reference will be happy with these few words of general thanks. Trade literature available from manufacturers of PVC resins and additives, and of processing machinery has been particularly valuable, and the more useful publications are listed among the references. Of individual sources the RAPRA Technical Review No. 31 by W.C. Geddes was especially useful in the writing of Chapter 5, and is still a useful source 27 years after its publication.

The author is still indebted to Bernard Dyer for providing the original for Plate 9.2, and to his ex-colleagues John Dando and Maurice Morgan for assisting in preparing the other plates.

Special thanks were also due to the late Morris Kaufman for permitting the author to read early drafts of relevant chapters of his book *The History of PVC* (Maclaren, 1969), and for himself reading the text of the monograph chapters at the draft stage. Thanks are also due to members of the then Plastics Institute's Monograph Sub-Committee, especially Professor P.D. Ritchie and the late Dr C.A. Redfarn, for their constructive criticisms of the original text.

In some ways translation from industry to the academic world in 1964 led to a widening of perspective, and introduced the author to a broader spectrum of individuals with interests in and information about PVC and plastics in general. Since retirement the author has maintained contact with many ex-colleagues and associates, and has become involved particularly with environmental matters relating to PVC and other materials, and hence has continued to benefit from the interchange of information with others.

As for the revisions and additions to this edition, special thanks are due to Ron Pearson and David Cadogan for helpful suggestions in relation to various aspects of plasticisers and plasticisation; and to Ron Jones and John Baldwin for detailed information relating to toxicity and environmental matters; to Marianne Gilbert for supplying reprints of papers on gelation, fusion, and other topics; to Edward J. Wickson for advice on the subject of formulation; and to the referees for helpful, constructive criticisms of the script in general.

Most of their suggestions have been incorporated into the text. A list of companies that have supplied up-to-date information on their products is to be found in the Appendix.

Finally thanks are due to my wife for tolerating the consequences of the commitment of so much of my time to the various activities involved in the preparation of the manuscript, leaving her with the full burden of gardening etc., and to our daughter Jane for dealing with the photocopying.

Preface

Comments made by academics and people in industry have emboldened the author to revise and update as a new book the first 17 chapters of his *Vinyl and Allied Polymers: Volume two*, published as a Plastics Institute monograph in 1972. At that time, although the UK and worldwide production and consumption of PVC surpassed all other polymers with the possible exception of polyethylene, no comprehensive treatise on the material appeared to have been previously published in English. Such books as had been published had either been devoted mainly to polymerisation and the fundamental properties of vinyl chloride polymers or were concerned mainly with trade products.

Since that time some other books have appeared, notably *Manufacture and Processing of PVC*, edited by R.H. Burgess, revisions of the late W.S. Penn's *PVC Technology* and W.V. Titow's *PVC Plastics*, but these are of somewhat different scope and approach to this author's 1972 book.

Most development since 1972 has been by way of improvements in process control and quality, and expansion in existing applications and into new ones. Although these developments are significant, they have not required many major alterations or additions to the original text. However, three topics have come into greater prominence during that period, and have consequently necessitated significant amendments and additions. These topics are (1) the carcinogenic nature of vinyl chloride and safeguards to avoid this hazard, (2) the structure of polymer particles and (3) environmental issues.

As pointed out in the Preface to the original volume, although a single author could scarcely hope to produce a work of magnitude comparable with the well-known *Polythene* (Iliffe, 1957) edited by Renfrew and Morgan from contributions by thirty-seven experts in their own particular fields, the eighteen chapters presented here do constitute an attempt to survey the whole field of PVC technology with particular regard to industrial practice and to the fundamental principles which underly it. Thus the basic variables in polymer properties have been related to the end-products and types of process for which the polymers may be destined. Similarly the selection of plasticisers, stabilisers and other additives has been related to requirements of processing and service use, and the processes and equipment used with PVC are discussed in terms of the end-products to be produced and the properties of the PVC feed-stock. Trade names have been excluded from the text as far as possible since their frequent use in the literature has often resulted in unnecessary confusion and obscurity.

While the contents include all that a student aiming at a Higher National Certificate, Advanced Certificate of Education, or a degree with polymer specialisation can reasonably be expected to know about PVC, the coverage goes well beyond that scope and is intended to be of value to all those concerned with production, development and research involving

PVC. At the time of the original edition there was a very extensive literature on PVC, and it has now become so large that it would be impossible to cover all but the more important. However, each chapter has a fairly extensive list of references which should be good sources for those wishing to investigate in detail the literature on any particular work. In addition the bibliography at the end of Chapter 1 includes a number of volumes which cover a wide range of PVC topics.

Wadebridge   G.M.
1993

1
Introduction

1.1 TERMINOLOGY

In the plastics industry the terms PVC or pvc are used indiscriminately to describe a wide variety of different products characterised only by the fact that they are based on polymers of vinyl chloride, or copolymers of vinyl chloride with minor proportions of other monomeric compounds. The same terms are used to describe the polymers themselves, as well as compositions derived from these polymers by combination with other materials such as stabilisers, plasticisers, and fillers. In the extreme case of vinyl floor tiles the vinyl chloride content can be as low as 15 per cent, yet these products are frequently referred to as PVC. It would probably be unrealistic at this late stage of development to attempt to obtain widespread restriction of the application of the terms PVC and pvc to the material for whose chemical name they are abbreviations, namely polyvinyl chloride. Moreover, both forms of abbreviation are so convenient that it would also be both unrealistic to try to ban them altogether as well as embarrassingly restrictive to have to avoid using them, even in written form as in a Monograph such as this. Users of the terms should, however, ensure that the context is such that no ambiguity arises, and listeners and readers should exercise caution whenever the terms are used in circumstances which do not make their meaning clear. In this Monograph the abbreviations will be used for convenience wherever this is possible without loss of meaning. The author has found it convenient to use the capital form 'PVC' whenever reference is being made to a composition containing polymer and other ingredients, restricting the lower case 'pvc' to polymer without additives.

A second controversial issue arises when considering the full name for homopolymers of vinyl chloride. Over two decades ago it was proposed that where the name of a monomer comprises more than one word (e.g. vinyl chloride, methyl methacrylate) the name of its homopolymer should consist of the prefix 'poly' followed by the full name of the monomer enclosed in brackets. Thus for vinyl chloride the homopolymer should be called 'poly(vinyl chloride)' rather than 'polyvinyl chloride'. The advantages of using the former nomenclature seem to the present author too marginal to support the inconvenience of writing it. In any case, the form with brackets has gained little acceptance, and the older form without brackets will be used throughout this Monograph. Where copolymers are concerned there is perhaps a greater case for using the brackets, but this leads to rather clumsy names, and it is preferred here to avoid the prefix 'poly' altogether and to use an alternative such as 'vinyl chloride/vinyl acetate copolymer'.

There is also confusion about names for other materials referred to herein. Thus when polymers of ethylene were first introduced by I.C.I. in the 1930s they were dubbed 'polythene' and produced by I.C.I. under the trade-name 'Alkathene'. Later other manufacturers regarded 'polythene' as a trade-name and preferred the term 'polyethylene'. Later still, when I.U.P.A.C. agreed on a new nomenclature for olefines 'ethylene' became 'ethene', so that 'polyethene' became the logical term for the polymer. However, although this latter term is used in schools, it has not obtained acceptance in industry, and the latest I.U.P.A.C. recommendation has reverted to 'polyethylene', which is therefore used throughout this volume.

Another word applied to vinyl chloride polymers, which has a slight aura of ambiguity, is 'resin'. This word was originally applied to some synthetic polymers to indicate their resemblance to already known natural resins, and has since passed into more general usage to distinguish those polymers which are not rubber-like from those which are. In the particular context of PVC the word 'resin' is sometimes useful in distinguishing polymers as such from compositions containing them. Moreover, expressions such as 'vinyl resin' can be used where it is desired to embrace a whole variety of homopolymers and copolymers.

1.2   DISCOVERY

The origins and development of vinyl chloride and its polymers have been described admirably in Morris Kaufman's book *The History of PVC*.[1]

Vinyl chloride itself appears to have been first produced by Liebig, although the first record of the preparation was in two papers by his young associate Regnault, which appeared in 1835.[2] In these Regnault described the formation of potassium chloride and a gas having an ether-like smell and the formula C_2H_3Cl, when s-dichloroethane and alcoholic potash were mixed.

The first recorded use of the name vinyl chloride appears to have been in 1854, in Kolbé's *Lehrbuch der Organischen Chemie*, although the origin of the term 'vinyl' is not discussed, and it was not until some years later that the structure of vinyl chloride was established.[3]

Regnault is frequently credited with having been the first to observe the formation of polyvinyl chloride as a white powder when vinyl chloride was exposed to sunlight, but the terminology of his paper[4] is a little obscure, and it now seems fairly certain that it was polyvinylidene chloride which he was describing. It seems that it was indeed Baumann who, following up work by Hofman on vinyl bromide,[5] was the first to produce polyvinyl chloride, in 1872. Baumann reported[6] that on exposing vinyl chloride to sunlight he obtained a white solid having a specific gravity of 1.406, which could be heated to 130°C without decomposition but which at higher temperatures melted to a black-brown mass, at the same time producing considerable amounts of acid vapours. It is doubtful, however, if Baumann or his contemporaries were aware of the polymeric nature of the product he had produced.

1.3   DEVELOPMENT

Apart from a few investigations of some academic relevance to the synthesis of vinyl chloride and other vinyl esters[7,8,9] nothing much relating to the development of PVC appears to have taken place until the early years of the twentieth century. By this time acetylene had become commercially available on a large scale in anticipation of the expected but, as it turned out, unrealised large demand for the gas for lighting.[10] This had stimulated investigations into the chemistry and particularly into other possible uses of acetylene. During the period around the years 1906–1913, Klatte and Rollett, and co-workers of the firm Griesheim Elektron, carried out extensive investigations in this field. Amongst other work they synthesised vinyl chloride and vinyl acetate by reaction of acetylene with hydrogen chloride and acetic acid respectively,[11] and also described the thermal polymerisation of vinyl chloride with the aid of initiators such as peroxides.[12]

About the same time Ostromuislenskii published details of a method of polymerising vinyl halides by exposure in solution to heat or ultra-violet light.[13] He postulated three different forms of polyvinyl chloride, but these were probably merely polymers having different molecular weight characteristics.

1.3.1 COPOLYMERS AND PLASTICISERS

The next stage was the introduction,[14] in 1928, of vinyl chloride copolymers which could be processed at lower temperatures than vinyl chloride homopolymer. Then, around 1929–1930, the possibility of plasticising polyvinyl chloride by using esters like tritolyl phosphate or dibutyl phthalate was discovered.[15]

1.3.2 PRODUCTION

By 1933 development quantities of PVC had been introduced in a variety of product forms in the USA and Germany, and full-scale production began in the latter country in 1937. Pastes were also developed in Germany around the same time. Vinyl chloride/vinyl acetate copolymers were introduced for proofing cloths, lacquers, and films during the period 1935–1937, and vinyl chloride/vinylidene chloride copolymers were also introduced in the late 1930s.

In the UK the earliest interest in PVC appears to have been in 1930,[16] but systematic study of the material was not undertaken until 1937–1938, and small-scale production of polymer, by emulsion polymerisation, only started in 1939. It is interesting to note that as late as 1941 a standard textbook of organic chemistry could still dismiss vinyl chloride by referring to the formation of resinous polymerisation products by exposing the gas or its solutions in organic solvents to ultra-violet radiation or ozone.

The 1939–1945 war undoubtedly stimulated development of many plastics, and other materials, both because of shortage of essential conventional materials and by virtue of demands for properties not possessed by available materials. In spite of the now evident fact that PVC has much to offer as a relatively cheap raw material in peace-time society, and that development had reached an advanced stage in Germany, USA and the UK, it is also certain that the tremendous increase in production of PVC during the 1950s and 1960s owed much to the war-time need to find a temporary replacement for rubber in cable insulation. It is therefore perhaps ironic that, as well as branching out on a large scale into other applications, PVC as now gone a long way to ousting rubber in cable production and some other applications.

1.3.3   DEGRADATION AND STABILISATION

In some ways the development of PVC has now gone full circle, in that, whereas the introduction of copolymers and plasticisers was necessary to permit PVC to be processed at temperatures attainable with available equipment without excessive decomposition, nowadays more PVC is processed in unplasticised than in plasticised form. The significance of copolymers and plasticisers in the development of satisfactory procedures for processing PVC has, perhaps, been over-emphasised. In addition to the temperature limitations of processing equipment during the early development, there have always been temperature limitations due to the tendency of PVC to decompose at the relatively high temperature required to convert the material into a plastic state suitable for processing in one way or another, and stabilisation to overcome this limitation has been a very significant factor. Documentation on early developments in the subject of degradation and stabilisation of PVC is somewhat sparse, apart from a large number of patents mostly of doubtful practical value. Stabilisation with ethylene oxide derivatives was patented in 1934,[17] and the use of metal soaps in 1937.[18] White lead, or basic lead carbonate, still used as a stabiliser, though to a much lesser extent than in the 1950s and 1960s, was undoubtedly in use at a very early stage in the development of PVC, and was already recognised as well established at the time of the first mention in print. This pre-eminence of white lead for such a long period was rather remarkable in view of the introduction of so many other stabilisers with better properties, such as other basic lead salts,[19] barium, cadmium, calcium, and zinc complex organic compounds,[20] and organotin compounds.[21]

1.3.4   POLYMERISATION

The major advances in PVC resins have been the introduction of emulsion polymerisation[22] and suspension or granular polymerisation systems,[23] both around 1933, and the development of the so-called 'easy-processing' or EP suspension resins. These were originally seen to be remarkable for their ability to absorb plasticiser without gelation, to form free-flowing powders known as plasticised dry blends,[24] but the use of the latter was rather slow to develop until the late 1960s. Nevertheless the EP-type resins were found to have other advantages over the then standard suspension resins, and have since become the general type of all 'general purpose' resins. More recently resins of quite different characteristics were developed specifically for unplasticised applications where EP-type resins have certain limitations.

Nowadays the terms 'easy-processing' and 'EP' have been discarded and the two types of polymer are differentiated by apparent density and plasticiser absorption.

Bulk or mass polymerisation methods using free-radical initiators were used on a small scale for vinyl chloride during the 1930s. By the mid-1950s large scale mass or bulk polymerisation plants began to appear, and by the 1960s it could be claimed that the resins they produced were of 'high quality', with higher bulk densities and greater uniformity of particle structure and size distribution.[25]

1.3.5   GROWTH

It will be clear that production before 1939 was small, even on a world-wide basis, but growth since then has been so rapid that most countries of any degree of industrialisation produce some vinyl resin. It is difficult to obtain accurate figures since those obtained from different sources frequently disagree, and it is not always clear whether quoted figures relate to vinyl chloride polymer usage or production, or to amounts of compound or finished products. For present purposes, however, it is sufficient to indicate that consumption in the UK rose from a figure of around 6600 t in 1947 to around 66 000 t by 1957; that is, consumption was approximately doubled every 3 years. Since then the percentage rate of increase has naturally fallen, although the numerical increase has risen steadily until in 1979 consumption was over 440 000 t. Consumption fell considerably during the early 1980s, but recovered to reach 615 000 t by 1990, and, in spite of the recession, 1992 consumption was estimated at around 570 000 t. Similar growth has occurred in other major industrial countries such as the USA, Japan, West Germany, Italy, and France. The precise pattern of growth has, of course, varied from one country to another, depending on a variety of factors such as availability of raw materials and of alternative polymers or other materials usable for the same applications as PVC. Thus, during the war and immediately afterwards, cable insulation and sheathing accounted for the major usage of available PVC. By 1960 calendered sheet and film was accounting for more PVC usage than cable, and other products such as general extrusions and coated fabrics were each using amounts of the same order as the 1950 total consumption. Other important markets for PVC have been in floor coverings, conveyor belts, articles moulded or dipped from pastes, gramophone records, and unplasticised extrusions and mouldings. The latter market deserves special mention because of its extremely rapid growth during the late 1960s and early 1970s.

As long ago as 1960 unplasticised applications accounted for more PVC than any other single type of product in Japan and Italy, and a similar situation was developing in Germany. In the UK usage of unplasticised PVC overtook that of plasticised PVC towards the end of the 1970s, and became the major outlet by a comfortable margin by 1990. This growth is largely due to penetration into some large-scale building and civil engineering markets, notably rainwater goods, drainage and water pipes, window frames, and expanded profiles for cladding, fascia boards and sills etc.

The rather spectacular growth of PVC consumption during the two or three decades following the 1939–1945 war was accompanied by an equally dramatic though less frequently appreciated fall in prices. The two are, of course, to some extent interdependent. Over the period 1960 to 1972 the UK price of polymer fell by over 35 per cent, so that it became the cheapest polymer on a price-per-unit-weight basis, although the significance of this fact is somewhat diminished by the relatively high density of PVC resin at 1.41 g/ml, and by the fact that the cost of a PVC composition depends markedly on formulation and the costs of ancillary ingredients.

The large usage of PVC is obviously dependent to some extent on its relatively low production cost, but also on the versatility of vinyl chloride polymers, in that they can, by appropriate formulation and processing, be converted into a wide range of different products of widely different properties, from flexible garden hose to rigid pipe for water and sewage conveyance, from flexible wire coverings to rigid gramophone records, from flexible sheet for raincoats to rigid sheet for packaging, from soft toys to upholstery. Physical and chemical properties of PVC have, of course, also been important, and these will be considered in some detail in later chapters.

REFERENCES

1. M. KAUFMAN, *The History of PVC*, Maclaren, London (1969)
2. H.V. REGNAULT, *Annls Pharm.*, **14**, 22 (1835): *Annls Chim. Phys.*, **58**, 307 (1835)
3. F.A. KEKULÉ and I. ZINCKE, *Ber. dt. chem. Ges.*, **3**, 130 (1870)
4. H.V. REGNAULT, *Annls Chim. Phys.*, **69** (2), 157 (1838)
5. A.W. HOFFMAN, *Justus Liebigs Annln Chem.*, **115**, 271 (1860)
6. E. BAUMANN, *Annln Chem. Pharm.*, **163**, 308 (1872)
7. S. ZIESEL, *Justus Liebigs Annln Chem.*, **191**, 366 (1878)
8. H. BILTZ, *Ber. dt. chem. Ges.*, **35**, 3525 (1902)
9. J.B. SENDERENS, *Bull. Soc. chim. Fr.*, **3** (4), 828 (1908)
10. M. KAUFMAN, *Trans. J. Plast. Inst.*, **35**, 115, 365 (1967)
11. F. KLATTE, *U.S. Pat.* 1 084 581; *D.R. Pat.* 204 883; 271 381; 278 249
12. F. KLATTE and A. ROLLET, *U.S. Pat.* 1 241 738; *D.R. Pat.* 281 687; 281 688
13. I. OSTROMUISLENSKII. *Chem. Zentral*, **1**, 1980 (1912); *Chemiker Ztg.*, **36**, 199 (1912); *Brit. Pat.* 6299/12; 255 837
14. E.W. REID and R.M. WILEY, *U.S. Pat.* 1 935 577; 2 160 931; 1 867 014; 2 012 177
15. *Brit. Pat.* 349 100; *Brit. pat.* 366 461; W.L. Semon, *Brit. Pat.* 398 091; *U.S. Pat.* 2 188 396
16. *Brit. Pat.* 363 009; 349 017; 366 897
17. G. MEYER, *Brit. Pat.* 418 230; *U.S. Pat.* 2 166 604; *D.R. Pat.* 656 133
18. F. GROFF and M.C. REED, *Brit. Pat.* 451 723; *U.S. Pat.* 2 057 543
19. A.K. DOOLITTLE, *Brit. Pat.* 450 856
20. *Brit. Pat.* 752 053; 815 875
21. V. YNGVE, *Brit. Pat.* 497 879; *U.S. Pat.* 2 219 463; 2 307 157; 2 344 002
22. H. FIKENTSCHER, *Brit. Pat.* 358 534; 410 132
23. J.W.C. CRAWFORD, *Brit. Pat.* 427 494; J.B. Morgan and W.M. Morgan, *Brit. Pat.* 444 257
24. S.R. DRESSER, *Wire & Wire Prod.*, **26**, 9, 904 (1951)
25. G.C. MARKS, 'Developments in PVC Technology', edited by J.H.L. Henson and A. Whelan, Applied Science Publishers (1973)

BIBLIOGRAPHY

M. KAUFMAN, *The First Century of Plastics*, The Plastics Institute, London (1963), (especially Chapter 10)
M. KAUFMAN, 'The Industrial Origins of Polyvinyl Chloride', *Trans. J. Plast. Inst.*, **35**, 115 (1967)
M. KAUFMAN, *The History of PVC*, Maclaren, London (1969)
D.G. OWEN, 'The Commercial Development of Polyvinyl Chloride', *Plastics Progress*, Iliffe, 77 (1961)
J.H.L. HENSON and A. WHELAN (eds), *Developments in PVC Technology*, Applied Science Publishers (1973).
A. WHELAN and J.L. CRAFT (eds), *Developments in PVC Production and Processing – 1*, Applied Science Publishers (1977)
R.H. BURGESS (ed), 'Manufacture and Processing of PVC', applied Science Publishers (1982)
W.V. TITOW, 'PVC Technology', Applied Science Publishers Ltd. (1984)
L.I. NASS and C.A. HEIBERGER, (eds), 'Encyclopedia of PVC', (3 volumes), Marcel Dekker (1986)
W.V. TITOW, *PVC Plastics*, Elsevier Applied Science (1990)
The Plastics and Rubber Institute, *PC Processing*, Proceedings of International Conferences (1978, 1983, 1987, and 1991)
Institute of Materials, *PVC '93, the Future*, Proceedings of International Conference (1993)
E.J. WICKSON (ed.), *Handbook of PVC Formulating*, John Wiley & Sons, Inc. (1993)

2

Production and properties of vinyl chloride and related monomers

2.1 VINYL CHLORIDE

2.1.1 LABORATORY PREPARATION

As was indicated in the Introduction, vinyl chloride was first produced by elimination of hydrogen chloride from s-dichloroethane, or ethylene dichloride, by means of alcoholic potash:

$$CH_2Cl.CH_2Cl + KOH \rightarrow CH_2:CHCl + KCl + H_2O$$

This is still a reasonable method for the laboratory,[1,2] although aqueous alkali can be used in an agitated pressure vessel at temperatures in the region of 420 K (150°C).

2.1.2 MANUFACTURE

2.1.2.1 From acetylene

The early development of vinyl chloride as an industrial chemical was dependent on the ready availability of acetylene, and for many years the most important method of manufacture was the addition of hydrogen chloride to acetylene:

$$CH:CH + HCl \rightarrow CH_2:CHCl$$

The reaction is carried out either in the gaseous state at temperatures around 420–270 K (150–200°C) in the presence of a catalyst such as mercuric chloride, or in aqueous medium at 293–298 K (20–25°C).

Since the acetylene is usually prepared by the action of water on calcium carbide, obtained by reaction of coke and lime in an electric furnace, the cost of vinyl chloride made by this process is very dependent on the availability and cost of electricity, and for this reason there have been considerable variations in the price of the monomer from one part of the world to another.

2.1.2.2 From ethylene

With the development of the petrochemical industry, the alternative route via ethylene has become more and more prominent and has now ousted the route from acetylene for large-scale production.[3, 29]

The ethylene is first converted to ethylene dichloride by addition of chlorine:

$$CH_2:CH_2 + Cl_2 \rightarrow CH_2Cl.CH_2Cl$$

After purification by washing and fractional distillation, the ethylene dichloride is pyrolysed at temperatures between 513 and 1273 K (240 and 1000°C), depending on the catalyst employed.[4] Conversion to vinyl chloride is about 50 per cent, but recycling produces overall yields as high as 95 per cent:

$$CH_2Cl.CH_2Cl \rightarrow CH_2:CHCl + HCl$$

The hydrogen chloride may be used to produce more vinyl chloride by reaction with acetylene, but the oxychlorination route is the preferred means for utilising all the chlorine, and has been employed in most plant construction since 1965.[3]

2.1.3   PROPERTIES

Vinyl chloride has a boiling point of 249.1 K (−13.9°C) at atmospheric pressure, and is therefore gaseous at normal atmospheric temperatures. For this reason it is normally polymerised under pressure. As with most other unsaturated compounds with a greater or lesser tendency to polymerise spontaneously, problems of storage and transport arise in handling vinyl chloride on a production scale, but these do not constitute a particularly major difficulty. Liquid vinyl chloride without stabiliser can be transported in tank cars. Spontaneous explosive polymerisation does not appear to be a hazard at normal temperatures, provided that a small amount of oxygen is present in the monomer. Nevertheless, over a considerable period, polymer can build up in storage tanks and pipe-lines, which therefore need to be inspected and cleaned at regular intervals. Apart from this tendency to polymerise, vinyl chloride is chemically quite stable.

Other technologically important properties of vinyl chloride monomer include vapour pressure, density, specific heat, latent heat of vaporisation, and heat of polymerisation. The first of these can be calculated from the formula:

$$\log P = 0.8420 + 1150/T + 1.75 \log T - 0.002\,415\,T$$

where P = pressure in atmospheres
and T = thermodynamic (absolute) temperature

The density of liquid vinyl chloride is about 0.9 g/ml 293 K (20°C) and the specific heat 0.38 cal/g°C (1.59 J/g°C). Latent heat of vaporisation falls from about 80 cal/g (335 kJ/kg) at 283 K (10°C) to about 72 cal/g (301 kJ/kg) at 323 K (50°C). Heat of polymerisation is 22 kcal/mol or 350 kcal/kg (92 kJ/mol or 1465 kJ/kg). Thermodynamic and other physical properties have been well characterised.[2, 5, 6] The solubility of vinyl chloride in water is very small,[7] but high in common organic solvents.

The heat of combustion of vinyl chloride is 286 kcal/mol (1197 kJ/mol) at 353 K (80°C), and the inflammability limits in air are 4–21.7 per cent by volume.

Until the early 1970s it was thought that vinyl chloride was non-toxic but slightly narcotic. At high concentrations at which its smell is still insufficient to offer a reliable warning, the gas can cause loss of consciousness: but then evidence began to accumulate that vinyl chloride monomer is carcinogenic. Naturally this was a cause for considerable concern both within and outside the industry, and urgent steps were taken to gather information, and to institute modifications to plant and procedure to overcome the problem. Special care is necessary in all monomer handling, polymerisation, and polymer isolation processes. This subject is discussed more fully in Chapter 17.[7]

2.2   VINYL ACETATE

2.2.1   LABORATORY PREPARATION

Vinyl acetate is formed by addition of acetylene to acetic acid, as noticed by Klatte in preparing ethylidene diacetate:[8] $CH:CH + CH_3CO_2H \rightarrow CH_3CO_2CH:CH_2$. Demand for the monomer for the production of synthetic resins around 1920 stimulated research into possible methods of forming vinyl acetate as the main product. Laboratory preparation is best carried out in aqueous medium, and a satisfactory though somewhat complicated method has been described by Groth and Johnnson.[9]

2.2.2   MANUFACTURE

Commercial production of vinyl acetate is also carried out in aqueous medium, using various mercury salts as catalysts,[10, 11] or in the vapour phase at 483–523 K (210–250°C) in the presence of cadmium or zinc acetate or phosphate.[12, 13] Since 1965 commercial synthesis by oxy-chlorination from ethylene and acetic acid in both liquid and vapour phases has been achieved.[3, 14]

2.2.3   PROPERTIES

The handling of vinyl acetate, b.p. 345.7 K (72.7°C), is rather simpler than that of vinyl chloride. Since, however, its main use is in copolymerisation with vinyl chloride, this does not markedly simplify polymerisation plant. With a specific gravity of 0.9338–0.9342 at 293 K/293 K (20°C/20°C) vinyl acetate is slightly denser than vinyl chloride. Its specific heat is 0.415cal/g°C (1.74 kJ/kg°C) and latent heat of vaporisation is 7.8 kcal/mol (32.7 kJ/mol) or 91 cal/g (381 kJ/kg).[9, 11] The vapour pressure at 20°C is 88.6mm Hg (11 812 N/m²).

The flash point (Tag open cup) is 266.3 K (−6.7°C), the explosive limits in air being from 2.65 to 38 per cent (vol/vol).

Vinyl acetate is slightly soluble in water [1–2.3% at 293 K (20°C)[11]], and this could have a marked effect on polymerisation in aqueous systems. It is miscible with organic solvents such as hydrocarbons, alcohol's, ketones, and esters. The following azeotropes are formed:

With water	92.7% vinyl acetate;	b.p. 359 K (86°C)
With methanol	63.4% vinyl acetate;	b.p. 331.8 K (58.8°C)
With isopropanol	77.6% vinyl acetate;	b.p. 343.8 K (10.8°C)
With cyclohexane	61.3% vinyl acetate;	b.p. 240.4 K (67.4°C)
With heptane	8.35% vinyl acetate;	b.p. 345 K (72°C)

The monomer does not polymerise readily in the absence of light, but is nevertheless usually stabilised for storage and transport. Small amounts (*c.* 0.01%) of stabiliser do not usually need to be removed before polymerisation, but larger amounts (e.g. 0.5%) must be. The heat of polymerisation is 21.3 kcal/mol (89 kJ/mol).

No important health hazards appear to arise in handling vinyl acetate. No systemic, chronic, or local irritant effects have been recorded.[11]

2.3   OTHER VINYL ESTERS

Vinyl esters other than the acetate are employed in a small scale as comonomers with vinyl chloride, but their value appears doubtful. The stearate has, for example, been proposed as a comonomer to produce internally plasticised or lubricated copolymers.

These esters are made by vinylation of the corresponding acid, or by transvinylation (acidolysis or vinyl interchange) using vinyl acetate. Many have been made and characterised.[11]

Table 2.1 gives some of the characterised properties of vinyl esters including vinyl acetate.

Table 2.1

	Acetate	Formate	Propionate	Butyrate	Stearate	Benzoate
Boiling point (°C/mmHg)	72.7	46.6	94.9	116.7	187–8/4.3	203
Melting point	180.2 K		186.2 K		308–309 K	
	(−92.8°C)		(−86.8°C)		(35–36°C)	
Density (d_{20}^{20})	0.9338	0.9651	0.9173	0.9022		1.0703
Refractive index (n_D^{25})	1.3940	1.3859	1.4038	1.4097		1.5266
Viscosity at 20°C (cP)	0.432	0.36	0.5	0.6		1.8
Solubility in water						
(% by weight at 20°C)	1–2.3	Soluble	0.82	0.3		0.01
Solubility of water						
(% by weight at 20°C)		Soluble	−0.6			0.32

2.4   VINYLIDENE CHLORIDE

After vinyl acetate, vinylidene chloride is the next most important monomer for copolymerisation with vinyl chloride. Vinylidene chloride (*as*-dichloroethylene) can be prepared by dehydrochlorination of trichloroethane by means of aqueous calcium hydroxide at about 333–363 K (60–90°C),[15] and is obtained as a by-product in the manufacture of trichloroethylene. It is manufactured by dehydrochlorination by catalytic cracking of trichloroethane produced by chlorination of ethylene or ethane obtained from petroleum.

At room temperature the monomer is a colourless liquid with a sweet odour, which becomes acidic on exposure to air and is a skin irritant. Some physical properties are listed below.

Boiling point   304.7 K (31.7°C)/760mm/Hg
Freezing point   150.5 K (−122.5°C)
Density (d_4^0)   1.2500
    (d_4^{20})   1.2129
Refractive index (n_D^{20})   1.4249

When exposed to air at room temperature, [16, 17] it forms an explosive white peroxide powder which readily decomposes to form phosgene and formaldehyde. It will initiate vinyl chloride and vinylidene chloride polymerisation. The vapour also forms explosive mixtures with air over a range of concentrations of 7–16% at 25°C. For these reasons handling of the monomer requires care. It is usually inhibited for storage and transport by means of triethanolamine, tributylamine, dimethyl sulphoxide, or *p*-methoxyphenol.[11]

2.5   OTHER MONOMERS

The only other monomers of importance for copolymerisation are vinyl ethers and maleic anhydride. The former can be prepared by a variety of laboratory methods,[18] but in large-scale production the Reppe process[19] is most common, in which acetylene is reacted under pressure at 120–180°C with an appropriate alcohol in strong alkali:

$$CH{:}CH + ROH \rightarrow CH_2{:}CH.OR$$

A fair number of vinyl ethers, including divinyl ether, have been made, and their general properties characterised.[11, 17]

Maleic anhydride is made by passing benzene vapour and air over vanadium catalyst at about 450°C:

It is also obtained as a by-product in the manufacture of phthalic acid and phthalic anhydride.

Many other vinyl compounds have been prepared and polymerised.[11, 20–28] These include vinyl bromide and fluoride, as well as vinyl ketones, thioethers, carbazoles, thiazoles, pyrrolidones, pyrazines, furfurans, lactams, and sulphonamides.

Other monomers that have been copolymerised with vinyl chloride include ethylene, propylene, and acrylic and methacrylic esters.[29]

REFERENCES

1. G.F. D'ALELIO, *Experimental Plastics and Synthetic Resins,* Wiley, New York (1946)
2. H. SHALIT, *Monomers* (E.R. Blout and H. Mark, eds), Interscience, New York (1949–51)
3. P.W. SHERWOOD, *Chem. Process.,* **14**, 11, 4 (1968)
4. *D.R. Pat,* 585 793
5. S.C. BANERJEE and L.K. DORAISWAMY, *Br. Chem. Engng,* 316 (1958)
6. L. BOLOGNA, *Poliplasti,* 4, 15, 7 (1956)
7. R. VILLARD, *Annls Chim. Phys.,* (7), 11, 387
8. *Brit. Pat.* 12 246/1913; *U.S. Pat.* 1 084 581; *D.R. Pat,* 271 383; 313 696
9. P. FRAM, *Monomers* (E.R. Blout and H. Mark, eds), Interscience, New York (1949–51)
10. *U.S. Pat.* 1 626 713; 1 710 197
11. S.A. MILLER, *Acetylene. Its Properties, Manufacture and Uses,* Vol. II, Ernest Benn Ltd., London (1966)
12. S.N. USHAKOV and J.M. FEINSTEIN, *Ind. Engng Chem. analyt. Edn,* **26**, 561 (1934)

13. *U.S. Pat.* 1 822 325; 2 336 208; 2 398 820
14. P.W. SHERWOOD, *Petro/Chem Engr,* **40**, 6, 36 (1968)
15. J.J.P. STAUDINGER, *Br. Plast.,* **19**, 229, 381 (1947)
16. R.C. REINHARDT, *Chem. Engng News,* **25**, 30, 2136 (1947)
17. K. MAYUMI, O. SHIBUYA and S. ICHINOSE, *J. chem. Soc. Japan,* **78**, 280 (1957)
18. C.E. SCHILDNECHT, *Monomers* (E.R. Blout and H. Mark, eds), Interscience, New York (1949–51)
19. *U.S. Pat.* 1 959 927
20. H. Gibello, *Plastiques,* **2**, 139 (1944); *C.A.,* **40**, 3641 (1946)
21. G.M. KLINE, *Mod. Plast.,* **24**, 157 (1946); **25**, 130 (1946)
22. H. DAVIDGE, *J. appl. Chem. Lond.,* **9**, 241, 553 (1959)
23. *Brit. Pat.* 739 438, *G. Pat.* 931 731
24. E.H. CORNISH, *Plastics,* **28**, 61 (1963)
25. *Ind. Chemist,* **29**, 122 (1953)
26. L. GREENFIELD, *Ind. Chemist,* **32**, 11 (1956)
27. E.S. HANSON, 118th A.C.S. Meeting, Abstract of papers 89N (1950)
28. L.F. SALISBURY, *U.S. Pat.* 2 532 573 (1946)
29. R.H. BURGESS, *Manufacture and Processing of PVC,* Applied Science Publishers, London (1982)

3

Production of vinyl chloride polymers and copolymers

3.1   INTRODUCTION

Baumann's original method of obtaining polyvinyl chloride by exposure of the monomer to sunlight is not really amenable to large scale development, although bulk or mass polymerisation is used on production scale by a few manufacturers. A major problem associated with this method is that the polymerisation is highly exothermic and, at about 50 per cent conversion to polymer, heat transfer through the reacting mass becomes so poor that the internal temperature tends to rise excessively above the nominal temperature of polymerisation, thus leading to heterogeneity throughout the batch and possibly to explosion hazards arising from the accompanying rise in pressure. For this reason the monomer is usually polymerised while dispersed as droplets in water. There are basically two different methods of operating this type of process, namely *emulsion polymerisation* and *suspension (or granular) polymerisation*, though some polymerisation systems which have been proposed are hybrids of these two methods.

The problems associated with mass polymerisation were overcome in the mid-1950s, and production plants were commissioned during the succeeding period up to the 1980s, when the technique lost favour due largely to the difficulty of reducing residual monomer content of the polymer to meet requirements imposed by the confirmation in the 1970s that vinyl chloride is a carcinogen.

Polyvinyl chloride is insoluble in its monomer and hence precipitates from the monomer phase in mass, emulsion, or suspension polymerisation. These systems are thus heterogeneous. If polymerisation is conducted in a solvent for vinyl chloride the system may be homogeneous or heterogeneous according to whether or not the polymer is also soluble in the solvent.

3.2   KINETICS AND MECHANISMS OF POLYMERISATION

Vinyl chloride is polymerised by a free radical mechanism,[1] usually activated by added peroxide or free radical azo initiators, and little success appears to have been achieved with ionic initiators. Most studies on vinyl chloride polymerisation have led to the conclusion that chain transfer to monomer is a major reaction, and is the main factor affecting the degree of polymerisation and molecular weight; but Talamini and Vidotto[2] have cast doubt on this.

Prat investigated the polymerisation at 298 K (25°C) and 323 K (50°C), and found that up to 40 per cent conversion the rate increased, then remained nearly constant up to 60 per cent, and then gradually diminished.[3] He suggested that this might be due to control of the initiation by formation of a monomer-initiator complex, followed by decomposition or re-arrangement of this complex, or its reaction with monomer, to produce active centres.[4, 5] Jenckel, Eckmans, and Rumbach also found an increase in polymerisation rate during the earlier part of the reaction.[6]

Bengough and Norrish studied the polymerisation of vinyl chloride initiated by benzoyl peroxide over the temperature range 286–348 K (33–75°C).[7] They, too, found an initial gradual acceleration in rate up to 30–40 per cent conversion, with benzoyl peroxide con-centrations of from 0.025 to 1 mol %, but this acceleration was absent if polymerisation was carried out in a solvent for the polymer. They demonstrated that added polymer, or polymer produced in the early stages of the polymerisation, exerts a catalytic effect on subsequent polymerisation in the presence of benzoyl peroxide, and that the average molecular weight was almost independent of initiator concentration up to about 0.5 mol %, being mainly de-termined by the temperature of polymerisation. The higher the polymerisation temperature, the lower is the resultant average molecular weight, and vice versa. The catalytic effect was found to be approximately proportional to the surface area of the polymer producing it, lead-ing to the conclusion that the effect occurs only at the interface between polymer and poly-merising monomer. Bengough and Norrish explained their observations by proposing that growing radicals can terminate by transfer with 'dead' polymer molecules, forming free valencies in the surface of the polymer. Nozaki had already suggested that chain transfer to monomer was likely to be a major feature of vinyl chloride polymerisation,[8] and this is supported by the approximate constancy of molecular weight over a wide range of initiator concentration observed by Bengough and Norrish, and by Mead and Fuoss.[9] These con-siderations led to the following scheme, where B represents a benzoyl peroxide molecule, R represents an initiator-free radical (i.e. phenyl or benzoate), R' represents a mobile grow-ing free radical, M a monomer molecule, P 'dead' polymer, and P* a radical in the polymer surface.

Initiation:

$$B \rightarrow 2 \text{ R (rate constant} = k_i)$$
$$\text{(e.g. Ph. CO.O.O.CO.Ph} \rightarrow \text{Ph}^\cdot + \text{Ph.CO.O}^\cdot + CO_2)$$

Propagation:

$$R + M \rightarrow R' \text{ (rate constant} = k_p)$$
$$\text{(e.g. } R + CH_2:CHCl \rightarrow R.CH_2.CHCl)$$

Chain transfer with monomer:

$$R' + M \rightarrow P + R' \text{ (rate constant} = k_{tr})$$
$$\text{(R}^\cdot CH_2 \dot{C}HCl + CH_2:CHCl \rightarrow R^\cdot CH_2CH_2Cl + CH_2:\dot{C}Cl}$$
$$\text{or } R^\cdot CH_2.CHCl_2 + CH_2:\dot{C}H$$
$$\text{or } R' \text{ CH:CHCl} + CH_3.\dot{C}HCl)$$

Transfer at polymer surface:

$$R' + P \rightarrow P^* + P \text{ (rate constant} = k_a)$$

$$(R' + R'CH_2CHCl.\dot{C}H_2- \rightarrow R'CH_2\dot{C}H.CH_2- + P$$
$$\text{or } R'CH_2CHCl.\dot{C}H - + P)$$

Propagation from the polymer surface:
$$P^* + M \rightarrow P^* \text{ (rate constant } = k_p)$$

Termination of surface radical by transfer:
$$P^* + M \rightarrow P + R' \text{ (rate constant } - k_{tr})$$

Termination of normal radicals:
$$R' + R' \rightarrow P \text{ (rate constant } = k_t)$$

(By disproportionation:

$$R'CH_2CHCl - + R'CH_2CHCl - \rightarrow R'CH:CHCl + R'CH_2CH_2Cl$$

By combination:

$$R'CH_2CHCl - + R'CH_2CHCl - \rightarrow R'CH_2CHCl.CHCl.CH_2R')$$

Assuming that P' radicals, which being part of the polymer conglomerates are not mobile like the other radicals, will terminate only by transfer to monomer but that their rates of propagation or chain transfer to monomer will not be affected so long as they are effectively in an environment of monomer, the rate of polymerisation becomes:

$$-dM/dt = K(M + K'P^{2/3} (B/M)^{1/2}, \text{ where } M = \text{moles of monomer}$$
$$B = \text{moles of benzoyl peroxide}$$
$$K = (k_p + K_{tr}) (k_i/k_t)^{1/2}$$
$$K' = k_a/k_{tr}$$

This expression agrees with the observation that at equal extents of polymerisation the rates are proportional to the square root of the initiator concentration. If removal of monomer and initiator during the first 30–40 per cent of reaction is such that the concentration of initiator remains approximately constant, the rate of polymerisation should be proportional to the $\frac{2}{3}$ power of the polymer concentration; this is also found to be approximately true.

The proposed mechanism also accounts for the observation that the average molecular weight is almost independent of initiator concentration up to 0.05 mol %.

However, this scheme assumes that the rate constants for reactions between initiator radicals and monomer and between the resultant radicals and further monomer are the same. It has been pointed out[58] that for some initiators, azo compounds for example, with very short half-lives for decomposition, the radical produced is so stable that initiation is inefficient. Consequently the propagation step in the above scheme should be considered as two, with different rate constants:

$$I' + M \rightarrow R' \text{ (rate constant } = k'_p)$$

$$R' + M \rightarrow R' \text{ (rate constant } = k_p)$$

Where I' represents an initiator radical, differentiated from a radical R' produced by reaction with monomer.

Other differences in more recent proposals take account of the fact that, although it is insoluble in monomer and is precipitated on formation, vinyl chloride polymer is swollen by monomer to form a 'gel' of approximately 25 per cent vinyl chloride that is presumed to contain a 'normal' concentration of initiator so that further polymerisation occurs within the 'gel'. Thus there are two regions of polymerisation, with different termination rate constants.

Most other investigators have also concluded that chain transfer is the main factor determining molecular weight [10-14] and that molecular weight does not change much during polymerisation.[15, 16] However, Bier and Kramer[17] have suggested that there are differences in molecular weights of polymers of below and above 90 per cent conversion, which they attributed to branching at high conversions.

Bamford *et al.*[18] accepted that the proposition of Bengough and Norrish that chain transfer to polymer occurs could be true, but suggested that radicals produced by this method would have relatively brief existence since they would be destroyed by transfer to monomer. They pointed out that there is no evidence to preclude the possibility that heterogeneous polymerisation of vinyl chloride might follow the mechanism worked out in some detail for acrylonitrile.[19-21] This involves the so-called 'occlusion theory', which supposes that a high proportion of growing polymer radicals is precipitated from the liquid phase, and that the radical ends tend to be occluded in the tightly coiled structures which would be expected to occur under such conditions. This, together with the added effect of coalescence with dead polymer, would lead to reduced radical activity. All velocity coefficients for radical reactions would thus be reduced under heterogeneous conditions, particularly for bimolecular termination; whereas propagation and transfer to monomer would be reduced only to the extent that monomer diffusion is obstructed by the barrier of polymer. Thus, provided that the occlusion was insufficient to reduce the rate of propagation, the main effect would be to reduce the rate of termination and consequently to increase the overall rate of polymerisation. The number of polymer particles would determine the rate of occlusion of radicals, whereas their weight would determine the magnitude of the polymer barrier to diffusion of monomer to radicals. Hence the termination coefficients would be expected to fall progressively, leading to a corresponding progressive increase in overall rate of reaction.

Mickley, Michaels, and Moore[22] studied the polymerisation, using dilatometric methods, and their results led them to propose a kinetic scheme only slightly different in detail from that of Bengough and Norrish.

From molecular weight studies on polyvinyl chloride produced by solution, bulk, suspension, and emulsion techniques, Talamini and Vidotto,[2] while accepting the that molecular weight is independent of conversion and concentration of initiator, did not consider this a reliable proof of a high transfer to monomer, and concluded that termination by combination is the reaction predominantly limiting chain length. They estimated that no more than 20 per cent of the polymer molecules are formed by disproportionation or chain transfer, and rejected the possibility of transfer to polymer and hence the formation of long side-chains. Their results, however, seem to be in some conflict with those of Freeman and Manning, based on molecular weight studies by light scattering, osometry, and solution viscometry.[23]

3.2.1   RATE OF POLYMERISATION

In general the rate of polymerisation depends on the purity of the monomer, the nature and concentration of initiator, and on the temperature. Since the last-named is determined by the particular average molecular weight required in the polymer, the initiator effectively constitutes the only available variable by which rate can be varied. For a given initiation system the usual kinetic laws apply to the dependence of rate on temperature, so that within the temperature range 40–60°C the polymerisation rate generally increases by a factor of about 2.5 for each 10°C rise in temperature.[24]

Neither the mechanism of Bengough and Norrish, nor that of Bamford and his colleagues, would be expected to yield a simple relationship between overall rate of reaction and initiation rate or initiator concentration. The catalyst exponent would be expected to be greater than the usual value of 0.5, although as a rough working rule the overall reaction rate may be taken as approximately proportional to the square root of the initiator concentration. For this reason polymerisation is usually continued to between 75 and 95 per cent conversion, and may be 'short-stopped' chemically.[69]

At any particular temperature the instantaneous rate of polymerisation increases with time up to a maximum at about 75 per cent conversion, and then falls with fall in monomer concentration. For a more detailed discussion the reader is directed to Reference 58.

3.2.2 TEMPERATURE OF POLYMERISATION

Since the molecular weight of vinyl chloride is determined mainly by the polymerisation temperature, the polymer producer has little choice in selecting the latter once the required molecular weight has been decided. In order to keep the spread of molecular weight to a minimum it is desirable to conduct as high a proportion of the polymerisation as possible at the selected temperature. Any part of the reaction that occurs during heating up, or during periods when inadequate control leads to temperature variations, will result in an unnecessary increase in molecular weight spread. Only a limited amount of work correlating temperature and molecular weight has been published, but Table 3.1 gives some indication of the magnitude of molecular weights obtained in production. ISO Viscosity Numbers (*see* Chapter 4) are also indicated.

Table 3.1 DEPENDENCE OF MOLECULAR WEIGHT OF POLYVINYL CHLORIDE ON POLYMERISATION TEMPERATURE

Polymerisation temperature		Average molecular weight		ISO Viscosity Number
(K)	*(°C)*	*Weight*	*Number*	
308	35	500 000	92 000	190
313	40	340 000	82 000	165
318	45	240 000	70 000	140
323	50	200 000	64 000	125
328	55	140 000	55 000	105
333	60	94 000	48 000	90
338	65	84 000	42 000	80
343	70	70 000	36 000	70

The values tabulated here differ slightly from but are of the same order of magnitude as those given by the relationships between molecular weight, K-value, and temperature presented by Burgess.[58]

Burgess[58] has shown that, if termination of polymer molecular growth is much more frequent by chain transfer to monomer than by disproportionation or combination, the degree of polymerisation is approximately equal to the ratio of the propagation and transfer rate constants, whence:

$$\text{Molecular weight} = 62.5 k_p/k_{tr}$$

The validity of this relationship increases with conversion. In commercial production conversion of monomer to polymer is usually carried out to over 70 per cent, and good agreement is found between experimental values[23] and those calculated from the above expression using Kuchanov and Bort's value for k_p/k_{tr}.[58, 59]

3.2.3   COPOLYMERISATION

The major proportion of vinyl chloride resins are homopolymers, but significant amounts of copolymers are manufactured, the most important comonomers being vinyl acetate and vinylidene chloride. Copolymerisation of vinyl chloride with other monomers appears to follow general theoretical principles. In the copolymerisation of two monomers the system may be considered as comprising two species of growing radicals, dependent on which of the monomers has contributed the reactive radical end, and two monomers. Hence there are four concomitant polymerisation reactions, each with its own rate constant, and the course of the copolymerisation reaction is determined by the activities of the growing radicals and of the monomers:

$$RM_1' + M_1 \longrightarrow RM_1' \text{ (rate constant} = k_{11})$$
$$RM_1' + M_2 \longrightarrow RM_2' \text{ (rate constant} = k_{12})$$
$$RM_2' + M_1 \longrightarrow RM_1' \text{ (rate constant} = k_{21})$$
$$RM_2' + M_2 \longrightarrow RM_2' \text{ (rate constant} = k_{22})$$

The distributions of the two monomer units along the polymer chains is governed by the reactivity ratios r_1 and r_2, where:

$$r_1 = k_{11}/k_{12} \text{ and } r_2 = k_{22}/k_{21}$$

Two other parameters have proved useful in predicting the course of copolymerisation. These are Q, which is related to monomer reactivity, and e, associated with the polar character of the monomer. Monomers in which the double bond is activated by conjugation, e.g. by a carbonyl, nitrile, or aryl group, have relatively high values of Q, whereas highly polar monomers have positive values of e. Typical values for the various copolymerisation parameters for a number of monomers are given in Table 3.2.[25, 26, 27, 30, 58, 62, 63, 64]

Vinyl acetate is relatively unreactive in copolymerisations, and generally retards the overall rate of polymerisation. Its relative reactivity with polyvinyl chloride radicals, as compared with vinyl chloride, has been quoted[25] as 0.5, the respective r_1 and r_2 values[26] being 0.23 and 1.68, and other work has produced results of similar magnitude,[27] so that in general the copolymer will contain a lower proportion of vinyl acetate than that present in the monomer mixture from which it is produced. Starting from a given composition, a polymer formed during the earlier stages will contain a higher proportion of vinyl chloride. As polymerisation proceeds the proportion of vinyl acetate incorporated into the growing chains will increase, until towards the end the monomer phase will be relatively rich in vinyl acetate. It is therefore extremely difficult to produce a truly random copolymer of vinyl

Table 3.2 COPOLYMERISATION PARAMETERS

Monomer	r_1	r_2	r_1/r_2	Q	e
Vinyl chloride				0.044	0.20
Acrylonitrile	0.02	3.28	0.07		
Allyl acetate	1.16	0			
Butadiene	0.03	9	0.27	2.4	−1.05
2-chloroallyl acetate	0.7	0			
Diethyl fumarate	0.12	0.47	0.06		
Diethyl maleate	0.77	0.009	0.007		
Ethyl vinyl ether				0.032	−1.17
Ethylene	2.8	0.05	0.14	0.015	−0.20
Isobutylene	2.05	0.08	0.16		
Isobutyl vinyl ether	2.0	0.02	0.04	0.023	−1.77
Isopropenyl acetate	2.2	0.025	0.6		
Maleic anhydride	0.29	0.008	0.0023	0.23	2.25
Methyl acrylate	0.083	9.0	0.75	0.42	0.60
Methyl methacrylate	0.1	10	1.0	0.74	0.40
Phenyl vinyl ether				0.082	−1.21
Propylene	5.6	0.006	0.034	0.002	−0.78
Styrene	0.07	35	2.5	1	−0.8
Vinyl acetate	1.68	0.23	0.39	0.026	−0.22
Vinylidene chloride	0.2	1.8	0.36	0.22	0.36

chloride and vinyl acetate; but an approach towards this ideal can be achieved by starting with a monomer mixture relatively rich in vinyl acetate and metering vinyl chloride into the reaction mixture at such a rate as to maintain the ratio of the two monomers constant.

On the other hand vinylidene chloride is more reactive than vinyl chloride when the two monomers are copolymerised, although the overall polymerisation rate is slower than homopolymerisation. Figure 3.1 shows a plot of the composition of the monomer phase in a

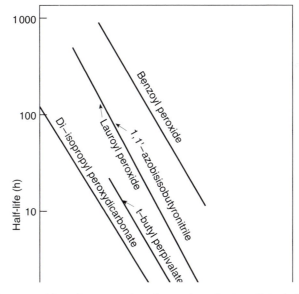

Figure 3.1. Variation in composition of monmer phase in copolymerisation of vinyl chloride with vinylidene chloride [28]

copolymerisation, starting with 17 per cent vinyl chloride and 83 per cent vinylidene chloride, against time of polymerisation.[28]

The relative reactivity of vinylidene chloride and vinyl chloride with polyvinyl chloride radicals has been quoted[25] as 5, and with polyvinylidene chloride radicals as 2. Konishi[29] determined the relative reactivities as 0.5 for vinyl chloride and 0.001 for vinylidene chloride copolymerised at 298 K (25°C). As with copolymerisation of vinyl chloride with vinyl acetate, homogeneous copolymers may be produced by maintaining constant the ratio of the two monomers in the reaction mixture; but this is rather simpler with vinylidene chloride, since the monomer employed in the lower concentration is the more reactive.

Other monomers used to a minor extent in commercial vinyl chloride copolymers include acrylonitrile, methyl acrylate, maleic anhydride, and vinyl ethers.

3.2.4   HEAT OF POLYMERISATION

As with other monomers the polymerisation of vinyl chloride is a highly exothermic reaction, but there is some doubt as to the precise value of the heat of polymerisation. By comparison with other monomers a value of 22 kcal/mol (92 kJ/mol) has been estimated[24] but more recent work[31, 34] has yielded a range of values, the consensus of evidence tending to suggest a figure around 30 kcal/mol (126 kJ/mol). Values for some other monomers are listed for comparisons in Table 3.4.

Table 3.4 SOME HEATS OF POLYMERISATION

Monomer	Heat of polymerisation			
	kcal/mol	cal/g	kJ/mol	kJ/kg
Vinyl chloride	30	480	126	2010
Vinyl acetate	21	240	88	1005
Vinylidene chloride	16	160	67	670
Maleic anhydride	14	140	59	586
Ethylene	24	860	101	3601
Propylene	20	480	84	2010
Styrene	17	170	71	712
Methyl methacrylate	13	130	55	436
Acrylonitrile	18	340	76	1424

The main practical significance of evolution of heat during polymerisation lies in its effect on temperature and, more particularly, temperature control. If unchecked, the rates of heat evolution and rise in temperature accelerate rapidly, since the rising temperature results in acceleration of polymerisation. In extreme cases this state of affairs can lead to 'run-away' reactions, possibly explosive, but in general practice this is avoided by the provision of adequate cooling facilities and safety devices. What requires rather more attention is the necessity to maintain the temperature constant during polymerisation in order to control molecular weight within as narrow limits as possible. If this is catered for adequately there is no danger of 'run-away' polymerisation unless the cooling arrangements fail.

The problem of temperature control is at its most acute in bulk or mass polymerisation, which is one reason for the development and widespread use of emulsion and suspension techniques. Here the aqueous phase fulfils a primary role as a 'heat sink', considerably reducing temperature variation for a given change in heat content of the system. Nevertheless, problems of heat transfer were solved in the 1950s, from which time a number of plants were commissioned until other problems arose leading to a decline from the 1980s onwards.

3.2.5 CHAIN-TRANSFER AGENTS

As indicated previously, the molecular weight of vinyl chloride polymer is mainly determined by the temperature of polymerisation, so that the usual means of obtaining polymers of relatively low molecular weights is to employ relatively high temperatures. It is not always convenient, however, to achieve a required low molecular weight by the employment of the appropriate temperature, because the pressures generated may be too high for the available autoclaves, because of the risk of gelation of polymerising droplets in suspension or emulsion polymerisation, or because of the dangerously high rate of polymerisation. In such circumstances chain-transfer agents may be added to assist in control of molecular weight.[35] Hydrogen chloride is an effective chain-transfer agent in mass polymerisation, but halogen-containing organic compounds and mercaptans are the most common. In addition to reducing molecular weight, chain-transfer agents usually retard polymerisation appreciably, and in high concentrations can inhibit polymerisation to an unacceptable degree. Moreover they constitute a potential source of impurity in the final polymer, and are therefore to be avoided if possible.

3.2.6 INITIATORS

While polymerisation of vinyl chloride was first observed to be initiated by sunlight, commercial processes usually employ free radical generators. These are mainly organic peroxides and azo compounds soluble in liquid vinyl chloride monomer, or water-soluble peroxy compounds such as hydrogen peroxide and alkali metal or ammonium persulphates. The most important monomer-soluble initiators about which details have been published include benzoyl peroxide, lauroyl peroxide, caproyl peroxide 1,1'-azobisisobutryronitrile, azobis (2,4-dimethylvaleronitrile); dimethyl 1,1'-azobisisobutyrate; di-isopropyl, dicetyl, and di(t-butylcyclohexyl) peroxydicarbonates; t-butyl perpivalate, t-butyl perneodecanoate, and acetylcyclohexylsulphonyl peroxide.[35–39, 58] Of the water-soluble initiators the most common are potassium, ammonium, and sodium persulphates, frequently activated by salts of reducing sulphur-containing acids, such as sodium sulphite, bisulphite, or thiosulphate, to give a reduction-oxidation or 'redox' system[·40] and sometimes further activated by small concentrations of metals and/or oxygen.[41]

The decomposition of most free radical generators in first order, i.e. the rate of reaction is proportional to the concentration of the substance, which may be expressed in the form:

$$dx/dt = k_d(a - x)$$

or

$$dx/(a - x) = k_d dt$$

whence

$$k_d = (1/t) \ln [a/(a - x)]$$

where dx/dt = reaction velocity
$\quad\quad a$ = initial concentration

x = decrease in concentration

k_d = velocity coefficient

A useful guide to the activity of an initiator at any particular temperature is its 'half-life'. This is related to the velocity coefficient by the relationship $t_{0.5}$ = (ln 2)/k, since $(a - x) = a/2$.

It does not follow that all free radicals formed by decomposition of an initiator will actually produce polymer radicals, and the efficiency with which such radicals are produced is obviously as much of interest as the rate of decomposition of the initiator. The efficiency of initiation is defined as half the ratio of the rate of initiation to the rate of decomposition of the initiator, since each molecule of initiator should give rise to two free radicals.

In experimental studies on the decomposition of benzoyl peroxide, the apparent order of the reaction is frequently higher than one. It is believed, nevertheless, that the first step in the decomposition of benzoyl peroxide is a first order homolytic fission into two benzoyloxy radicals, followed by a reaction of higher order involving decomposition of the peroxide by reaction with benzoyloxy or derived radicals. Thus, in the presence of free radical acceptors such as vinyl monomers, iodine, trinitrobenzene, or α,α'-diphenyl-β-picrylhydrazyl,[42, 43] decomposition rates tend to become independent of solvent and the order of reaction becomes close to unity.

Of the published data, those of Bawn and Mellish[43] seem most applicable to the behaviour of benzoyl peroxide in initiation of heterogeneous polymerisation of vinyl chloride. Their results lead to the following relationship for the variation of the velocity coefficient of decomposition with temperature:[18]

$$k_d = 3.0 \times 10^{13} e^{-29600/RT} (s^{-1})$$

Half-lives (hours) calculated from the same data may be obtained from the following relationship:

$$\log_{10} t_{0.5} = 6.52 \times 10^3/T - 17.5.$$

This line is plotted in Figure 3.2, and a plot of log k_d versus reciprocal temperature in Figure 3.3, together with similar plots for a few other initiators of activity appropriate to vinyl chloride polymerisation.

Half-life data may be presented in a variety of ways; namely in the form of an equation as above; graphically as in Figure 3.2; as values of half-lives at various temperatures of interest;[58, 64] or as temperatures at which particular half-lives are obtained.[68] Where correlations between different sources can be attempted agreement is not particularly good, and the data can be taken only as a guide to the selection of appropriate initiators. This may be because half-life can be very dependent on the nature of the solvent in which determinations are carried out.[68] For example, the temperature for a 10 hour half-life of diisopropyl peroxydicarbonate was raised from 18–35°C on changing from methanol to benzene, and to 50°C on changing to trichloroethene.[68]

Lauroyl peroxide is more active than benzoyl peroxide at any particular temperature. The data available indicate that its activity is very similar to that of 1,1′-azobisisobutyronitrile, which has been studied in some detail. As well as allowing greater reaction velocities than benzoyl peroxide, lauroyl peroxide appears to yield polymers superior in colour, and in heat and light stability.[24] Caproyl peroxide seems to be very similar to lauroyl peroxide, and has been used as an alternative.

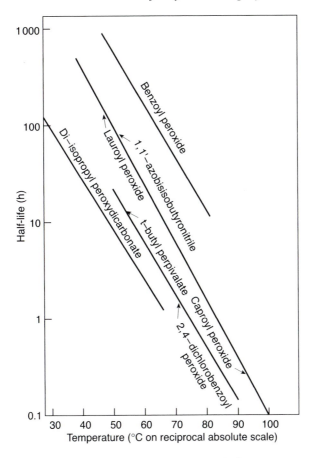

Figure 3.2. Half-lives of free-radical indicators

Other peroxy compounds that have activities appropriate for vinyl chloride, and are generally more reactive than acyl peroxides, include *t*-butyl perpivalate and various per-oxydicarbonates (e.g. dibutyl, diisopropyl, di(2-ethylhexyl), and di(t-butylhexl) per-oxydicarbonates), which latter appear to be preferred to other peroxides, at least for polymerisation at the lower temperatures.[58] Published data leads to the following approximate relationship between half-life and temperature:

t-butyl perpivalate $\qquad\qquad\qquad \log_{10}t_{0.5} = 6.48 \times 10^3/T - 18.8$
di-isopropyl peroxydicarbonate $\quad \log_{10}t_{0.5} = 6.23 \times 10^3/T - 18.45$

Substitution of chlorine in the aromatic ring of benzoyl peroxide increases reactivity. For example, *p*-chlorobenzoyl peroxide appears to behave similarly to lauroyl and caproyl peroxide, and 2,4-dichlorobenzoyl peroxide is rather similar to *t*-butyl perpivalate, but there is no indication that these aromatic peroxides are of any particular interest for vinyl chloride polymerisation.

Of non-peroxy compounds, the only initiators of any interest are aliphatic azo compounds, of which 1,1'-azobisisobutyronitrile is by far the most common and most studied. Bamford *et al.* have pointed out[18] that the published values of rate constants for the decomposition of this compound in a variety of aromatic solvents are in good agreement with the expression:

$$k_d = 1.0 \times 10^{15} e^{-30450/RT} \, (s^{-1})$$

This leads to the following relationship between half-life (in hours) and temperature:

$$\log_{10} t_{0.5} = 6.61 \times 10^3/T - 18.7$$

Values calculated from this equation are in reasonable agreement with half-life data published more recently.[58] It has already been stated that lauroyl peroxide has similar activity to 1,1'-azobisisobutyronitrile, and these two initiators can be used to yield similar polymerisation rates under similar conditions. The low molecular weight of the azo compound gives the theoretical advantage over the common peroxides that it leads to a lower level of contamination of the polymer, but there appears to be no practical evidence to suggest that this has any significance. Also, it has been suggested that azo initiators are advantageous as compared with peroxides since they do not contribute to oxidative reactions occurring during photodegration; but conclusive supporting evidence has not been published.

There has been some controversy over the mode of decomposition of 1,1'-azobisisobutyronitrile, but the details need not concern us here. Some of the possible decomposition

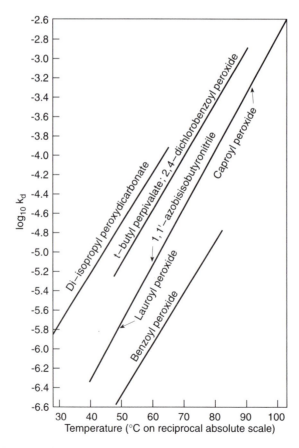

Figure 3.3. decomposition rate constants of free-radicals initiators

products are undoubtedly toxic, but investigations have shown that their concentration in the polymer is usually negligibly small, even for foodstuff packaging applications.

Efficiency of initiation by 1,1′-azobisisobutyronitrile has been determined for a few monomers, but no data appear to be available for vinyl chloride. A value of 0.58 seems to be accepted for styrene but values as low as 0.5 for methyl methacrylate and as high as 0.79 for acrylonitrile have been quoted.

Other azo initiators of interest for vinyl chloride polymerisation include dimethyl-1,1′-azobisisobutyrate and azobis(2,4-dimethylvaleronitrile). The former is slightly less reactive than the azonitrile but of low toxicity. It is, however, more expensive, since it is prepared from the nitrile. Azobis(2,4-dimethylvaleronitrile) decomposes somewhat more slowly than the peroxydicarbonates at any particular temperature, but considerably faster than 1,1′-azobisisobutyronitrile or the acyl peroxides.

Because their radicals are very energetic, peroxides generally have high efficiency, while azo compounds, which produce relatively less energetic radicals, are less efficient.[58] For example, 2,2′-azobis-(4-methoxy-2,4-valeronitrile) produces such stable radicals that it would be effective only at undesirable and uneconomic concentrations.[58]

Because the molecular weight of the polymer is determined by the temperature of polymerisation, no single initiator is appropriate where, as is the norm, a range of polymers of different molecular weights is produced. In some cases mixtures of initiators may be advantageous.

Other factors important in selecting an initiator system for commercial polymerisation include stability and storage behaviour, which may require special storage and handling facilities involving extra costs.

3.3   POLYMERISATION PROCEDURES

From the time of Baumann's observation of the effect of sunlight on vinyl chloride, workers continued to concentrate on photopolymerisation for over fifty years, and even Klatte's first patent[44] describing thermal and photoactivated polymerisation seemed to favour the latter, although an associated patent[45] discloses the activation of polymerisation by 'catalysts' such as organic peroxides, ozonides, organic acid anhydrides in combination with oxygen, or oxygen-producing compounds such as perborates and percarbonates.

The inhibiting effect of oxygen was observed at a fairly early stage,[46] and Schoenfield[47] proposed the purging of polymerisation vessels with nitrogen, carbon dioxide, or vinyl chloride itself.

Polymerisation of vinyl chloride in solution was first investigated by Ostromuislenskii[46] and developed by Plotnikow,[48] but this method has now been largely superseded by polymerisation in aqueous media. This latter process has the advantage over mass polymerisation (Section 3.3.1) of easier temperature control, and over solution polymerisation of the avoidance of the use of relatively expensive solvents; but it has the disadvantage that the high specific heat of water leads to the necessity for relatively large amounts of heat in raising the system to the required temperature and to increase in overall cycle time because of the time required to heat the aqueous phase and to cool it after polymerisation.

There are thus four distinct separate methods available for the production of vinyl chloride polymers:

(1) Mass or bulk polymerisation
(2) Solution polymerisation
(3) Emulsion polymerisation
(4) Suspension polymerisation

Of these by far the most important commercially at the present time is suspension poly-merisation, followed by emulsion and mass polymerisation. Solution polymerisation is re-served for a few speciality copolymers where for some reason the other techniques are objectionable, or where the application makes it appropriate, as in solution coatings.

3.3.1   MASS OR BULK POLYMERISATION

Bulk polymerisation methods using free radical initiators were used on a small scale for vinyl chloride during the 1930s, but development to large scale production by bulk or mass polymerisation did not make much progress until the mid 1950s. Outlines of feasible pro-cesses were published in 1967,[49] followed later by more detailed desriptions.[59, 60, 68] Owing to the highly exothermic nature of the polymerisation reaction and the nature of the poly-merised material, conversion is limited to around 50 to 60 per cent, the residual monomer being flashed off from the polymer and recycled. The process resembles that used for some other polymers, in which monomer is pumped under pressure through a series of narrow tubular reactors in parallel, maintained at the appropriate temperature. Pumping rates can be adjusted so that discharge from the ends of the reactors corresponds to about 50 to 60 per cent conversion, the discharge being into a decompression chamber to flash off monomer from the liquid monomer and return it to the pumping system, possibly with intermediate purification by fractional distillation. Separation of polymer presents no special problems, and for homopolymer there is no difficulty in removing most of the residual monomer. However, reduction of monomer content to concentrations now required on health and en-vironmental grounds has necessitated the introduction of additional procedures. (See Chapter 17.)

The one-stage mass or bulk polymerisation process produces resins of rather low bulk density and plasticisers absorption. Nowadays bulk polymerisation of vinyl chloride is a two-stage process which can be adapted to produce resin particles of different character-istics, including porous particles for plasticised and high density particles for unplasticised dry-blend processing.[49, 59, 60, 68]

In the first or 'prepolymerisation' stage about 50 per cent of the monomer with initiator is charged to a conventional autoclave with a paddle stirrer producing very turbulent agita-tion. This is necessary in order to produce beads of high density. This stage is sometimes known as the 'bead formation' stage. Polymerisation is carried out in the liquid phase to 7–10 per cent conversion.

The 'prepolymerised' product is passed to a second-stage autoclave, together with the rest of the monomer and initiator, where polymerisation is continued to 65–85 per cent conver-sion. Originally the second-stage autoclave was a horizontal reactor, but vertical reactors are now preferred, apparently for agitation considerations.

Thomas[49] has presented a brief theoretical explanation of the formation of particles dur-ing bulk polymerisation, referring to work by the Russians Bort, Rylov, and Okladov, and Clark[69] has shown how polymerisation conditions can be varied to produce different types of particle.

The number of manufacturers operating commercial bulk polymerisation of vinyl chloride is relatively small, though the sizes of individual plants may be comparable with those of the more common suspension and emulsion processes. Precise production figures are difficult to obtain, but it is doubtful if bulk processes account for more than about 10 per cent of total production of vinyl chloride resins. By 1982 the world capacity of plants for bulk production of PVC was estimated to be of the order of one million kilotonnes per annum.[59]

As discussed in the next chapter, bulk polymers have the advantage over other types that the necessary additives are at a minimum, and hence the residual impurities also; but the resultant property advantages over the best suspension polymers are only marginal. Moreover, removal of the last traces of residual monomer from bulk polymer appears to be more difficult than with suspension polymers, and requires special degassing arrangements.

Fuller details of bulk polymerisation plant and procedures have been presented elsewhere.[59, 60, 68]

3.3.2   EMULSION POLYMERISATION

Emulsion polymerisation was the first method developed industrially for vinyl chloride and was the only method used industrially in the UK up to 1944. The earliest British patent on emulsion polymerisation of vinyl chloride appeared in 1933.[51] It was based on a German application of the previous year.

In this method of polymerisation the monomer is dispersed as a relatively stable dispersion in water and the resultant product is a stable suspension or latex of very small polymer particles in the water. Dispersion of monomer in the aqueous phase is achieved by agitation with a dissolved emulsifying agent. Although much information about recipes for emulsion polymerisation has been published it has been mainly in the form of patent specifications and rather non-committal about specific details and procedures actually used by manufacturers. Thus, although much work has been carried out and published to show the effects of various initiators and activators on polymerisation rates, little precise information has been disclosed about initiation systems actually in use. Still less has been released about specific emulsifying agents and reaction conditions. It is certain that the choice of initiation system has a profound effect on polymerisation rates and cycle times, and therefore on outputs from autoclaves of given size. Similarly, the agitation conditions, and emulsifying agents and other ingredients in the aqueous media, can markedly affect the size of polymer particles and the stability of the latex obtained.

3.3.2.1   Recipes

The proportion of monomer to water employed is limited by heat transfer considerations, since the higher this ratio the greater will be the amount of heat evolved by the polymerisation, and the greater the amount of heat to be extracted from the reaction system through the walls of the autoclave if a specific reaction temperature is to be maintained. Another limitation on the proportion of monomer employed is the necessity to produce a stable aqueous polymer latex, which imposes an upper limit on the polymer and thus the monomer concentration. This limit appears to be in the region of a concentration of monomer of about 95 per cent of the aqueous phase.

In emulsion polymerisation the initiator employed is usually water-soluble, and for this purpose alkali metal or ammonium persulphates are common, possibly activated by a water-soluble reducing agent such as sulphur dioxide, sodium sulphite, sodium bisulphite, sodium hydro-sulphite, sodium thiosulphate, or sodium formaldehyde sulphoxylate, to form a reduction-activation system.[52] Other initiators described as suitable include hydrogen per-oxide and alkali metal azodisulphonates. The choice of initiation system and its concentration determine the rate of polymerisation at the selected reaction temperature, and thus the cycle time and throughput of the polymerisation plant.

Anionic surface-active agents are most commonly used as emulsifying agents, and other materials described as suitable include alkali metal or ammonium salts of long-chain fatty acids, sulphonates and sulphates, such as sodium and ammonium oleate, palmitate, and stearate; sodium cetyl sulphate and similar compounds; salts of dialkyl sulphosuccinates and of alkane- and alkylbenzene-sulphonic acids; ammonium dinonyl citrate; sodium di-alkyl phosphites and phosphates; saponified vinyl alkyl ether/vinyl acetate copolymers; glyceryl octadecyl ether; and soluble urea-formaldehyde resins. The emulsifying agent needs to be present in a concentration exceeding the critical micelle concentration. Over the range normally employed the particle size decreases with increase in concentration of sur-face-active agent.[65] Where the emulsion polymer is destined for 'melt' processing the primary particle size is generally less than 0.3 μm, but paste polymers require a wider range of particle size and some greater than 1 μm may be desirable. A typical recipe might be as follows (parts by weight):

Water	100
Emulsifying agent	0.0–1
Buffer salt	0.05–0.1
Ammonium persulphate	0.050–0.25
Sodium hydrosulphite	0–0.2
Vinyl chloride monomer	55–90

3.3.2.2   Procedures

In batch processes the aqueous phase of emulsifying agent and initiator, possibly with al-kaline and buffer salts to control pH, dissolved in the appropriate amount of water, is first charged to a pressure autoclave. All or most of the air is removed from the autoclave, by flushing with nitrogen or monomer, and the vinyl chloride is then charged under pressure with agitation. The temperature of the reaction mixture is then raised to the selected tem-perature by circulating water though the jacket of the autoclave. This will raise the pressure from about 2–10 atmospheres (200–1000 kN/m^2) consequent upon the charging of the monomer, to about 5–15 atmospheres (500–1500 kN/m^2). As polymerisation proceeds, however, there is a drop in pressure due to the reduction of the amount of vinyl chloride monomer present, and the pressure changes can be used to follow the progress of the poly-merisation. Conversion is taken well towards completion, around 85 to 95 per cent conver-sion; the time required depends on the temperature, and the concentration and nature of the initiator system, but may be as short as six hours or less under favourable conditions. At the selected time the autoclave is vented to recover unpolymerised monomer. The autoclave may then be heated under vacuum to boil the latex to reduce the residual monomer content to a minimum, but this stripping may be deferred to a later spray-drying stage in an evacu-ated chamber against a counter-current of steam.[66] The latex is then discharged, possible after cooling by circulating cold water through the autoclave jacket.

When operating continuously a number of possible alternatives exist, but in any case the aqueous phase (containing emulsifying agent, initiator, buffers) and the monomer are usually charged separately into the plant. This latter can be in a number of stages, dispersion of monomer in the aqueous phase being effected in the first. The dispersed mixture can then be transferred to one or more reaction vessels. These might be a number of pressure autoclaves in succession, although a single continuous tube has been employed.

As already indicated, emulsion polymerisation systems are selected to produce stable polymer latices. A typical commercial latex might have a solid content of between 35 and 45 per cent, consisting mainly of polymer particles having diameters in the region of 0.05–0.25 μm. The stability of a latex is critical for handling, storage, and processing, and is strongly dependent on the emulsifying system and the pH. The later is usually well on the alkaline side, typically about 10–11. Marked changes towards acidity are liable to reduce latex stability and cause coagulation. Latex stability is sometimes increased by adding further emulsifying agent. The more dilute latices may be concentrated to a solids content of about 50 per cent by a creaming process analogous to that used for concentrating rubber latices. This involves adding latex to a mixture of a small quantity of the latex with up to about 1 per cent of its total weight of a thickening agent such as Calasec MAH. Over a period of 1–2 days the mixture separates into a lower (concentrated) and an upper (dilute) layer, which can be separated by conventional means.

An alternative method of producing PVC latices, which can be regarded as 'microsuspension' polymerisation, uses monomer-soluble initiators, and the monomer is emulsified in the aqueous medium by mechanical homogenisation by means of a high-speed pump, colloid mill, or high-speed stirrer[29, 66] prior to charging to the autoclave. Originally [29] recipes of the following type were proposed:

Water	100
Fatty sulphonate	0.4–0.6
Fatty alcohol	0.4–0.6
Lauroyl peroxide	0.4–0.5
Vinyl chloride	40–50

but the lauroyl peroxide has now been replaced by peroxydicarbonates, preferably together with a long-chain hydrocarbon to reduce build-up of polymer on the autoclave surfaces.[66]

The dispersions obtained spray-dry readily, and the stability of the dried resins is comparable with that of many suspension polymers, but it has been difficult to obtain good paste resins. This difficult is reduced by 'seeding' the system with a mixture of conventional emulsion latex and a microsuspension latex containing a high concentration of initiator.[66] Paste resins made by microsuspension are now offered alongside those made by emulsion polymerisation.

Plasticisers can be incorporated into PVC latices by homogenising in a colloid mill or similar machine.

Latices of PVC, plasticised or unplasticised, can be used directly for some processes, as discussed more fully in Chapter 15, but most latex is further processed to produce dry polymer in powder form.

3.3.2.3   *Isolation of polymer from latices*

Coagulation by addition of electrolytes, followed by washing, filtering, and drying is to a limited extent employed for this purpose, but this method is not very convenient on a large scale, and spray-drying is more commonly used. The equipment is essentially similar to that used for the drying of melamine–formaldehyde resins and the production of dehydrated milk.[53] It usually consists (Figure 3.4) of a chamber of essentially inverted conical shape,

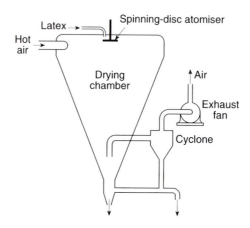

Figure 3.4. Spray-drier

into the top of which latex is introduced and atomised by a disc rotting at high speed. At the same time heated air is introduced into the chamber so that the droplets of latex are dried as they pass downwards to the base. During the drying process the primary particles of polymer within each droplet agglomerate to form larger particles which become cenospherical as a result of expansion due to pressure of the water entrapped within the droplet. The dried particles, now having an average diameter of about 30–50 μm, pass out through the base of the drier and are conveyed to a packaging point. Excessively fine particle are carried out in the air stream, from which they are recovered as fine dust by means of a cyclone separator. Commercial polymers will usually be found to contain a proportion of particles of indeterminate shape, which are fragments of cenospheres broken during conveyance from one part of the plant to another. The mechanical strength of the cenospherical particles depends on the heating they have received, which will determine the extent to which their cenospherical nature is retained.

Emulsion polymers are in part used for the production of plastisols or pastes (Chapter 14), and the polymers so used are commonly known as plastisol, paste, or dispersion polymers or resins. They may consist of unmodified dried emulsion polymers or may have been submitted to a grinding process to produce particles of a size required for particular paste viscosity characteristics.

3.3.3.   SUSPENSION POLYMERISATION

Large-scale suspension polymerisation of vinyl chloride appears to have developed from the granular or bead polymerisations of methyl methacrylate and styrene, but uncommercialised experimental work dates back to the early 1930s. The patent position on suspension poly-merisation was confused, but the most important patents were probably *Brit. Pat.* 427 494 (1933), which described the use of organic colloids such as starch and gelatine, and *Brit. Pat.* 444 257 (1934), which related to the use of synthetic polymeric colloids such as polyvinyl alcohol. Neither specification, however, mentioned vinyl chloride specifically, the nearest reference being to 'vinyl esters'.

The process was used on a small scale (about 2000 t/a) in Germany,[54] and on a somewhat larger scale in the US, during the 1939–45 war.

Suspension polymerisation bears a superficial resemblance to emulsion polymerisation in that the monomer is polymerised while dispersed as droplets throughout an aqueous phase. In suspension polymerisation, however, dispersion is maintained by a combination of stirring with the so-called 'protective' action of a colloid dissolved, or a finely divided inorganic powder suspended, in the aqueous phase. In addition, polymerisation is initiated by a monomer-soluble initiator, so that in effect the process may be considered as bulk polymerisation within each of the dispersed droplets. Marx studied the progress of both emulsion and suspension polymerisation of vinyl chloride using optical and electron miscroscopy.[55] He found that the surfaces of suspension polymer particles showed inversion due to volume contraction during polymerisation, and that cavities appear for the same reason. Hollow spheres also appeared to be formed owing to entrapment of heat within par-ticles, the outside skin being difficultly soluble and the interior relatively easily. These ex-planations do not seem adequate to explain the mechanisms of particle formation in all types of suspension polymerisation, particularly in view of the more recently introduced non-porous particle polymers.

Allsopp[61] and colleagues, and Clark,[69] studied the mechanism of particle formation dur-ing suspension polymerisation in detail, and showed how its complexity accounts for the variety of types of particle that can be produced by variations in recipe and polymerisation conditions. Because of its importance in determining major polymer properties, this subject is discussed in more detail in the next chapter.

Industrial suspension polymerisation has been described fairly fully.[24, 58, 68]

3.3.3.1   RECIPES

Suitable initiators include benzoyl, lauroyl, caproyl, dodecyl, *p*-chloro-benzoyl, acetylcy-clohexanesulphonyl, and 3,5,5-trimethylhexanoyl peroxides; polypropylene hydroperoxide; diethyl and di-isopropyl peroxy-dicarbonates; and azo compounds such as 1,1′-azobi-sisobutyronitrile and dimethyl 1,1′-azobisisobutyrate. Historically the first of these was probably most common in the early manufacture, but was largely replaced by lauroyl per-oxide and some other aliphatic peroxy compounds, owing largely to the superior colour and stability which these latter impart to the polymer. More recently peroxydicarbonates appear to be most favoured. As stated previously, the choice is mainly dependent on the tempera-ture of polymerisation to be employed. The concentration depends on the chemical nature of the particular initiator, the polymerisation temperature, and the required rate of reaction, but is typically in the range of 0.01–0.1 per cent.

Owing to the insolubility of its polymer in monomer, polymerising vinyl chloride does not pass through the 'sticky' stage characteristic of styrene and methyl methacrylate, and there have been claims to the use of very low concentrations (less than the critical micelle concentration) of suspension agent, reducing the size of the particle,[56] and even claims to suspension polymerisation without dispersing agents.[57]

Suspension agents are usually water-soluble colloids such as polyvinyl alcohol (more correctly, hydrolysed polyvinyl acetate), gelatin and natural proteins, water-soluble cellulose derivatives such as methyl and carboxy-methyl cellulose, dextran, starch, sodium alginate and various gums, acrylic acid/ester copolymers, acrylate/maleate copolymers, salts and partial esters of maleic acid or anhydride/styrene copolymers, and partially esterified maleic anhydride/ethylene copolymers. Many other materials have been described.[56]

Finely divided water-insoluble inorganic powders have also been used as suspension agents, presumably acting by interposing a physical barrier between approaching droplets. Such materials have included talc, kaolin, bentonite, titania, barium sulphate, and aluminium hydroxide.

Of the various possible suspension agents, polyvinyl alcohol is probably the most commonly used commercially. It is available in a variety of degrees of hydrolysis (of polyvinyl acetate) and molecular weights, and these affect the protective action and the type, size, and shape of monomer droplets and polymer particles obtained. In addition to the nature of the suspension agent, its concentration and the intensity and rate of stirring will also affect the nature of the droplets and polymer particles. By manipulation of these variables, different types of polymer can be produced but appropriate systems have usually to be discovered by trial and error. The concentration employed is generally between 0.05 and 0.5 per cent by weight of the monomer.

It has been suggested that water-soluble colloids act as suspension agents by their effect on the surface tension at the monomer/water interface, but effective colloids are generally poor emulsifying agents and are used in much smaller concentrations than are emulsifying agents in emulsion polymerisation. Moreover, suspension agents do not induce dispersion of monomer in water to form an emulsion. Bologna[24] has described a simple demonstration of the effect of suspension agents. Vinyl chloride monomer is sealed in a Carius tube with about four times its own volume of a 0.05 per cent solution of polyvinyl alcohol in pure water. The liquid vinyl chloride floats on the aqueous layer, from which it is separated by a fairly flat meniscus. On agitation a dispersion of clearly discernible droplets is formed. On stopping the agitation these droplets rise to the surface and retain their separate nature for several hours. In the absence of the polyvinyl alcohol at rest the meniscus between the vinyl chloride and water phases is highly curved, the aqueous surface being concave. Even violent agitation will only disperse the monomer into large droplets, and these reform to a single phase very quickly on stopping the agitation.

It has also been suggested that, in addition to a conventional effect on interfacial tension, the molecules of the suspension agent form an external coating on the droplets, thus forming a barrier to coalescence. It has also been suggested that an important factor is the increase in viscosity of the aqueous phase arising from the presence of high molecular weight colloid in solution; but, while this may be true for high concentrations, where small spherical droplets and ultimately polymer particles are produced, the increase in viscosity for the smaller concentrations employed for the production of suspension polymers is unlikely to be significant.

In addition to suspension agents, buffer salts such as disodium hydrogen phosphate or borax are sometimes used to prevent any tendency for the pH of the aqueous phase to fall during polymerisation. Sometimes, too, 'anti-foam' materials such as octyl alcohol, polyethylene silicate, or a suitable silicone are added before or after polymerisation in order to reduce frothing during discharge of residual monomer at the end of the polymerisation cycle.

A typical recipe might be as follows (parts by weight):

Water	100
Protective colloid	0.1–0.5
Buffer salt	0–0.1
Initiator	0.05–0.3
Antifoam	0–0.002
Vinyl chloride monomer	50–70

The water need not be as pure as is usually necessary for emulsion polymerisation, but any dissolved or suspended impurities are potential sources of contamination of the polymer and may affect the colour, clarity, and electrical resistance of products. It has been suggested that hard water leads to polymers more stable than those produced with water of high purity,[19] but it is doubtful if the difference is of practical significance, and in general the highest level of purity commensurate with economical operation would be preferred.

3.3.3.2 Procedures

Suspension polymerisation is effected batchwise in autoclaves. As in emulsion polymerisation, the aqueous phase is charged to the autoclave first, the suspension agent and buffer being dissolved therein or having previously been dissolved in a separate solution vessel. Inert gas or monomer is then used to remove air, and the vinyl chloride and initiator are then charged under pressure, the agitation being such as to break down the liquid monomer phase into droplets of appropriate size. Heating and cooling arrangements are similar to those of emulsion polymerisation, as also are temperatures and pressure cycles. Within the autoclave, however, the course of events is usually markedly different.

Suspension polymerisation is essential a series of small-scale bulk polymerisations conducted within the aqueous phase, and the mechanisms and kinetics are the same as those of bulk polymerisation. Moreover, the droplets formed initially do not necessarily retain their identity throughout the process. Depending on the nature and concentration of the suspension agent and on the agitation, droplets may coalesce with others and also re-disperse. The final polymer particles may be formed from single droplets or a number of droplets which have coalesced at some stage of the process without re-dispersion.

On completion of the process, the product consists of an unstable slurry of solid polymer particles suspended by agitation in the aqueous phase and containing residual monomer. Cessation of agitation leads to deposition of the particles; agitation is therefore continued until the slurry has been discharged from the vessel, usually into a stirred storage tank. The size and shape of the individual polymer particle depends mainly on the suspension system and agitation, and can be varied over a wide range to suit particular requirements (Chapter 4). However, the details have still to be derived by experience and trial and error. No theoretical approach is available which would permit design of recipe and plant to produce a particular type of polymer, though the nature of polymer particle structures and the mechanisms of particle formation are now well understood.[61]

Residual monomer left in the resin amounts to 10–20% of the original charge. The major

proportion of this is recovered by gasification and liquefaction, but because of its carcinogenic nature it is necessary to reduce monomer content still further. This is most conveniently achieved by heating the polymer/water slurry to 80–110°C with steam, which is
a carrier for the monomer from which it is separated in a condenser,[70] as discussed in
Section 17.2.

3.3.3.3   Isolation of polymer from slurries

On a small scale, suspension polymers can be filtered from the aqueous phase, washed, and
dried by conventional means; but such a procedure is not convenient in large-scale production. Commonly slurry is stocked in stirred storage tanks capable of holding several
batches, and is pumped from these to continuous centrifuges which remove most of the
water to leave a 'wet cake' of polymer with about 10–15 per cent of water. Simultaneous
washing with additional water is usually possible. Final drying is effected in conventional
equipment, usually a rotary drier of some sort. The design must be such that the polymer
does not stagnate in any position, that corrosion cannot occur, and that the drying air is free
from suspended contaminants. Dried polymer is freed from excessively large agglomerates
by coarse screening on issuing from the drier, and most of the polymer is then either bagged
directly, or conveyed to a storage hopper for subsequent packing, usually by a pneumatic
system. Flash and fluid bed driers are also used for suspension polymers.[67]

   While some commercial polymer is still packed and delivered in paper sacks, bins of various kinds and bulk storage with tanker delivery are now most common. Sacks usually hold
about 25 kg, whereas bins are available to hold a ton or more, and tankers can carry much
larger quantities. Quite apart from the convenience of bulk deliveries, contamination by
paper and stitching from paper sacks has always been a problem, which can be eliminated
with bulk equipment by due care in design and operation. The higher initial capital investment in bulk handling equipment can be set against the saving in non-reusable paper sacks,
and handling equipment, both of which account for an important part of the final cost of
polymer.

3.3.4   POLYMERISATION PLANT

Some details of plant for bulk or mass polymerisation have been published, and the reader
is referred to the references[59, 60] for further information.

   A typical plant for the production of vinyl chloride polymers and copolymers by the suspension method is illustrated diagrammatically in Figure 3.5. Emulsion polymerisation
plant would be essentially similar, apart from the substitution of a spray-drier for the cen-
trifuge/rotary drier system. Indeed, it is possible for both methods of polymerisation to be
carried out in the same vessels.

   A considerable part of the plant is involved with receipt, storage, preparation, conveyance, and measurement of raw materials. This includes, for example, tanks for vinyl
chloride and conomoner storage, water storage, mixing vessels for preparation of the aqueous phase and initiator solution, weighting machines and burettes, pumps, and pipework.

   Polymerisation vessels or autoclaves have to be capable of withstanding pressures of 15
atmospheres (1500 kN/m²), or appreciably above if low molecular weight polymers or

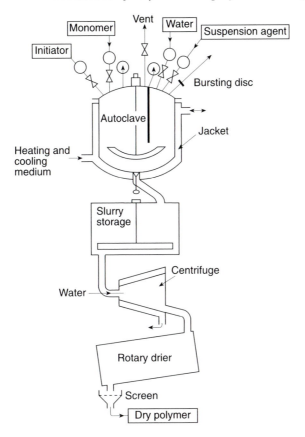

Figure 3.5. Plant for suspension polymerisation

copolymers are to be produced, since such pressures are developed at the temperatures necessary to obtain the lowest molecular weights of commercial importance, unless added chain transfer agents are employed. Even at higher molecular weights and correspondingly lower temperatures, pressures of about 5–10 atmospheres (500–1000 kN/m²) are encountered. Autoclaves and most other parts of the plant must be made of stainless steel, or ordinary steel lined with stainless steel, since iron contaminates the polymer sufficiently to impair colour and stability, and must be avoided at all costs. The internal surfaces of the autoclave or any items of equipment located in the polymerisation medium constitute regions where polymer is likely to deposit and form hard lumps and skin which can be difficult to remove, and moreover act as nuclei for the growth of further deposits of polymer which is usually of very high molecular weight. All such surfaces should therefore be maintained in as clean and smooth a condition as possible. It is also desirable to keep items of equipment inside the autoclaves to a minimum so as to avoid pockets and recesses where polymer might be deposited. Fouling by deposited polymer may also be suppressed by chemical means.[58]

A factor of importance in autoclave design is the provision of facilities for cleaning, and in particular of removal of build-up of polymer on the autoclave walls, baffles, and stirrer. At one time this was done manually by operatives entering the vessels between polymerisations, but, as well as the obvious inefficiency of such a procedure, it is a highly hazardous operation, and not only because of the now recognised carcinogenic nature of vinyl chloride.[70] Nowadays high pressure jets of water propelled from rotary heads are employed to remove the build-up, and, as mentioned earlier, chemical means are used to suppress it.[58]

The autoclaves have pipe connections or ports, usually fitted to the lids, for the introduction of water, monomer, suspension agents, initiator, and nitrogen, as well as a safety pressure release valve or bursting disc and a pressure release line for removal of residual monomer at the end of the reaction. In addition there are pressure and temperature recorders and an agitator or stirrer, usually of the paddle or anchor type. Each autoclave is fitted with a jacket through which cooling or heating fluid, usually water, can be circulated to control temperature, possibly controlled by temperature recorders. At the base of each autoclave is a valve usually of the mushroom type, through which the product can be discharged into a storage tank.

With reasonable recipes and consistent quality of monomer and initiator, combined with consistent operation, the polymerisation is unlikely to get out of control, but bursting discs and other pressure-relieving attachments must be carefully maintained a clean and efficient condition since the violence of an uncontrolled polymerisation can be high and dangerous. A more likely source of trouble is adventitious failure of the stirrer motor, which leads to a major reduction in heat transfer from the system to the jacket and thus to excessive temperature rise and incrase in polymerisation rate. In thse circumstances it is usually necessary to release the pressure and lose the batch.

For a particular recipe the shape of the autoclave, baffles, and other fixed items of equipment in the polymerisation system, and the shape and speed of rotation of the stirrer, are critical in determining the form of the final polymer particles, and indeed the success of the whole operation. Attempts to establish theoretical equations for the design of stirred autoclaves have not been adequate to permit design without recourse to the lessons of past experience, and to the methods of trial and error where insufficient past experience is available. Precise details of design, dimensions, etc., are consequently preserved as secret by manufacturing companies and no details are published. Ideally all regions of the polymerising mass should be in equal turbulence, but this is rarely achievable in practice. Consequently, rapid circulation of material between zones of relatively low and high turbulence is necessary. A complication of stirred systems is that centrifugal separation of the two phases can occur where there is density difference. In vinyl chloride polymerisation the monomer droplet density starts at about 0.9, but during polymerisation increases towards the polymer density of about 1.4. There is the further complication that in the initial stages the agitation system should prevent the monomer from floating on the surface of the aqueous phase, while towards the end of the polymerisation it has to prevent the relatively dense monomer/polymer particles from settling to the bottom. Baffles constitute regions where polymer is liable to deposit, and for this reason they may be water-cooled, and should be kept to a minimum. Consequently stirrer speeds are often quite high (e.g. up to 200–300 rev/min) and efficient bearings, glands, and lubricating systems are essential in order to avoid contamination of the product. It has been suggested[24] that eccentrically mounted stirrers are more efficient than those mounted symmetrically, in that their use avoids the necessity for baffles. It has also been suggested that a stirrer system mounted at the base of the autoclave is advantageous as the stuffing-box works in a water and not at vapour environment. It is probable, however, that the majority of autoclaves used are fitted with conventional centrally-mounted stirrer shafts operated by motors above.

Sizes of autoclaves vary widely, and have naturally tended to increase as new plants have been built to meet the ever-increasing demand for polymer. The smallest production unit is probably about 1 m^3 in capacity but much larger sizes are common (e.g. 10 m^3, or even up to 100 m^3). Owing to reduction in the ratio of surface area to volume as volume increases, the proportional area available for heat transfer between batch and jacket decreases as autoclave volume increases. Also increased volume of autoclave involves increased thickness of autoclave walls in order to meet pressure rating requirements. The resultant reduced heat transfer capability can be compensated by increasing the water/monomer ratio. Consequently the larger the autoclave the smaller is the proportion of its volume occupied by monomer, i.e. batch size increases more slowly than increase in autoclave size. To limit this effect it is common practice to design larger autoclaves by increasing height rather than diameter relative to smaller autoclaves. This brings with it problems of agitation.

Even without these limitations it would be undesirable to instal one autoclave and associated plant to cope with the total output required. With only one autoclave each item of plant would be operated intermittently with long periods of down-time. Moreover a shutdown of the autoclave would stop production completely. By having a minimum of four autoclaves the ancillary plant can be operated more or less continuously, and breakdown of one autoclave is not so disastrous.

An autoclave of about 3 m^3 capacity will accommodate a monomer charge of about 1 ton (1.016 t), and will therefore produce a like quantity of polymer. Larger autoclaves will not produce proportionally more, but something rather less than this, so that an autoclave of about 10 m^3 capacity will produce something like 3 tons (3.05 t) per batch.

Usually autoclaves are arranged in groups discharging to a common storage or stock tank, through monomer stripping plant where the stripping is not performed in the autoclaves. By this arrangement slurry or latex from the stock tank can be pumped continuously to the drying plant. This arrangement also serves to homogenise the products from a number of batches, thus removing any small batch to batch variations. Fairly vigorous stirring is necessary to maintain suspension of polymer in slurry, but only mild stirring is required to homogenise emulsion polymer latices.

Spray-dryers for latices have already been described (Figure 3.4). With suspension polymers slurry is usually pumped from the stock tanks to continuous centrifuges, usually of the horizontal type, in which the slurry is fed into the centre of a filter basket spinning on an horizontal axis. Water passes through the basket to drain or recirculate in the normal way, the solid polymer being held on the basket and transferred along it to a discharge outlet whence it issues with a retained water content of about 10–15 per cent. If desired a limited amount of washing with water can be carried out during the separation. The resultant so-called 'wet-cake' is finally dried, usually in a conventional rotary drier.

Dried polymer can be transported by means of conveyer belts or vibrating conveyers, but pneumatic pipe systems are common.

Until the mid-1960s, polymerisation plants were commonly controlled manually, but since then operation from a remote control room and control by computer have become increasingly common, reducing process labour costs and hazards, and improving consistency.

3.4   CONCLUSION

It will be seen that three main methods are employed for the production of vinyl chloride polymers and copolymers, namely bulk, emulsion, and suspension polymerisation, of which the last-named is easily the most important in that it is employed for a very high proportion of world manufacture. By any of the methods it is possible to produce homopolymer and copolymers containing a range of conomoner content, and to produce these with a wide range of useful molecular weights. Furthermore, variations in detail of production methods and conditions can result in polymers of widely differing particle shape and size. Thus by varying these factors a large number of polymers of different properties can be produced, and a high proportion of these are indeed produced on a manufacturing scale.

The importance of these different properties, and in particular, their effect on the behaviour of polymers is discussed in the next chapter.

REFERENCES

1. P.J. FLORY, *J. Am. chem. soc.*, **59**, 241 (1937)
2. G. TALAMINI and G. VIDOTTO, *Chimica Ind., Milano*, **46**, 1, 16 (1964)
3. J. PRAT, *Méml Servs chim. État*, **32**, 319 (1945)
4. G. GEE and E.K. RIDEAL, *Trans. Faraday Soc.*, **36**, 656 (1936)
5. A.C. CUTHBERTSON, G. GEE and E.K. RIDEAL, *Proc. R. Sol.*, **A170**, 300 (1939)
6. E. JENCKEL, H. ECKMANS and B. RUMBACH, *Makromolek. Chem.*, **4**, 15 (1949)
7. W.I. BENGOUGH and R.G.W. NORRISH, *Proc. R. Soc.*, **A200**, 301 (1950)
8. K. NOZAKI, *Discuss Faraday Soc.*, **2**, 337 (1947)
9. K.H. MEAD and R.M. FUOSS, *J. Am. chem. Soc.*, **64**, 277 (1942)
10. J.W. BREITENBACH and A. SCHINDLER, *Mh. Chem.*, **80**, 429 (1949)
11. F. DANUSSO and G. PERUGINI, *Chimica Ind., Milano*, **35**, 881 (1953)
12. F. DANUSSO and D. SIANESI, *Chimica Ind., Milano*, **37**, 695 (1955)
13. J.W. BREITENBACH, *Makromolek. Chem.*, **8**, 147 (1956)
14. F. DANUSSO, G. PAJARO and D. SIANESI, *Chimico Ind., Milano*, **41**, 1170 (1959)
15. F. KRASOVEC, *Rep. Josef Stefan Inst.*, **3**, 203 (1956)
16. G. PEZZIN, G. TALAMINI and G. VIDOTTO, *Makromolek. Chem.*, **43**, 12 (1961)
17. G. BIER and H. KRAMER, *Makromolek. Chem.*, **19**, 151 (1956)
18. C.H. BAMFORD, W.G. BARB, A.D. JENKINS and P.F. ONYON, *The Kinetics of Vinyl Polymerisation by Radical Mechanism*, Butterworths, London, 117, 220 (1958)
19. C.H. BAMFORD, W.G. BARB, A.D. JENKINS and P.F. ONYON, *The Kinetics of Vinyl Polymerisations by Radical Mechanisms*, Butterworths, London, 111 (1958)
20. C.H. BAMFORD and A.D. JENKINS, *Proc. R. Soc.*, **A216**, 515 (1953); **228**, 220 (1955) and *J. Polym. Sci.*, **14**, 511 (1954); **20**, 405 (1956)
21. C.H. BAMFORD, D.J.E. INGRAM, A.D. JENKINS and M.C.R. SYMONS, *Nature, Lond.*, **175**, 894 (1955)
22. H.S. MICKLEY, A.S. MICHAELS and A.L. MOORE, *J. Polym. Sci.*, **60**, 121 (1962).
23. M. FREEMAN and P.P. MANNING, *J. polym. Sci. A*, **2**, 2017 (1964)
24. L. BOLOGNA, *Poliplasti*, **4**, 15, 7 (1956)
25. F.R. MAYO and C. WALLING, *Chem. Rev.*, **46**, 191 (1950)
26. P.J. Flory, *Principles of Polymer Chemistry*, Cornell University Press (1953)
27. G.M. BURNETT and W.W. WRIGHT, *Proc. R. Soc.*, **A221**, 41 (1954)
28. M.S. WELLING, *Adhes. Resins*, **5**, 105 (1957)
29. A. KONISHI, *Bull. chem. Soc. Japan*, **35**, 395 (1962)
30. J. BRANDRUP and E.H. IMMERGUT, (eds.), *Polymer Handbook*, Interscience, New York (1966)
31. D.E. ROBERTS, *J. Res. natn Bur. Stand.*, **44**, 221 (1950)
32. G.C. SINKE and D.R. STULL, *J. Phys. Chem., Ithaca*, **62**, 397 (1958)
33. J.R. LACHER, H.B. GOTTLIEB and J.D. PARK, *Trans. Faraday Soc.*, **58**, 2348 (1962)

34. J.R. LACHER, E.E. MERZ, E. BOHMFALK and J.D. PARK, *J. Phys. Chem., Ithaca,* **60**, 492 (1956)
35. W.M. SMITH, *Manufacture of Plastics*, Reinhold, New York (1964)
36. F.H. DICKEY, J.H. RALEY, F.F. RUST, R.S. TRESEDER and W.E. VAUGHAN, *Ind. Engng Chem. analyt. Edn,* **41**, 8, 1673 (1949)
37. F.M. LEWIS and M.S. MATHESON, *J. Am. chem. soc.,* **71**, 747 (1949)
38. *Brit. Pat.* 626 155; 649 934; 651 315
39. R.E. KIRK and D. F. OTHMER, *Encyclopaedia of Chemical Technology*, Vol. 14, Interscience, New York (1947–60)
40. *Brit. Pat.* 573 366; 574 482
41. *Brit. Pat.* 589 264; 598 777
42. C.G. SWAIN, W.H. STOCKMAYER and J.T. CLARKE, *Am. chem. Soc.,* **72**, 5426 (1950)
43. C.E.H. BAWN and S.F. MELLISH, *Trans. Faraday Soc.,* **47**, 1216 (1951)
44. *D.R. Pat.* 281 688
45. *Brit. Pat.* 319 588
46. I. OSTROMUISLENSKII, *Zh. russk. fiz.-khim. Obshch.,* **44**, 204 (1912); *Brit. Pat.* 6299/1912
47. F.K. SCHOENFIELD, *U.S. Pat.* 2168 808
48. J. PLOTNIKOW, *Z. wiss. Photogr.,* **21**, 133 (1922)
49. J.C.THOMAS, *S.P.E.Jl,* **61**, Oct. (1967)
50. *Plast. Rubb. wkly,* 1st Nov. (1968)
51. *Brit. Pat.* 410 132 (1933)
52. R.G.R. BACON, *Trans. Faraday Soc.,* **42**, 170 (1946)
53. B.B. FOGLER and R.V. KLEINSCHMIDT, *Ind. Engng Chem. ind. Edn,* **30**, 12, 1372 (1938)
54. BIOS Final Report No. 811
55. T. MARX, *Plaste Kautsch.,* **1**, 211 (1954)
56. S.A. MILLER, *Acetylene: Its Properties, Manufacture and Uses*, Ernest Benn Ltd, London (1966)
57. *U.S. Pat.* 2625 539; *G. Pat.* 1 065 612
58. R.H. BURGESS, *Manufacture and Processing of PVC*, Applied Science Publishers Ltd., London (1982) Chapter 1
59. M.W. ALLSOPP, *Manufacture and Processing of PVC*, R.H. Burgess, ed., Applied Science Publishers Ltd., London (1982) Chapter 2
60. G.C. MARKS, *Developments in PVC Technology*, J.. Henson and A. Whelan, eds. Applied Science Publishers Ltd., London (1973) Chapter 2
61. M.W. ALLSOPP, *Manufacture and Processing of PVC*, R.H. Burgess, ed., Applied Science Publishers Ltd., London (1982) Chapter 7
62. G. FLEISCHER and F. KELLER, *Plaste u. Kaute.,* **20**, 7 and 10 (1973)
63. G. SIELAFF and D.G. HUMMEL, *Makromol. Chemie.,* **175**, 1561 (1974)
64. H-G. ELIAS, *Macromolecules-2*, John Wiley, London, 779 (1977)
65. P. RANGNES and O. PALMGREN, *J. Polym. Sci., Pt.C.,* **33**, 181 (1971)
66. D.E.M. EVANS, *Manufacture and Processing of PVC*, R.H. Burgess, ed., Applied Science Publishers Ltd., London (1982) Chapter 3
67. V.G. LOVELOCK, *Manufacture and Processing of PVC*, R.H. Burgess, ed., Applied Science Publishers Ltd., London (1982) Chapter 6.
68. M. LANGSAM, *Encyclopedia of PVC*, L.I. Nass and C.A. Heiberger, eds., Marcel Dekker, New York (1985) Volume 1, Chapter 3
69. M. CLARK, *Particulate Nature of PVC*, G. Butters, ed., Applied Science Publishers Ltd., London (1982) Chapter 1.
70. R.H. Burgess, *Manufacture and Processing of PVC*, Applied Science Publishers Ltd. (1982) Chapter 5.

4

Properties of vinyl chloride polymers and copolymers

4.1 INTRODUCTION

By virtue of the large number of variants possible in details of vinyl chloride polymerisation systems, as described in the previous chapter, many variations are possible in the properties and combinations of properties of the resultant polymers. At least 400 different grades of resin have been offered on the world market over the years, and whilst some of these are undoubtedly intended to match competitive products, or products of associated companies, it is almost certain that no two of these grades of resin are identical. Indeed, one of the problems encountered in the manufacture of PVC has been the difficulty of maintaining constant properties during product over a period of time, and even where nominally identical plants and recipes have been employed, as when there has been an exchange of 'know-how' between associates or licensers and licensees, some minor differences have arisen. Improvements in control systems have now resulted in such a high level of consistency, especially within any one plant, that these problems have largely disappeared.

The physical and mechanical properties of a finished PVC product are dependent on its composition, including the particular resin selected, and on the processing treatment used to make the product. The important properties of the resin in this respect are average molecular weight and possibly molecular weight distribution, the presence or otherwise of comonomer, and for some properties (e.g. electrical resitivity) the presence of impurities. All these properties are important also in affecting processing behaviour of resins, and in addition the processing behaviour is affected by the type of particle, e.g. porosity and shape, as well as the particle size and size distribution, and by the bulk and packing densities of the resins.

Chemical properties of vinyl chloride homopolymers are of course determined to a large extent by the basic chemical structure of the polymer molecule, but it is almost certain that some chemical properties (e.g. thermal stability) are influenced out of all proportion by the presence of structural irregularities, such as side-chains and unsaturation. Thermal stability is certainly influenced markedly by the nature and concentration of residual and adventitious impurities, and also by previous thermal history.

Introduction of comonomers can affect chemical properties very materially, depending on the nature and proportion of the particular monomers involved.

4.2   STRUCTURE OF VINYL CHLORIDE POLYMERS AND COPOLYMERS

The structure of a polymer is naturally intimately bound up wtih the structure of the monomer or monomers from which it was derived, and with the kinetics and mechanisms of the polymerisation process by which it was formed; these have already been discussed.

Polyvinyl chloride has a mainly 'head-to-tail' main chain structure, in which a chlorine atom is situated on each alternative carbon atom in the polymer chains:

$$\text{ww}\ CH_2.CHCl.CH_2.CHCl.CH_2.CHCl\ \text{ww}$$

This was demonstrated by Marvel et al.,[1-4] who removed chlorine from the polymer by refluxing in dilute dioxan solution with zinc. The residue was found to contain between 13 and 16 per cent chlorine, close to the expected value of 13.53 per cent for a regular 'head-to-tail' structure, compared with 18.4 per cent which would be expected from a 'head-to-head', 'tail-to-tail' arrangement. This view of the structure of polyvinyl chloride is supported by the similarity of its ultra-violet spectrum to that of 2,4-dichloropentane and its dissimilarity from that of 2,3-dichloropentane.[6]

Evidence on linearity or chain-branching of polyvinyl chloride is somewhat sparse. Infra-red studies by Grisenthwaite et al.[7, 8] of polymers before and after reduction with lithium aluminium hydride indicated that there are differences in structure, probably in chain-branching, between products obtained by polymerisation at different temperatures, the amount of branching tending to increase with increasing temperature of polymerisation over the range 233–318 K ($-40 - +45°C$). On the other hand, investigations by Freeman and Manning[9] indicate that chain-branching increases with increasing molecular weight, at least for commercial polymers, which are usually made by polymerisation at temperatures within the range 308–343 K (35–70°C), according to the molecular weight-required. They esti-mated that the commercial polymers they examined contained between 5 and 10 per cent of long-chain branched molecules. Other estimates[10] have suggested that commercial poly-mers contain about 16 branches per molecule. Krasovec[11] and Cotman[12] have also found evidence of chain-branching. From an analysis of molecular weight distribution curves, Talamini and Vidotto[13] concluded that polyvinyl chloride could not include long side-chains but that short side-chains might arise, as postulated by Bovey and Tiers.[14] It is doubtful, however, if one would on this basis be justified in denying the existence of the small amount of long-chain branching suggested by Freeman and Manning. Baker and co-workers con-cluded that commercial polymers contain about five branches per 1000 carbon atoms.[73] Abbas and co-workers found that the branches were mainly methyl groups amounting to about four per 1000 carbon atoms.[74] In different studies[75] they suggested that the frequency was within the range 1.9 to 2.8 per 1000 carbon atoms, and also found evidence for about one per 1000 carbon atoms of branches containing six or more carbon atoms. The same group of workers found chlormethyl groups to be present.[76]

There has been a mild controversy on the question of crystallinity. It is generally agreed that commercial polymers are mainly amorphous, containing around 5 to 10 per cent of crystalline regions,[15-20, 77] higher molecular weight commercial polymers tending to have a higher proportion of crystallinity than lower molecular weight polymers.[78]

Work by Fuller[21] and Natta and Corradini[22] indicated that the crystalline regions have a syndiotactic structure, the repeat distance being 5.1 Å (0.51 nm). Kockott[85] regarded stereo-regular PVC as a copolymer of crystallisable syndiotactic parts and uncrystallisable isotactic parts, and calculated the melting temperature of 100% crystallised syndiotactic PVC to be 273°C, with an enthalpy of fusion of 2.7 kcal/mol. The levels of crystallinity in commercial polymers parallel the proportions of syndiotacticity, which have been calculated from NMR studies at about 0.56 for polymer of ISO Viscosity No. 165, and 0.54 for polymer of ISO Viscosity No. 70.[77, 79] It has been suggested that the crystallinity is in the form of cylindrical crystallites of about 4 nm diameter,[77] with a melting range of 165 to 260°C or above,[78, 82, 83] so that the extent to which the original crystalline structures are retained during processing depends on processing temperatures, though crystallinity may return on cooling. (Section 9.3.)

Bawn considered the level of crystallinity in PVC to be unexpectedly low, and attributed this to chemical non-conformities arising from some 'head-to-head' structures or chain-branching.[23] On the other hand later workers have found the observed level of crystallinity to be considerably in excess of what is anticipated.[77, 85, 86] On the assumption that syndiotactic sequences of ten to twelve units are required to permit crystallisation Juijn concluded that crystallinity of commercial polymers should not exceed a few tenths of one per cent.[80] Even if sequences as short as six units are sufficient, as suggested by Gray and Gilbert,[81] the estimated level of crystallisation is still less than that observed.[77]

Considerable work has been carried out in attempts to produce commercially stereoregular (syndiotactic) polyvinyl chloride, but a number of processing problems are still to be solved.[24] The usual Ziegler–Natta catalysts cannot be used because both monomer and polymer react with them, but radical initiators active considerably below 273 K (0°C) have been effective. More highly crystalline polyvinyl chloride has been made by polymerisation at temperatures down to about 203 K (−70°C) using for initiation y-radiation or very active substances such as alkyl boranes. White[25] prepared what appears to have been a completely syndiotactic polymer by X-ray irradiation of a vinyl chloride-urea canal complex, at 195 K (−78°C).

Most crystalline polyvinyl chloride produced commercially or on a development basis has been obtained by polymerisation at relatively low temperatures, e.g. 233 K (−40°C). The degree of crystallinity achieved in an end-product depends on the maximum obtainable inherent in the structure of the polymer and also on the processing conditions used for specimen preparation; the maximum observed is usually between 20 and 25 per cent, or occasionally somewhat higher where polymerisation temperature has been relatively very low (cf. Section 4.7.2.4).

Increase in crystallinity reduces solubility in normal solvents for the polymer, and, as would be expected, raises the softening temperatures. This naturally raises the temperatures required for thermoplastic processing. This, coupled with the affects that the highly crystalline polymers yield relatively brittle products, and are rather expensive, has precluded any appreciable commercial development to date.

While the gross structural features of polyvinyl chloride are reasonably well understood, a number of minor features about which rather less is known[5, 12, 16] may nevertheless have important effects on chemical properties, particularly stability. It is fairly certain that most if not all free radical initiators will leave residues, such as carbonyl, hydroperoxide, or aldehyde groups,[78] attached to the polymer chains which they initiated. For example, with acyl peroxide initiation end groups of the following type would be expected:

$$R.CO_2.CH_2.CHCl \sim\!\!\sim\!\!\sim$$

Points of branching, too, provide variations from the idealised structures:

$$CH_2.CCl.CH_2.CHCl \text{ ⟋⟍}$$
$$|$$
$$CH_2.CHCl \text{ ⟋⟍}$$

Perhaps the most important variation from ideality is, however, the occurrence of unsaturation. If, as most workers suggest, 'chain-transfer' is significant in vinyl chloride polymerisation, terminal unsaturation involving groups $CH_2{:}CH$— or $CH_2{:}CCl$— are bound to occur, and the possibility of unsaturation in a chain, i.e. —$CH{:}CH$—, due to loss of hydrogen chloride, cannot be completely precluded. Points of unsaturation constitute centres of relative instability and may have disproportionate effects on the stability of the whole polymer chains. That terminal groups can constitute points of weakness is indicated by the observation that thermal stability of a series of similar polymers decreases with molecular weight.[27, 28] Baum and Wartman[29] demonstrated that terminal unsaturation of polymer molecules constitutes the main source of instability and suggested that tertiary chlorine atoms have only a secondary effect.

Finally, there is also a possible introduction of extraneous groups by reaction to growing polymer chains with molecules of polymerisation additives, particularly chain-transfer agents used to control molecular weight, although other types of molecules would not be excluded.

Structure is clearly more complicated for copolymers than for homopolymers,[30, 31] and will depend on the relative reactivities of vinyl chloride and the comonomer or comonomers, and on the method of polymerisation. Truly random copolymers will only result from an isolated mixture of monomers when the reactivities are equal, but controlled addition of one or more of the monomers so as to maintain constant the proportions of unreacted monomers can be used to increase the uniformity of distribution of the monomer units in the polymer. It is extremely difficult to assess the success of such procedures, except in qualitative terms, by comparison of properties of copolymers produced by different techniques.

Spectrographic and chemical techniques available for investigations of PVC structure, and conclusion to be drawn from these, have been reviewed.[67]

4.3   MOLECULAR WEIGHT AND MOLECULAR WEIGHT DISTRIBUTION

As with other synthetic addition polymers, those of vinyl chloride consist of mixtures of polymeric chains of different lengths and molecular weights. The actual molecular weights and molecular weight distributions of vinyl chloride polymers have been studied,[9, 11–14, 31–36] but for everyday practical purposes it is usual to employ some measure of average molecular weight only.

Such studies have been numerous, and some have been applied in attempts to elucidate mechanisms of polymerisation.[13] The main methods used have been osomometry, solution, viscometry, and light-scattering, and attempts have been made to correlate the results obtained by the different methods.[9] The molecular weights and molecular weight distributions do not appear to vary with the degree of conversion of monomer to polymer, but there is some disagreement over the conclusions to be drawn from this observation. Since commercial emulsion and suspension polymers are usually made at high conversions (i.e. 75 to 95%) the point is only of academic interest, although it could be relevant where comparisons are being drawn between these polymers and those made by mass or bulk methods, where conversion may be somewhat lower. It has also been suggested [46] that the molecular weight distribution of polymers produced by the emulsion technique is wider than that of polymers produced by the suspension process. This suggestion has been disputed, and there is certainly much evidence to suggest that the difference is small.

The molecular weight in the absence of aided chain-transfer agent is almost entirely dependent on polymerisation temperature. It is thus clear that the molecular weight distribution of the final polymer is very dependent on the extent to which polymerisation occurs at temperatures different to that nominally set for the process. If all the components of a polymerisation recipe are mixed at room temperature, therefore, the time required to reach the nominal polymerisation temperature can be important in contributing a proportion of high molecular weight polymer. This problem can become particularly acute in scaling-up from smaller to larger polymerisation autoclaves.

The range of weight-average molecular weight encountered in commerical vinyl chloride polymers is 40 000–480 000, with a corresponding range of 20 000–91 500 for number-average molecular weight,[9, 47] the ratio of weight- to number-average molecular weight thus increasing from 2 at the low end to 5.25 at the high. The majority of commercial polymers come within the range 100 000–200 000 and 45 500–64 000 for weight-average and number-average molecular weights respectively.[47]

A number of different methods are used for the purposes of production control and general comparisons.[47] All the methods depend on measurement of viscosities in solution, but they differ in the solvent, concentration, and temperature of determination used. Most methods involve measurement of relative viscosities of solutions compared with pure solvents, but the results may be expressed as relative viscosities, specific viscosities, inherent viscosities,[48] viscosity numbers,[49] or *K* values.[34, 50, 51] This subject is discussed in more detail in Chapter 18, where a table correlating the commoner methods of expressing PVC molecular weights will be found.[47] It should be noted that the *K*-values quoted by many resin manufacturers are not necessarily on the same scale. A particular sample of resin can be allocated different *K*-values by different laboratories according to the method of test. It is therefore important to specify this method where numerical values are quoted. The widespread use of different methods of expressing the molecular weight characteristics can lead to confusion and even technical problems. As long ago as 1963, in an attempt to remove this confusion (at that time at least sixteen different methods were in use world-wide), because manufacturers were generally reluctant to abandon their own in favour of a single standard method, it was proposed that they should quote an ISO Viscosity Number alongside their own *K*-value or viscosity figures, and a table of equivalent values was published to enable this to be done quite simply.[47] (See Chapter 18.) This proposal has been adopted by a number of manufacturers, and some have adopted the ISO standard method (ISO1628–2) in place of their own. For present purposes it is proposed to use ISO Viscosity Numbers[49] where numerical values are necessary and the term *K*-value will generally only be used qualitatively.

Table 4.1 is intended merely to indicate the correlation between molecular weight and the most common methods of characterisation, for the purpose of subsequent discussion.

Table 4.1 MOLECULAR WEIGHT CHARACTERISATION OF PVC

Average molecular weight		ISO 1628–2	K-value			ASTM D 1243–58T		JIS K6721 PD
Weight	Number	Viscosity number	0.5% in Ethylene dichloride	0.5% in Cyclo-hexanone	0.4% in Nitro-benzene	Method A Inherent viscosity	Method B Specific viscosity	
54 000	26 000	57	45	48	52	0.49	0.19	380
70 000	36 000	70	50	54	57.5	0.62	0.24	560
100 000	45 500	87	55	60	62.5	0.75	0.3	760
140 000	55 000	105	60	65	68	0.88	0.36	975
200 000	64 000	125	65	70	70.5	1.01	0.41	1230
260 000	73 000	145	70	76	78	1.13	0.47	1570
340 000	82 000	165	75	78.5		1.26	0.54	1980

4.4 PARTICLE SIZE AND SHAPE

On handling a number of different commercial PVC resins the most obvious difference immediately noticed is in particle sizes. Some polymers are fine and dusty, spreading the evidence of their presence on every available surface nearby, while others are relatively coarse and handle more like sand. Paste-making, microsuspension, and emulsion polymers are generally of the former type, as are some of the older suspension resins, whereas suspension resins are of the latter type.

Quite apart from handling behaviour, the size, size distribution, porosity, and shape of PVC resin particles have important effects on their processing behaviour.

Particle size and size distributions are usually determined by sieve analysis using a stack of standard sieves in the range 12–300 BS mesh sizes (Chapter 18). This method is not suitable for very fine particle paste resins, and laborious microscope counts are used whenever it is necessary to charactersise such resins. More refined methods, such as the Malvern laser diffraction particle sizer, are occasionally used for special purposes. Some typical sieve analyses of commercial resins are set out in Table 4.2.

Particle shape and structure are conveniently studied by optimal microscopy. Vertical illumination can reveal surface features of particles, while transmitted light is useful in revealing not only external shape but also internal structure. During the 1970s more sophisti-

Table 4.2 PARTICLE SIZE DISTRIBUTIONS OF SOME TYPICAL COMMERCIAL POLYMERS

	Weight % retained on						Weight % passing
BS Mesh Nos.	36	52	72	100	150	200	200
Nominal size in microns	420	295	210	150	105	76	76
EP-type suspension polymer I	0	1	8	30	40	16	5
EP-type suspension polymer II	0	0	0	20	60	15	5
EP-type suspension polymer III	0	0	0	2	28	30	40
Early-type suspension polymer	0	0	1	1	1	7	90
Spray-dried emulsion polymer	0	0	1	4	10	15	70
Coagulated emulsion polymer	0	0	0	0	0	0	100
Non-porous suspension polymer	0	0	2	28	52	15	3

cated techniques, especially electron microscopy, began to be used, and the fine structure of PVC resin particles is now understood in considerable detail.[60, 61, 63]

Examination under a microscope using transmitted light (60–300 diameters magnification is convenient, according to the particle size and the amount of detail required), with the sample preferably suspended in plasticiser or a non-volatile non-solvent, reveals the fact that a wide variety of particle types is represented by the different resins available. Microphotographs of some typical samples of commercial resins are shown in Plates 4.1–4.4. The particles of paste-producing resins are usually too small for optical microscopy to reveal details of structure, and electron microscopy is necessary if these are required.

The sizes of primary particles of polymer suspended in a latex produced by emulsion polymerisation depend on the ingredients of the polymerisation recipe, particularly the chemical nature of the emulsifying agent and the pH of the aqueous phase, the agitation of the system, and the temperature cycle, but are typically in the region of 0.01–1 μm in diameter. The particle size of dried polymer depends, however, not only on the size of the primary particles in the latex but also on the method and conditions of isolation and drying. Thus, the particles of a spray-dried emulsion polymer consist of agglomerates of primary particles and may be as large as 80 μm or even more in diameter. Spray-dried emulsion polymer particles are basically 'cenospherical' in shape – that is to say, each particle consists of a hollow sphere with a hole in the surface. These particles are usually fragile, however, and a proportion may become broken during transportation. As will be seen from the table of sizes and size distributions, most of the particles in a commercial spray-dried emulsion polymer are usually less than 76 μm in diameter, but a few are frequently as large as 150–200 μm; whereas emulsion polymers isolated by coagulation usually have particles less than 76 μm in diameter.

The shapes and size distributions of paste-making resins vary with the types of resins, particularly the range of paste viscosities which they are designed to yield[53, 54] (Chapter 14). A polymer yielding pastes of relatively low viscosity for a given plasticiser concentration is likely to consist of a number of species of different particle sizes, for example averaging about 0.06, 0.25, and 1.1 μm. A 'high-viscosity' paste polymer will usually be found to comprise irregular particles of relatively large size, e.g. of about 1–4 μm. The pastes produced from such a polymer will usually be dilatant, i.e. their viscosities will increase with increase in shear rate. Pastes whose viscosities decrease with increasing shear can be obtained by using polymers comprising particles of regular and uniform shape and size, the latter usually being in the region of 0.4 μm.

The suspension resins originally developed were of relatively fine particles size, mostly below 100 μm in diameter, similar to spray-dried emulsion polymers, and usually consisted of a mixture of porous and non-porous particles. Most modern resins, on the other hand, are relatively coarse, a high proportion frequently being larger than 150 μm in diameter, and often of irregular shape. These tend to fall into two different types. One of these, originally known as 'easy-processing' or 'EP' resins, has particles with a porous structure. These resins are particularly valuable because the porosity allows ready absorption of plasticiser to form 'plasticised dry blends'. The other type comprises particles of low or zero porosity, specifically intended for unplasticised compositions. Shape and size distribution vary considerably from one grade of resin to another, as exemplified by the three sieve analyses tabulated earlier, but the major proportion is usually in the 75–150 μm range. Other types of suspension polymer also exist, such as one in which the particles are non-porous and substantially spherical. The particle sizes and size distributions of commercial resins of this type also vary widely from one grade to another, but the average diameter is usually above 100 μm.

At an early stage in the development of 'easy-processing' or 'EP' polymers it was recognised that the particles or granules are agglomerates of very much smaller particles.[61] It is now realised that the particle or grain structure is much more complex,[60, 61, 63] and at least six morphological species have been identified. Although the nomenclature for these seems now to be reasonably established,[64] it will not get everybody's total approval because the terms used do not immediately call to mind the nature of each feature; not that it is an easy matter to think of a more descriptive scheme of terms. The proposed terminology is used for the sake of consistency. The reason for the complexity is that during the course of suspen-

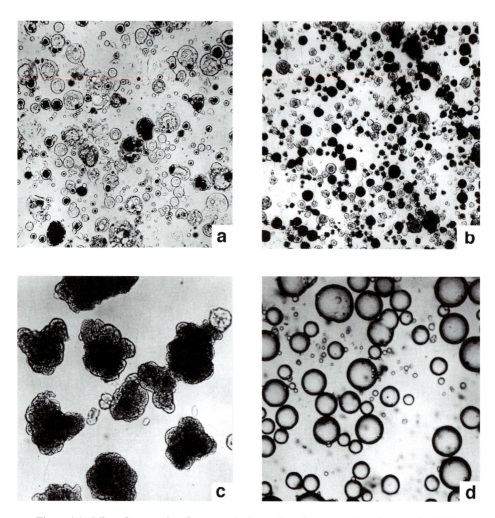

Figure 4.1. Microphotographs of some typical samples of commercial resin taken in DIOP.
(a) Spray-dried emulsion polymer; (b) original type of suspension polymer; (c) 'easy-processing' (EP) suspension polymer; (d) non-porous type of suspension polymer (approx. ×85)

sion polymerisation, changes occur concurrently on the macroscopic scale and on a microscopic scale.

At the start of suspension polymerisation the agitation of the autoclave stirrer combined with the effect of the protective colloid produces a suspension of monomer droplets in the aqueous phase, the droplet size ranging from 5 to 150 μm. Very early in the process the droplets are found to be surrounded by a membrane formed from a graft copolymer of vinyl chloride and the protective colloid. With a polyvinyl alcohol/acetate colloid the membrane consists of a polyvinyl acetate/polyvinyl chloride graft. The droplets now coalesce to a degree dependent on the particular recipe used, up to a conversion of about 30 per cent, by which time the membrane is strong enough to prevent further coalescence. As polymerisation proceeds monomer of relative density 0.85 is converted to polymer of relative density 1.41, corresponding to a contraction of nearly 40 per cent. If the contraction is taken up uniformly the outside dimensions are progressively reduced and the resultant polymer particles or 'grains' are almost non-porous, as is the case with some commercial polymers designed for unplasticised PVC processing. However, with so-called 'easy-processing; or 'EP' resins this reduction in external contraction is accommodated by the formation of pores or voids within the particles, and by disruption of the regular droplet shape.

At the same time physical changes at a much smaller scale are occurring within each droplet. In the earliest stages of polymerisation polymer coagulates to form aggregates of about 50 molecules, and by the time conversion has reached 2 per cent, particles of 0.1 to 0.2 μm comprising about 1000 of these aggregates have been formed. In Geil's terminology,[60, 61, 64] these two structural features are known as 'micro-domains' and 'domains', respectively. The domains grow as polymerisation proceeds, and by 10 per cent conversion have agglomerated to form 'primary particles' of about 0.7 μm in diameter, which then coalesce further to form 'agglomerates', the spatial packing of which appears to determine the degree of porosity in the final product. Subsequent polymerisation results in increase in size of these agglomerates to between 2 and 10 μm. The precise distribution and sizes of these structural features depend on the polymerisation recipe and conditions. In general, porosity in the final 'grains' increases with increasing stirrer speed and decreases with increase in temperature.[61]

Bulk or mass polymers appear to have similar morphological features to those of suspension polymers.

For convenience Allsopp[61] and Geil's[64] scheme of nomenclature is summarized in Table 4.3.

Table 4.3 MORPHOLOGICAL FEATURES OF PVC RESINS

Term	Approximate size (μm)		
	Range	Average	
Micro-domain	0.01–0.02	0.02	Aggregate of ~50 polymer molecules
Domain	0.1–0.2	0.2	Primary particule nucleus containing ~1000 micro-domains
Primary particle	0.6–0.8	0.7	Grow from domains
Agglomerate	2–10	5	Agglomerate of primary particles
Sub-grain	10–150	40	Polymerised monomer droplet
Grain	50–250	130	The visible polymer powder

These features are of more than academic interest in determining the polymer properties. They may retain their identity to varying degrees in subsequent processing, affecting processing and gelation behaviour, and the extent of their retention can sometimes be related to the physical proprieties of the end-product. Some workers[63] prefer to consider the particle sub-structure in three stages, namely Stage I particles of approximately 0.01 μm diameter, corresponding to 'micro-domains'; Stage II particles of approximately 0.1 to 2 μm diameter; and Stage II particles corresponding to 'grains'. This subject is discussed more fully in Chapter 9.

The size and size distribution, shape, and nature of the particles of a resin are reflected in other important properties, such as surface area, which largely determines capacity for plasticiser absorption, and bulk and packing densities (which will affect packaging a well as batch sizes and throughputs in some forms of processing equipment), and powder flow. Bulk and packing densities vary widely from one grade of polymer to another. The bulk density of an emulsion polymer can be as low as 250 g/l with corresponding packing density a little over 300 g/l, but the values are more commonly in the regions of 400–500 and 500–700, respectively. Porous particle suspension polymers tend to have lower densities than the older types of suspension polymer, and, indeed, there appears to be an approximate correlation with plasticiser absorptive capacity, which tends to increase with decrease in bulk and packing densities. A non-porous suspension polymer is likely to have a bulk density in the region of 500–600 g/l with a corresponding packing density between 600 and 750 g/l, whereas the values for porous types of polymer are more usually 400–500 and 480–580, respectively.

The particle type and size distribution of pvc resins have profound effects on processing behaviour. Thus, at equal polymer/plasticiser concentrations, a polymer having porous and relatively large particles will yield dryer, less sticky mixtures with plasticisers than a polymer with small, non-porous particles. Moreover, heating the former while mixing with up to about 70 per cent of its weight of plasticisers, results in absorption of the plasticiser to form a dry 'plasticised powder', without gelation or agglomeration, provided that the temperature is not allowed to rise much above 110°C. Other types of polymer gel to weak cheese-like masses if treated similarly. Furthermore, when compounded in internal mixers of the Banbury type (see Chapter 10) plasticised premixes based on porous particle polymers tend to gel and also to attain substantially complete dispersion of the particles more rapidly than those based on other types. On the other hand, in the extrusion of unplasticised dry blends (see Chapter 11) entrapment of air in the pores of porous particle polymers can lead to porosity in extrudates, and in such cases polymers with non-porous particles and high packing densities are preferable.

4.5   PURITY

Commercial polymers differ not only in particle size and structure, and molecular weight, but also in other relatively minor respects. Thus the polymers will always contain small proportions of residues from the ancillary ingredients of the polymerisation recipes used. Some of the polymer chains will contain chemical groupings from the polymerisation initiators or catalysts, and the nature of these groupings will clearly depend on the particular polymerisation initiation systems employed. Suspension polymers contain traces of residual suspension agents used to aid maintenance of dispersion during polymerisation, and emulsion polymers usually contain somewhat larger proportions of residual emulsifying agents. These residues affect the colour and clarity of articles produced from the polymers and also affect heat stability. In some cases it is possible that processing behaviour is also affected by residues of emulsifying agent.

The presence of volatiles, especially residual monomer, is obviously undesirable, and production procedures are designed to reduce volatile content to a minimum. In particular steps have been taken to remove vinyl chloride monomer as far as possible, because of its carcinogenic effect (see Chapter 17). In general the volatile content of homopolymers is less than 0.3 per cent, but values for vinyl acetate copolymers are slightly higher, presumably due to the relative inactivity and lower volatility of the acetate.

4.6   COPOLYMERS

The other factor in the constitution of polymers which has a profound effect on properties is the presence or absence of comonomers. The most common of these in commercial use are vinyl acetate and vinylidene chloride, although other monomers such as maleic anhydride and maleic esters, vinyl alcohol (from hydrolysis of vinyl acetate in the polymer), fumaric acid and esters, styrene, acrylonitrile, methyl and ethyl acrylates, and vinyl ethers, are also employed for special effects. The proportions of comonomers used in vinyl chloride copolymers varies over practically the whole theoretically possible range, but in the resins here concerned, the range is generally from 2 to 20 per cent.

Copolymers of vinyl chloride and propylene were introduced in the USA around 1970, but they have found only a small market. Like other copolymers, they are claimed to process at lower temperatures than homopolymers and also to lubricate and increase stability. They do, however, appear to produce less rigid products than do homopolymers.

4.7   PHYSICAL PROPERTIES

As already indicated, the processing behaviour of a polymer is very dependant on its particle size and particle structure, and these factors are discussed more fully in Chapters 9, 10, and 14. The molecular weight of a polymer will also affect processing behaviour and, in addition, will determine the physical properties of any given composition.

Only a small proportion of vinyl chloride polymer is processed to finished product without the incorporation of additives such as stabilisers (Chapter 5), lubricants (Chapter 7), and plasticisers (Chapter 6), and these additives affect the processing behaviour of a PVC composition and also the physical properties of the processed composition. However, with any particular formulation the physical properties of PVC are dependent on the molecular weight of the resin, just as is found with other polymeric materials. Thus properties like softening temperature, tensile strength, modulus, elongation at break, tear strength of film, hardness, and impact strength of unplasticised compositions all increase in value with increase in molecular weight. In short, strength increases with molecular weight. Realisation of this effect is sometimes obscured because, since softening temperature increases with molecular weight, temperatures required for proper processing are also raised and the full potential of physical properties is not obtained if a composition is not submitted to adequate processing conditions.

It will be seen that it is difficult enough to consider the physical properties of polyvinyl chloride, let alone determine them, since so much depends on other ingredients and the way in which test specimens are prepared. Even unplasticised PVC usually contains between 2 and 10 per cent of additives, and it is questionable whether discussion of physical properties of the unmodified polymer is meaningful. Some further discussion of properties appears in the considerations of applications (Chapter 16).

4.7.1   DENSITY

The density of polyvinyl chloride is variously quoted at between 1.39 and 1.43 g/cm^3 , and a value of 1.41 can probably be taken to be as valid as any. Introduction of vinyl acetate into the polymer reduces density. At concentrations of 10 and 16 per cent of vinyl acetate, density has fallen to values of about 1.38 and 1.36 respectively. Vinylidene chloride, on the other hand, increases density, but the small portions (2–3 per cent) used in most PVC applications do not lead to differences measurable by routine methods.

In a shaped product the density depends not only on that of the polymer, but also on the densities and concentrations of the various additives. Thus the very dense lead stabilisers and fillers will naturally tend to increase the density of the overall composition, whereas plasticisers of low density will tend to decrease it. This is a very important factor to consider in formulating PVC compositions, since in many applications overall cost depends on cost per unit volume.

4.7.2.   THERMAL PROPERTIES

4.7.2.1   *Specific heat*
Values ranging from 0.83 to 1.15 J/g/degC are to be found in the literature, but most of these are based on measurements on material compounded from polymer with other materials, and a value of 1.0 is probably sufficiently accurate for most practical purposes.

4.7.2.2   *Thermal expansion*
A range of values between 5 and 7×10^{-5} m/m/K has been published, and again, an intermediate value of 6×10^{-5} is probably adequate for practical calculations.

4.7.2.3   *Thermal conductivity*
This appears to be between values of 1449 and 1674×10^{-4} W/m/degC.

4.7.2.4   *Temperature effects*
As with other amorphous thermoplastic polymers, the softening/melting behaviour of polyvinyl chloride is much more ill-defined than that of crystalline polymers like isotactic polypropylene, nylon 66 or even polyethylene, and there is no clearly defined melting point. Provided decomposition is inhibited or kept to a low level, the polymer slowly softens as the temperature rises, and is usually noticeably soft by the time 150°C or thereabouts is reached. If decomposition did not take place it would undoubtedly be possible to continue to raise the temperature to such a point that the polymer became a viscous liquid, but, even

so, the visual transition from solid to liquid would be diffuse. The second order transition temperature of polyvinyl chloride has been quoted[20, 55] as 348 K (75°C). Values for softening temperature depend on the composition and the method of test. Thus Vicat softening points of 363–368 K (90–95°C), BS 2782 softening points of 350–351 K (77–78°C), and ASTM heat distortion temperatures of 343–344 K (70–71°C) (264 psi) (18.2 kN/m²), are to be found in the published literature. Incorporation of plasticiser lowers these values dramatically and it is rarely considered worth while quoting softening temperatures of plasticised PVC. The values probably all imply inadequate performance due to distortion at temperatures well below 373 K (100°C), e.g. about 333 K (60°C), but at least one application ('high temperature' electrical cable), where there is support for the polymeric material, involves the use of plasticised PVC running at a temperature above this (105°C).

A heat of fusion of 10.5 cal/g has been quoted.[55] The precise significance of this figure is not clear, since, because crystalline regions account for only a small proportion of the whole, the transition of solid PVC to melt is diffuse, and the temperature range over which 'fusion' is deemed to occur needs to be defined. Thermal analysis indicates a melting range from about 100 to 250°C.

At the other end of the scale, the polymer becomes stiffer and more brittle as the temperature falls, so that brittle failure under impact or shock may limit the use of PVC in low temperature applications. Thus, unplasticised PVC is generally not suitable for applications where impact shock is likely at temperatures below about 243 K (−30°C). The presence of plasticiser at concentrations above about 25 per cent considerably lowers the onset temperature of brittleness (Chapter 6).

Between the extremes of relatively high and low temperatures, most mechanical properties are temperature-dependent, but this is more appropriately considered in discussion of mechanical properties specifically.

So-called crystalline PVC rarely attains a level of crystallinity above 25 per cent, though predominantly syndiotactic polymer with 84 per cent crystallinity can be produced by the use of special initiator systems at −75°C.[68-70] Even at 25 per cent the full level of crystallinity is only reached by processing at relatively high temperatures, e.g. 200°C. In unplasticised compositions the crystalline polymer yields Vicat softening points some 20°C higher than those obtained with the usual suspension polymers, and rods or rubes of the material remain rigid up to about 393–403 K (120–130°C). In plasticised compositions the plasticiser is the overriding factor, and the Vicat softening points are similar.

Somewhat surprisingly, crystalline PVC also has low-temperature flexibility some 20°C lower than that of the usual suspension polymers, but products based on unplasticised crystalline PVC are generally brittle at normal temperatures.

Vinyl chloride copolymers have lower softening points than the homopolymer of the same molecular weight, and under equivalent processing conditions yield melts of lower viscosity. Thus for every 1 per cent of vinyl acetate introduced the softening temperature is likely to fall about 0.5°C.

4.7.3   OPTICAL PROPERTIES

Polyvinyl chloride appears to be essentially transparent and colourless, but additives have to be selected with care if high clarity and low colours are to be retained, though light transmissions of over 80 per cent through a thickness of 2 mm are not particularly difficult to achieve.

Refractive index n_D^{20} is quoted as 1.537, and 16 per cent of vinyl acetate appears to lower this to 1.53.

In applications where transparency is particularly important, and where any inadequately dispersed particles would be objectionable, such as in medical products, bottles, and windscreens, it may be desirable to test and select polymer specially to ensure that particles that are difficult to disperse are absent.

Translucent and opaque effects can be obtained by the incorporation of appropriate additives.

4.7.4   ELECTRICAL PROPERTIES

Polyvinyl chloride itself is a typical organic polymer insulator, and although the insulation properties can be largely retained in the presence of additives, great care is required in selection, if the maximum retention is to be achieved.

4.7.4.1   *Volume and surface resistivities*

For a shaped product, these are very dependent on the polymer selected, as well as on the nature and concentrations of any additives. Unmodified bulk-polymerised and suspension polymers can yield volume resistivities greater than 10^{15} Ω m at 296 K (23°C). Volume resistivity is markedly temperature-dependent; for example, tests on the same samples at 296 K (23°C) would be expected to yield values of about $\frac{2}{3}$ of the values observed at 293 K (20°C).

Surface resistivity can be maintained at levels of above 10^{14} Ω at 293 K (20°C).

Introduction of other ingredients usually lowers both volume and surface resistivities, so that plasticised compositions generally yield lower values, roughly in proportion to the amount of plasticiser involved, but very dependent on its nature and purity. Indeed, the small amount of residues present in spray-dried emulsion polymers is sufficient to reduce volume and surface resistivities dramatically. Thus 1 per cent of an emulsion polymer in a specific plasticised composition based on a suspension polymer, reduced the volume resistivity of the latter from a value of 2×10^{12} to about 1×10^{11} Ω m.[56]

4.7.4.2   *Permittivity and power factor*

These are both dependent on the frequency of the applied test voltage, as well as on the temperature and the formulation of the particular material under examination.

Unplasticised PVC generally has a permittivity of about 3 at a frequency of 1 kHz and 296 K (23°C), although values as low as 2.2 have been quoted.[57] At normal temperatures, permittivity of unplasticised PVC is not very dependent on frequency, but marked frequency dependence develops at temperatures of 353 K (80°C) and above.[57] Introduction of plasticiser increases permittivity, and very soft PVC compositions may reach values as high as 10 and over.

The power factor is high relative to other dielectric polymers like polyethylene and polystyrene, and although lower values have been quoted, unplasticised PVC usually exhibits a value in the region of 0.02. This rises rapidly with temperature, values between 0.1 and 0.18

being obtained at 378 K (105°C). Power factor also rises with the introduction of plasticiser, reaching values in the region of 0.1–0.12 at plasticiser contents around 30 per cent.

4.7.4.3   *Dialectric strength and other properties*

For unplasticised PVC a dielectric strength of 1300 V/mil (51.18 kV/mm) has been quoted,[55] but values of 500 (19.69) are more commonly encountered in commercial materials. Introduction of plasticiser reduces dielectric strength slightly. Both unplasticised and plasticised PVC are regarded as track- and arc-resistant.[57]

4.7.5   MECHANICAL PROPERTIES

It is almost meaningless to talk about mechanical properties of the vinyl chloride polymers or resins themselves, but the mechanical properties of a finished PVC product are very dependent on the resin as well as on the other ingredients from which it was derived. The most important resin parameters from this point of view are molecular weight, molecular weight distribution, and comonomer constitution. In commercial practice the second of these is dependent mainly on the polymerisation temperature cycle, and this is largely determined by heating arrangements and the heat transfer characteristics of the polymerisation vessels. Thus little can usually be done deliberately to vary molecular weight distribution, and it is probable that variations between different commercial resins of similar average molecular weights are of no significance.

The effects of varying the average molecular weight of the polymer in a PVC composition are in general similar to those encountered with other thermoplastics. Precise values of mechanical properties will depend on the nature and amounts of other ingredients, but as an indication of the magnitude of effects likely to be encountered the following typical cases are quoted.

A composition containing 50 phr (parts per hundred of resin) of DOP was found to yield a tensile strength of 21 374 kN/m² (~ 21.4 MPa) and an elongation at break of 290 per cent, when based on a polymer having an ISO Viscosity No. of 125; sheeting produced from the composition had a tear strength of 6250 kg/m. When based on a polymer with an ISO Viscosity No. of 90, the values of these properties fell to 15 168 kN/m² (~ 15.2 MPa), 215 per cent, and 3750 kg/m respectively. Similarly, a composition containing 30 phr of DOP developed a tensile strength of 25 511 kN/m² (25.5 MPa) and an elongation at break of 235 per cent when based on a polymer of ISO Viscosity No. 125, the corresponding values with a polymer of ISO Viscosity No. 90, being 20 684 kN/m² (20.7 MPa), and 210 per cent, respectively. It should be noted that reduction in molecular weight of the polymer also causes slight increases in softness and flexibility. If these changes are compensated by appropriate decrease in plasticiser content (e.g. about 3–8 per cent), the fall in tensile strength can be considerably reduced but the elongation at break falls still further.

While there is no doubt that the intrinsic values of all mechanical properties increases with increase in molecular weight as is usual with thermoplastic polymers there is sometimes confusion when it comes to correlating these dependent variables. This is partly because ingredients of a composition other than the polymer itself often have a major effect, but also because the observed properties are very dependent on how the test-piece has been made. Indeed the properties of a finished product are very dependent on the processing it has received, and especially on the level of gelation it has attained.[71, 72] (Section 9.3.2.) In a given formulation increasing molecular weight leads to increase in melt viscosity at the same processing temperature or requires increase in processing temperature if melt viscosity is to be maintained constant. The latter is often precluded or limited by the onset of excessive decomposition, so that in practice the full potential mechanical properties of a composition are undeveloped owing to lack of homogenisation. This factor has often led to the at first sight surprising observation of a particular polymer yielding mechanical properties superior to those of a polymer of higher molecular weight.

Because of the dependence of mechanical properties on additives to the resin itself, fuller discussion is deferred to those chapters dealing with the additives in question. At this point it need only be stated that the tensile strength of unplasticisd PVC will usually be found to lie between 49 600 and 69 000 kN/m^2 ($\sim$ 50 and 69 MPa), and tensile modulus between 2.7 $\times$ 10^6 and 3.5 $\times$ 10^6 kN/m^2 (2.7 and 3.5 GPa). Both these properties diminish with the introduction of plasticiser, the former to a value of about 6550 kN/m^2 ($\sim$ 6.6 MPa) and the latter to about 2.1$\times$ 10^6 kN/m^2 (2.1 GPa) at a BS Softness of 145. Over the same range, elongation at break increases from a few per cent to over 400 per cent.

In the unplasticised form, therefore, PVC is a fairly stiff hard material, but the introduction of plasticiser yields a very wide range of flexibility and softness. Copolymerisation also generally results in reduced stiffness, though the effect is never as marked as that produced by plasticiser.

4.7.6 PHYSICAL RESPONSE TO OTHER MEDIA

PVC itself is essentially a relatively inert material, although it is softened, dissolved, or degraded by some other media, these effects usually increasing with the introduction of plasticiser. The behaviour of unplasticised PVC has been reported in some detail in trade literature and elsewhere.[57, 58] Apart from a small absorption, usually well under 1 per cent, unplasticised PVC is unaffected by liquid water or dilute aqueous solutions. The water vapour permeability of unplasticised PVC is low, usually around 250 g/ μm/24h/m^2 at 298 K (25°C) and 75 per cent RH, and 635 g/ μm/24 h/m^2 at 311 K (38°C) and 90 per cent RH;[57] but these values increase with the introduction of plasticiser to reach somewhere around 4570 g/ μm/24 h/m^2 at 311 K (38°C) and 90 per cent RH. These values may be compared with a value of 480 g/ μm/25h/m^2 at 311 K (38°C) and 90 per cent RH for a polyethylene of density 0.92 g/cm^3 and MFI 7.

Unplasticised PVC is less permeable to nitrogen, oxygen, and carbon dioxide than polyethylene, and the same is also generally true for plasticised PVC, although the presence of plasticiser increases permeability. Table 4.4 gives typical values for measurements at 293 K (20°C), the units being cm^3/m^{2}cm Hg mm 24h.[57] These values should be divided by a factor of 1.333 to convert them to units of cm^3/m^{2}kN/m^{2}mm 24 h.

Table 4.4

	Unplasticised PVC	Polyethylene (density 0.92 g/ml)
Carbon dioxide	0.5	8.64
Hydrogen	1.5	6.2
Oxygen	0.1	2.75
Air	1.0	

Unplasticised PVC is affected by only a relatively small number of organic liquids. It is essentially unaffected by alcohols, many organic acids, and aliphatic hydrocarbons; but some acids and anhydrides, esters, aromatic compounds, chloro-compounds, and ketones are absorbed, and may even dissolve the polymer, although the solubility is never very high by normal solvent solution standards. The most effective solvents are ethylene dichloride, nitrobenzene, ketones (such as cyclohexanone), and tetrahydrofuran; and these solvents are used for solution viscometry. Absorption of solvents by PVC leads to a plasticising effect with resultant swelling, lowering of softening point, and increase in flexibility.

In general, plasticisers will be extracted from PVC by solvents, though this can often be reduced to a low level by selecting plasticisers appropriate to the solvent involved (Chapter 6).

Copolymerisation of vinyl chloride with other monomers also markedly affects resistance to solvents. Thus, low molecular weight copolymers with about 10–15 per cent acetate or 30–50 per cent vinylidene chloride are sufficiently soluble in a number of solvents and solvent mixtures to be suitable for lacquers (Chapter 15).

4.8   CHEMICAL PROPERTIES

Heating PVC at temperatures above 373 K (100°C) leads to decomposition at rates which increase with temperature. The immediately observed effects are the development of colour (usually yellow, orange, brown, and eventually black), and the evolution of hydrogen chloride. Ultimately a charred mass is left, which can be consumed by strong ignition. As estimated by rate of colour development, which is commonly the most important aspect, emulsion polymers usually decompose more rapidly than suspension polymers, although there is considerable variation amongst the various grades of the latter. This is due to the level of impurities present in the different polymers. Bulk or mass produced polymers are the most heat-stable, though only marginally better than the best of suspension polymers. There is some doubt as to how much of the decomposition is intrinsic to the basic structural unit of polyvinyl chloride, $—(CH_2.CHCl)_n—$, and how much may be due to irregularities in the polymer structure, such as unsaturated groupings. The subject of decomposition is discussed more fully in the next chapter.

Oxygen and ozone do not appear to attack PVC at room temperature, but severe oxidising systems do cause some breakdown. Thus, concentrated solutions of oxidising agents such as potassium permanganate attack PVC superficially. Hydrogen peroxide, however, is ineffective at any concentration.

Chlorine has little effect on PVC, but bromine and fluorine attack it even at room temperature.

With the exception of sulphuric acid above 90 per cent and nitric acid above 50 per cent, PVC is resistant to acids and aqueous alkalis up to 333 K (60°C), but above this temperature the stronger acids attack the polymer.

PVC can be reduced by strong reducing agents like lithium aluminium hydride, and the chlorine can be removed by refluxing in dioxan solution with zinc. These reactions were used in elucidation of the head-to-tail structure of polyvinyl chloride.[1–5, 7, 8] Presumably the chlorine reacts as in other chlorine-containing aliphatic compounds, but little if any investigation of this appears to have been carried out.

Copolymerisation introduces the chemical behaviour characteristic of the structural groups contributed by the comonomer. Both vinyl acetate and vinylidene chloride reduce heat stability as compared with homopolymer of similar molecular weight.

4.9   CRYSTALLINE POLYMERS

As indicated above (Sections 4.2 and 4.7.2.4), so-called crystalline PVC rarely attains a level of crystallinity above 25 per cent, and it is not to be expected that this amount would have a very large effect on properties. It appears that even this proportion of crystallinity can only be obtained with polymers of appropriate syndiotacity by processing at relatively high temperatures, e.g. 473 K (200°C) or above, depending on the molecular weight. Some evidence exists to suggest that the polymer is slightly more crystalline in plasticised than in unplasticised compositions, as observed with 'normal' pvc resins. Thus increase in crystallinity in PVC increase brittleness, which may be one of the reasons why small additions of plasticiser yield products more brittle than unplasticised PVC.

Tensile strengths of unplasticised PVC with 85 per cent crystallinity are lower than those of corresponding 'normal' PVC, values of about 41 400 kN/m^2 (41.4 MPa) being typical, and the more crystalline material also fails in a brittle manner without yielding.

The main advantage of this level of crystallinity lies in increased softening temperatures, which are at least 20°C higher than those of corresponding unplasticised PVC. Even in plasticised compositions the more crystalline PVC yields weaker products than normal polymer, though the onset of brittleness with fall in temperature is decreased by some 20°C.

As might also be expected, increase in crystallinity reduces solubility; so much so that molecular weight determinations by solution viscometry can be difficult.

One other effect of increased crystallinity in polyvinyl chloride is to yield in a given plasticised composition a harder and stiffer product than that obtained with normal polymer.

The structure and properties of 25 per cent crystalline polymer have been reviewed.[20]

4.10   SELECTING POLYMERS

Selection of a polymer for a given application depends on a number of factors, most of which are discussed in detail in other chapters. It is sufficient here to note that the properties required in the finished product will determine within fairly narrow limits the permissible formulations in terms of proportions of polymer, plasticiser, and filler. The shape of the finished product will limit the forming technique to be employed, and the combination of formulation and shape, together with processing equipment and process, will impose limitations on the permissible processing behaviours of the selected composition. This may limit the choice of polymer in a given formulation.

From the point of view of high strength and related properties, the higher the molecular weight of the polymer employed the better. However, ease of processing, in terms of processing temperatures, melt viscosities, and power requirements, decreases with increase in molecular weight, and in practice a compromise has usually to be accepted. The majority of commercial polymers offered for general purpose plasticised PVC applications have viscosity numbers in the region of 125. Polymers with viscosity numbers as high as 145, 165, or even 185, have been produced, but these are not yet generally accepted for anything but speciality applications where the greatest possible strength is required and where the composition is such that the inclusion of high molecular weight resin does not render processing unduly difficult. The physical properties obtainable with polymers of viscosity numbers around 125 are adequate for most applications. Nowadays, too, suspension polymers are usually preferred to emulsion polymers, because the former are cheaper and generally yield products of better colour and clarity as a result of their relative freedom from polymerisation recipe residues and their superior heat stability.

In selecting a resin for a given application, therefore, a suspension polymer having a viscosity number of approximately 105–125 usually constitutes the first choice. Rejection of polymers in this category only arises when processing is impossible or difficult in the selected equipment, or where some property required in the finished product necessitates the use of an alternative resin. Thus difficulties can arise where the processing temperatures with the equipment available are insufficient to achieve adequate fluxing of the selected composition. This was common in the early days of calendering of PVC sheeting when converted rubber calenders were pressed into service with heating facilities inadequate for PVC. Difficulties can also arise where the processing temperatures required to achieve adequate fluxing are so high as to initiate excessive decomposition. With hard or unplasticised compositions, particularly where thin sections are being formed, the melt viscosity under processing conditions may be so high that excessive heat is generated by frictional work, and control of temperature is difficult. In all these cases alternative possibilities usually exist, namely selection of an emulsion polymer, a copolymer, or a homopolymer or copolymer of lower molecular weight. The best choice is not always obvious, and in fact more or less equivalent alternatives can often be found. Nowadays it is common practice to avoid these problems by incorporation of processing aids. (Section 7.5.)

The difficulties just discussed are at their greatest with unplasticised PVC compositions. It was for this reason that until the 1960s, it was common practice for unplasticised PVC to be processed either from emulsion polymers of viscosity numbers in the region of 85, or from copolymers containing appreciable proportions (e.g. 10 per cent) of comonomer such as vinyl acetate. Improvements in polymers and equipment, as well as in stabilisers and lubricants, have permitted the use of suspension polymers of similar and even higher molecular weight, and the use of suspension polymers having viscosity numbers as high as 105 or even 125 is now quite common. As an indication of the way molecular weight is matched to processing requirements, a viscosity number of about 105 might be selected for the extrusion of unplasticised PVC pipe or small sections, viscosity numbers of about 85 to 95 for the production of rigid sheet, and of about 70 for injection moulding of rigid objects. Where an easier flow is required as in the moulding of gramophone records, or in the calendering of vinyl floor tiles, it is usual to reduce the viscosity number still further, for example to about 70, and to use a copolymer containing between about 10 per cent and 20 per cent of comonomer. The same procedure is followed in the production of sheet for vacuum-forming, although it is usually not necessary to reduce the viscosity number below about 70.

As already indicated, vinyl acetate is copolymerised with vinyl chloride usually for the purposes of reducing softening temperature or increasing solubility. The effect is often referred to as 'internal plasticisation'; but this is something of a misnomer, as the lowering of softening temperatures can be achieved with much less consequent increase in flexibility than is effected by the use of conventional external plasticisers. The reduction in softening temperature obtained by copolymerisation with vinyl acetate permits the use of lower processing temperatures, e.g. in extrusion and calendering, and, at appropriate *K*-value and vinyl acetate concentrations, permits the production of sheet which can be shaped by vacuum-forming. Introduction of vinyl acetate into vinyl chloride polymer chains also reduces the heat-stability, and this, coupled with increased costs arising from relatively higher prices of vinyl acetate in most if not all parts of the world, usually limits the use of copolymers to those applications where the required technical effects are not achievable by other means. Copolymerisation of vinyl chloride with vinylidene chloride is also used for similar purposes, although the results are not identical with those obtained with vinyl acetate. Furthermore, vinyl chloride/vinylidene chloride copolymers have good barrier properties, and on this account are sometimes used for coatings. The other comonomers mentioned earlier are mainly used in copolymers for solution applications where improved adhesion is required.

The preceding discussion is intended to outline the basic principles of selection of PVC resins for various applications. Other factors arise in a number of processes and applications, and are discussed in the appropriate chapters.

In practical formulation one is of course concerned with selecting from the wide range of polymers commercially available. For details of these the reader is referred to the trade literature offered by the suppliers, which is usually sufficiently informative, and to the late W.S. Penn's book on *PVC Technology*,[59] and its revisions by Titow and Lanham.[65, 66]

REFERENCES

1. C.S. MARVEL, J.H. SAMPLE and M.F. ROY, *J. Am. chem. Soc.*, **61**, 3241 (1939)
2. C.S. MARVEL, G.D. JONES, T.W. MARTIN and G.L. SCHERTZ, *J. Am. chem. Soc.*, **64**, 2356 (1942)
3. C.S. MARVEL, *An Introduction to the Organic Chemistry of High Polymers*, Wiley, New York (1959)
4. C.S. MARVEL and C. HORNING, *Synth. Polymers*, **1**. 754 (1943)
5. F. CHEVASSUS and R. DE BROUTELLES, trans. C.J.R. Eichlorn and E.E. Sarmiento, *The Stabilization of Polyvinyl Chloride*. Arnold, London (1962)
6. E.C.A. HORNER, *Chemy Ind.*, **12**, 308 (1960)
7. R.J. GRISENTHWAITE and R.F. HUNTER, *Chemy Ind.*, No. 23, 719 (1958)
8. M.H. GEORGE, R.J. GRISENTHWAITE and R.F. HUNTER, *Chemy Ind.*, No. 34, 1114 (1958)
9. M. FREEMAN and P.P. MANNING, *J. Polym. Sci.*, A, **2**, 5, 1017–2024 (1964)
10. C.A. BRIGHTON, 'Factors Affecting the Properties of Vinyl Chloride Polymers and Copolymers', Chapter 1 of *Advances in PVC Compounding and Processing* (M. Kaufman, ed.), Maclaren, London (1962)
11. F. KRASOVEC, *Rep. Josef Stefan Inst.*, **3**, 203 (1956)
12. J.D. COTMAN, *Ann. N.Y. Acad. Sci.*, **57**, 417 (1953)
13. G. TALAMINI and G. VIDOTTO, *Chimica Ind., Milano*, **46**, 1, 16 (1964)
14. F.A. BOVEY and G.D.V. TIERS, *Chemy. Ind.*, **42**, 1826 (1962)
15. T. ALFREY, N. WIEDERHORN, R. STEIN and A. TOBOLSKY, *Ind. Engng Chem., analyt. Edn*, **41**, 701 (1949); *J. Colloid Sci.*, **4**, 211 (1949)
16. H. WILSKI, *Kolloidzeitschrift*, **210**, 37 (1966)
17. G. GARBUGLIO, A. RODELLA, G.C. BORSINI and E. GALLINELLA, *Chimica Ind., Milano*, **46**, 166 (1964)
18. A. NAHAJIMA, H. HAMADA and S. HAYASHI, *Makromolek. Chem.*, **95**, 40 (1966)
19. G. TALAMINI and G. VIDOTTO, *Makromolek. Chem.*, **100**, 48 (1967)
20. G. PEZZIN, *Plast. Polym.*, **37**, 130, 295 (1969)
21. C.S. FULLER, *Chem. Rev.*, **26**, 162 (1940)
22. G. NATTA and P. CARRADINI, *Polym. Sci.*, **20**, 262 (1956)

23. C.E.H. Bawn, *The Chemistry of High Polymers*, Butterworths, London (1948)
24. O.C. Bockman, *Br. Plast.*, **38** 6, 364 (1955)
25. D.M. White, *J. Am. chem. Soc.*, **82**, 21, 5678 (1960)
26. J.D. Cotman, *J. Am. chem. Soc.*, **77**, 2790 (1955)
27. Minoru Imoto, and Tajayaki Otayu, *J. Inst. Polytech., Osaka, Cy Univ.*, **4**, 1, Series C, 124 (1953)
28. S.G. Bankoff and R.N. Shreve, *Ind. Engng Chem., analyt. Edn*, **45** (2), 270 (1953)
29. B. Baum and L.H. Wartman, *J. Polym. Sci.*, **28**, 537 (1958)
30. F.T. Wall, *J. Am. chem. Soc.*, **62**, 803 (1940); **63**, 812 (1941)
31. R. Simha, *J. Am. chem. Soc.*, **63**, 1497 (1941)
32. P. Doty, H. Wagner and S. Singer, *J. phys. Colloid Chem.*, **51**, 32 (1947)
33. Hengstenberg, *Agnew. Chem.*, **62**, 26 (1950)
34. P. Platzek, *Plastica*, **5**, 19 (1952)
35. J.W. Brettenbach, E.L. Forster and A.J. Renner, *Kolloidzeitschrift*, **127**, 1 (1952)
36. F. Danusso, G. Moraglio and A. Gazzera, *Chimica Ind., Milano*, **36**, 883 (1954)
37. A. Peterlin, IUPAC *Symp. on Macromolecular Chem.*, Milan, 1954; *Ricerca Scient. Suppl.*, **25B**, 533 (1955)
38. Z. Mencik, *Chemické Listy*, **49**, 1598 (1955); *Chem. abstr.*, **50**, 650i, 1360H (1955)
39. G. Bier and H. Kramer, *Makromolek. Chem.*, **19**, 151 (1956)
40. H. Batzer and A. Nisch, *Makromolek. Chem.*, **22**, 131 (1957)
41. D. Laker, *Polym. Sci.*, **25**, 122 (1957)
42. G.M. Guzman and J.M.G. Fatou, *An. R. Soc. esp. Fis. Quim*, **54**, 601 (1958)
43. G. Pezzin, G. Talamini and G. Vidotto, *Makromolek. Chem.*, **43**, 12 (1961)
44. R. Endo, *Chemy high Polym.*, **18**, 143; **18**, 447 (1961)
45. A. Guyot and J.P. Benevise, *Industrie Plast. Mod.*, **13**, 37 (1961)
46. J.J.P. Staudinger, Lecture to Swedish Plastics Federation, November (1949)
47. G.A.R. Matthews and R.B. Pearson, *Plastics, Lond.*, **28**, 307, 98 (1963); **29**, 317, 99 (1964)
48. ASTM D 1243–58T
49. ISO 1628–2
50. H. Fikentscher, *Cellulose-Chem.*, **13**, 58 (1932)
51. H. Fikentscher, *Kolloidzitschrift*, **53**, 34 (1931)
52. JIS K 6721–1959
53. R. Hammond, *Trans. J. Plast. Inst., Lond.*, **26**, 49 (1958)
54. B.S. Dyer, *Trans. J. Plast. Inst., Lond.*, **27**, 84 (1959)
55. W.M. Smith, *Manufacture of Plastics*, Reinhold, New York (1964)
56. G.A.R. Matthews, 'Compounding Techniques', Chapter 5 of *Advances in PVC Compounding and Processing* (M. Kaufman, ed.), Maclaren, London (1962)
57. J.D.D. Morgan, *Plastics Progress*, Iliffe, London (1951)
58. Imperial Chemical Industries Ltd., Plastics Division I.S. Notes 736 and 1031.
59. W.S. Penn, *PVC Technology*, Maclaren, London (1966)
60. M.W. Allsopp, *Pure Appl. Chem.*, **53**, 449 (1981)
61. M.W. Allsopp, *Manufacture and Processing of PVC*, edited by R.H. Burgess, Applied Science Publishers Ltd. (1982) Chapter 7
62. A. Cittadini and R. Paolillo, *Mat. Plastiche*, **11**, 974 (1959)
63. G. Butters, ed., *Particulate Nature of PVC*, Applied Science Publishers Ltd. (1982)
64. P.H. Geil, *Macromol. Sci-Phys.*, **B14**, (1) 171 (1977)
65. W.V. Titow and B.J. Lanham, *PVC Technology*, Applied Science Publishers Ltd. (1971)
66. W.V. Titow, *PVC Technology*, Applied Science Publishers Ltd. (1984)
67. W.F. Maddams, *Degradation and Stabilisation of PVC* (E.D. Owen, ed.), Elsevier Applied Science Publishers, London (1984) Chapter 4
68. G. Natta and P. Corradini, *J. Polym. Sci.*, **20**, 251 (1956)
69. A. Nakajima, H. Hamada and S. Hayashi, *Makromol. Chem.*, **95**, 40 (1966)
70. A.W. Coaker and R.W. Wypart, *Handbook of PVC Formulating*, (E.J. Wickson, ed.), John Wiley & Sons, Inc., New York (1993) page 18
71. S.V. Pavel and M. Gilbert, *Plast. Rubb. Proc. Appn.*, **5**, 85 (1985)
72. J.A. Covas, M. Gilbert and D.E. Marshall, *Plast. Rubb. Proc. Appn.*, **9**, 107 (1988)
73. C. Baker, W.F. Maddams, G.S. Park and B. Robertson, *Makromol. Chem.*, **165**, 321 (1973)
74. K.B. Abbas, F.A. Bovey and F.C. Schilling, *Makromol. Chem. Suppl. 1*, 227 (June 1975)

75. W.H. STARNES, F.C. SHILLING, F.C. ABBAS, I.M. PLITZ, R.L. HARTLESS and F.A. BOVEY, *Macromolecules,* **12**, 13 (1979)

76. F.A. BOVEY, K.B. ABBAS, F.C. SCHILLING and W.H. STARNES, *Macromolecules,* **8**, 437 (1975)

77. W.F. MADDAMS, *Particulate Nature of PVC*, (G. Butters, ed.) Applied Science Publishers, London (1982) Chapter 3

78. A.W. COAKER and R.W. WYPART, *Handbook of PVC Formulating*, (E.J. Wickson, ed.), John Wiley & Sons, Inc., New York (1993) Chapter 2

79. Q.T. PHAM, I. MILLAN and E.L. MADRUGA, *Makromol. Chem.*, **175**, 945 (1974)

80. J.A. JUIJN, *Crystallinity in Atactic Polyvinyl Chloride*, (J.H. Pasmans, ed.) 'S-Gravenhage, Holland, (1972)

81. A. GRAY and M. GILBERT, *Polymer*, **17**, 44 (1976)

82. D.E. WITENHAFFER, *J. Macromol. Sci. Phys.,* **B4** (2), 915 (1970)

83. H. MUNSTEDT, *J. Macromol. Sci. Phys.,* **B14** (2), 195 (1977)

84. O.P. OBANDE and M. GILBERT, *J. Appl. Polym. Sci.*, **37**, 1713 (1989)

85. V.D. KOCKOTT, Kolloid-Z.u.Z., *Polym.*, **198**, 17 (1964)

86. A. KELLER, *Polym. Sci. & Tech.*, **22**, 25 (1983)

5

Degradation and stabilisation

5.1 INTRODUCTION

At room temperature polyvinyl chloride is inherently stable. Decomposition by ultra-violet light is very dependent on additives to the polymer, and also, apparently, on the degree to which the polymer has previously been decomposed during processing, the rate of decomposition by ultra-violet light being generally increased by increased prior decomposition. In general PVC may be considered as one of the more stable plastic materials to the influence of ultra-violet light, particularly where the composition as been specifically formulated for outdoor exposure.

Increase in temperature to above about 373 K (100°C) leads to the decomposition for which polyvinyl chloride is notorious, with evolution of hydrogen chloride and development of colour in the polymer and eventual charring (see Plate 5.1).

Dehydrochlorination is substantially complete[1] in 30 min at 573 K (300C). The rate of decomposition at processing temperatures is usually too great to be ignored, and steps have to be taken to reduce decomposition to an acceptable level, usually by the incorporation of stabilisers. The colour changes or actual processing difficulties are generally the main concern, because, if not masked by colorants, there is considerable darkening in colour before there is any significant change in physical properties; but deterioration in physical properties can sometimes be important.

For many years the reputation of PVC for decomposition led to some hesitancy on the part of processors to handle the material, particularly in those forms which might be regarded as particularly difficult (e.g. extrusion of thin 'non-toxic' unplasticised film), but given proper formulation, equipment, and control, decomposition need not be a matter of great concern.

5.1.1 THERMAL DECOMPOSITION

While decomposition of polyvinyl chloride can be induced by other forms of energy, thermal decomposition is by far the most important in practice. As decomposition proceeds the substantially colourless polymer usually becomes yellowish, then brown, and finally black, although considerable variations in hue are encountered between different commercial resins, probably due to variations in residual impurities. Thus pink, red, and even purple colours are not unknown. When decomposition occurs after the polymer has been compounded with additives (e.g. stabilisers), colour development can be very patchy, dark coloured regions often developing rapidly in an otherwise unchanged background. This is presumably due to uneven distribution of stabiliser in the polymer matrix. As a consequence of improvements in quality during the past few decades these problems are now rarely encountered.

The only volatile product of decomposition below 473 K (200°C), as shown by mass spectrometric methods, is hydrogen chloride.[2] Above this temperature small amounts of aliphatic, unsaturated, and aromatic hydrocarbons are formed, but free chlorine or hydrogen have not been detected.[1,3,4] The residue is generally insoluble in common solvents.

The rates of decomposition and colour development are temperature-dependent, increasing with increase in the latter. There are considerable differences in performance between different commercial resins, and it is thus common to speak of the 'heat-stability' of a particular resin and to compare the 'heat-stabilities' of different resins. Likewise, different stabilisers vary in their ability to reduce decomposition, and it is common practice to compare 'heat-stabilities' of different compositions.

As an approximate generalisation it will usually be found that heat-stability improves with increasing molecular weight in a series of otherwise similar commercial polymers. Mass-polymerised polymers have marginally better heat-stability than suspension polymers, most of which are more heat-stable than emulsion polymers, though there is a considerable variation amongst the suspension polymers available commercially. Contrary to statements in the literature, it is a practical observation that commercial vinyl chloride/vinyl acetate copolymers are less stable to both heat and light than homopolymers.

In comparing different compositions it is found that not only are there differences in rate of colour development, but that rates for individual compositions may change with time of heating, and that the variations in rate are not the same for all compositions. Thus some compositions exhibit a fairly steady development of colour, while others may retain a low level of discoloration for a period of time until they suddenly and rapidly deteriorate to dark shades. Other compositions may commence with some discoloration but not deteriorate on heating for long periods. These facts complicate the evaluation of heat-stabilisers and compassion of PVC compositions.

5.1.2   TECHNIQUES FOR STUDIES OF THERMAL DECOMPOSITION

There is voluminous literature concerning studies on the decomposition and stabilisation of vinyl chloride polymers and copolymers, and no attempt will be made here to list all the relevant references. For more exhaustive compilations the reader is referred to the numerous reviews of the subject,[5-12, 59, 60] particularly the comprehensive review of work up to 1965 by Geddes,[12] and the latter one by Hjertberg and Sorvik.[95]

Since decomposition of polyvinyl chloride involves evolution of hydrogen chloride it might be thought that studies of the kinetics by measurement of the rate of evolution of this compound should be simple. Indeed, the majority of workers have used this approach, usually absorbing volatile decomposition products in an appropriate liquid medium and measuring the hydrogen chloride absorbed by acid-base titration,[13-17] conductimetry,[18-22] measurement of pH,[23, 24] potentiometry,[25-29] or argentimetry,[30-38] colour change in an acid indicator,[90] Raman spectroscopy, and mass spectrometry.[101] Thermogravimetric methods are not sufficiently sensitive in decomposition of polyvinyl chloride to be of much value, and determination of volatile by pressure is little better.[39]

Development of colour was used at an early date to follow decomposition of PVC,[40] and this method is the most commonly employed for practical assessments of stability and stabilisation of polymers, stabilisers, compounds, etc., but absorption spectra, sufficiently resolved to permit analysis of sequences of unsaturation, can now be obtained.[32, 41–47] Techniques using visible light measurements linked to a computer have been developed, which are very sensitive in detecting yellowing, for example.

Other methods which have been used to follow the decomposition of polyvinyl chloride, usually in the form of compounds with additives, include observations of changes in melt viscosity using a Brabender Plastograph[48, 49] or a melt rheometer,[50] and observations of changes in mechanical properties, such as tensile strength and elongation at break,[50] but these methods are not in themselves of much value in determining the kinetics of any chemical process involved in the decomposition.

A 1967 paper[51] described the use of optical and X-ray diffraction to study the behaviour of crystalline stabilisers like basic lead carbonate in PVC, but while interesting and valuable in a few cases, possible application of these techniques seems rather limited.

One of the problems associated with investigations of polyvinyl chloride is that of obtaining completely pure samples of the polymers to be studied; and if this objective is attained there remains the probability that the impurities which have been so assiduously removed may provide an important contribution to the overall stability of the polymer. In addition it is virtually impossible to characterize the few anomalous groupings actually in the polymer molecules, which almost certainly also have a disproportionately major influence on stability.

5.2   KINETICS OF THERMAL DECOMPOSITION

While there has been considerable controversy, it is now generally agreed that the decomposition of polyvinyl chloride in air or oxygen is markedly different from that in an inert atmosphere or vacuum.[52] The rate of evolution of hydrogen chloride is more rapid and accelerating in the presence of air than it is in the absence of air under similar conditions, suggesting that hydrogen chloride has an autocatalytic or pseudo-catalytic effect in the presence of oxygen. In the absence of oxygen, hydrogen chloride has generally been thought to be non-catalytic, at least at low concentrations, but it has been suggested that there is a catalytic effect if the concentration of hydrogen chloride is sufficiently high.[5] Normally the evolution of hydrogen chloride from polyvinyl chloride in the absence of oxygen, at constant temperature, is linear with time.

Early work [53, 54] suggested that in the temperature range 481–496 K (208–223°C) the rate of decomposition increases linearly with extent of reaction or time; but more recent observations at higher temperatures[1] suggested a much more complex reaction with an order of 3/2. However, it must be noted that most of the work has been at temperatures above those normally employed in commercial processing, and it is not inconceivable that the kinetics, and indeed the mechanisms, should vary with temperature.

Fundamental study of the kinetics of polyvinyl chloride decomposition are also complicated by the evident fact that minor differences between different samples of polymer can profoundly affect rates of decomposition. Thus polymers prepared with 1,1′-azobisisobutyronitrile initiator have been shown to degrade more rapidly than those prepared with lauroyl peroxide initiator,[23] though it is doubtful if the extension to peroxide and nitrogen compound initiators is generally valid. Small proportions of impurities, such as metal ions (e.g. iron, copper, and zinc), or residues from emulsifying and suspension agents, can also markedly increase the rates of degradation, at least as evidenced by colour formation. As a rough guide, rate of decomposition increases with decrease in molecular weight, in accordance with the practical observation that, in a series of similar commercial polymers differing intention-ally in molecular weight only, heat-stability decreases with decreasing molecular weight. This agrees with a number of more theoretical studies,[18, 20, 55] although there is evidence of a levelling-off of rate of decomposition in the higher molecular weight range.[56]

A simple mathematical model of PVC degradation has been proposed.[102]

Even if precise kinetic relationships between rate of decomposition and temperature were available it would be difficult to translate these to practical processing situations, owing to the fact that processing of PVC (particularly harder grades) develops considerable frictional heat, so that the temperature usually varies through a flowing melt of material, and is in any case difficult to measure. Certainly little can be made of the usual set machine temperatures.

5.3   MECHANISMS OF THERMAL DEGRADATION

In spite of the considerable number of workers and the very large volume of work on the subject the mechanism of decomposition of polyvinyl chloride is still not completely understood, established, or agreed.[91, 92] A full critical review would be much too lengthy for a monograph such as this, and the seeker after more detailed consideration is referred to the numerous reviews already mentioned, particularly the excellent one by Geddes.[12] All that will be attempted here is a brief survey of the more important theories and typical supporting evidence.

Thermal decomposition of unstabilised polyvinyl chloride produces intense discoloration by the time as little as 0.1 per cent of hydrogen chloride is produced.[2] It is generally assumed that the colour arises from conjugated polyene structures arising from elimination of hydrogen chloride from consecutive vinyl chloride units of the polymer chain. The colours obtained require sequences of not less than seven consecutive double bonds.[18, 57] Moreover complete random elimination of hydrogen chloride should leave 13 per cent of chlorine in the product,[58] whereas only traces are found in practice.[59] This suggest that hydrogen chloride molecules are eliminated in succession along polymer chains, possibly by a so-called 'zipper' reaction.[60] However, it has also been suggested that the colour arising from decomposition of polyvinyl chloride is due to the formation of fulvene structures,[61] and formation of carbon has been blamed for discoloration in photodegradation.[62]

It has been stated[12] that cross-linking has been well established as a feature of the thermal decomposition of polyvinyl chloride,[6, 13, 18, 28, 32] but in normal industrial practice compositions are usually so well stabilised that any cross-linking which does occur must be very limited, since there is rarely if ever an appreciable change in mechanical properties which could be attributed to such reaction.

The various schemes which have been proposed for the thermal decomposition of polyvinyl chloride may be classified as involving unimolecular elimination, free-radical chain, or ionic mechanisms.

Studies of the decomposition of simple aliphatic compounds containing chlorine have been inconclusive. While some of the evidence[63, 64] is claimed to support a unimolecular elimination mechanism, the fact that polyvinyl chloride decomposes more readily than simply alkyl chlorides which are known to decompose by unimolecular processes, and much more readily than 2,4-dichloropentane[65] and 1,3,5-trichlorohexane, [66, 67] which have structures similar to the polymer, agrees better with a radical chain mechanism.

5.3.1 UNIMOLECULAR EXPLUSION THEORY

The suggestion that decomposition of polyvinyl chloride follows a mechanism involving unimolecular elimination of hydrogen chloride[54] followed earlier work on the decomposition of polyvinyl acetate,[53] a 'zipper' expulsion being explained on the basis of allylic activation of the chlorine atoms attached to carbon atoms adjacent to double bonds resulting from initial elimination of hydrogen chloride molecules. Apart from a small amount of supporting kinetic evidence[54] this theory largely stands on the basis of negative evidence which opposes other theories.

5.3.2 RADICAL-CHAIN THEORIES

The kinetics observed[1] in decomposition of polyvinyl chloride at 508 K (235°C) and above match a free-radical scheme, such as the following:

Initiation: $\sim\sim CH_2.CHCl.CH_2.CHCl \sim\sim \longrightarrow$

$\sim\sim CH_2.CHCl.\dot{C}H_2.CH \sim\sim + Cl^{\cdot}$

Propagation: $\sim\sim CH_2.CHCl.CH_2.CHCl \sim\sim + Cl^{\cdot} \longrightarrow$

$\sim\sim CH_2.CHCl.\dot{C}H.CHCl \sim\sim + HCl$

$\sim\sim CH_2.CHCl.\dot{C}H.CHCl \sim\sim \longrightarrow$

$\sim\sim CH_2.CH:CH.CHCl \sim\sim + Cl^{\cdot}$

Termination: $Cl^{\cdot} + Cl^{\cdot} \longrightarrow Cl_2$

$R^{\cdot} + R^{\cdot} \longrightarrow RR$

$R^{\cdot} + Cl^{\cdot} \longrightarrow RCl$

Most theories involving free-radical mechanisms involve the formation of free chlorine atoms as chain-carriers; but no chlorine has been detected. It may be, however, that the amounts present at any time are too small to detect. Also, there is no general agreement that a chlorine atom would extract a β- rather than an α-hydrogen atom. It has also been pointed out that chain propagation would not occur if one of the products had an inhibiting allylic structure.

However, free-radicals have been detected in degraded PVC by electron spin resonance.[4] Furthermore, some sources of free radicals increase the rate of decomposition,[18, 68] and some stabilisers (but not all) interfere with radical chain processes.[60] It has also been argued[1] that steric considerations oppose the suggestion that hydrogen chloride can be expelled from polyvinyl chloride in the way that acetic acid can be expelled from polyvinyl acetate.[53]

5.3.3   IONIC THEORIES OF DECOMPOSITION

Polyvinyl chloride is dehydrochlorinated by a variety of bases, and if the substitution reactions which normally occur simultaneously are inhibited, long polyene chains yielding red and violet coloured residues, according to the length of the polyene chains, result. The reactions involved are thought to be similar the reaction of alkyl chlorides with bases,[44] where elimination of hydrogen chloride involves an ionic mechanism:

$$
\begin{array}{c}
\qquad\qquad Cl \\
\qquad\qquad | \\
-CH-CH+B:\longrightarrow -CH:CH-+BH^{+}+Cl^{-} \\
\quad\ | \qquad | \\
\quad\ H
\end{array}
$$

It has been suggested[69] that thermal decomposition of polyvinyl chloride also proceeds by an ionic dehydrochlorination with a somewhat similar mechanism, involving a charge separation of the carbon–chlorine bond; but there is little support for the idea that appropriate conditions might be present normally.[39] However, more recently there has been increased support for the proposal that an ionic mechanism might be responsible for the dehydrochlorination of PVC.[93-95]

5.3.4   EFFECTS OF STRUCTURAL IRREGULARITIES

By analogy from the fact that low molecular weight compounds such as 1,3,5-trichlorohexane have good thermal stability,[66, 67] it may be argued that a polyvinyl chloride having the ideal structure $CH_3(CH_2.CHCl)_nCH_2.CH_2Cl$ would be unlikely to lose hydrogen chloride below 573 K (300°C), suggesting that initiation, if not propagation, of the decomposition of polyvinyl chloride might well arise from departures from the idealised structure.

5.3.4.1   *Irregularities arising from initiators*

Commercial production involves initiation by means of peroxy or azo compounds. Each chain initiation of this type involves the incorporation on to one chain-end of a grouping deriving from the particular initiating species. Thus acyl peroxides are likely to leave an acyloxy, acyl, or phenyl group at the initiation end of each polymer chain. Peroxydicarbonates may leave an ester group or an alkyl carbonate group. With azonitriles, the group attached to the polyvinyl chloride structure of each initiated chain is a substituted nitrile, e.g. $(CH_3)_2(CN)C$—from the common 1,1′-azobisisobutyronitrile. This latter structure appears to lead to more rapid thermal decomposition than that derived from organic peroxides.[23, 24] Qualitative observation also suggests that in polymer initiated by organic peroxides the nature of the particular peroxide is important. Thus benzoyl peroxide appears to yield less stable polymers than the aliphatic lauroyl and caproyl peroxides.

The generally more rapid decomposition of emulsion polymers when compared with suspension polymers might suggest that the end-grouping arising from persulphate initiation, such as $-SO_4M$ (where M is a metal dependent on the constitution of the initiator system), is more liable to lead to decomposition than those derived from organic peroxide initiation; but the difficulties of purifying emulsion polymers from residues from other ingredients of the recipe, such as emulsifying agents and buffers, make it difficult to be sure on this point. It has also been shown that polymerisation initiated by γ-radiation yields polymers which are thermally more stable than those initiated by chemical free radical generators.[1] It may be of some interest that somewhat similar correlations have been observed between colour in polymethyl α-chloroacrylate and the initiator systems used, though azo initiators were here found to be superior to benzoyl peroxide. It is clear that these initiator fragments, present in a high proportion of the polymer molecules, contribute to thermal instability, but it is not clear whether this can account for most of the decomposition observed under practical conditions, or to what extent stability would be increased if these fragments could be replaced by more stable groupings.

The precise mechanism by which an initiator residue on a polymer molecule leads to the usual dehydrochlorination process might vary from one initiator to another, but probably starts with dissociation into a small radical and a macroradical. This latter might activate unimolecular elimination from the allyl grouping, followed by rearrangement and repetition.[12, 18, 60]

5.3.4.2   Unsaturation

Chain-transfer involving monomer molecules is generally agreed to be an important feature of vinyl chloride polymerisation, leading to an unsaturated group in the molecule. Similar unsaturation would arise as a result of termination by disproportionation. Unsaturation along a polymer chain could arise from elimination of hydrogen chloride due to the action of nucleophilic reagents if any are present during polymerisation or subsequently. Small amounts of unsaturation have been reported in polyvinyl chloride.[2, 19, 28, 70] Chloro-olefinic compounds are generally more readily decomposed than their saturated analogues, and activation energies are of the same order as values calculated for polyvinyl chloride.[25, 26, 55, 56]

Unsaturation of this type would be prone to oxidation and peroxidation, which could account at least in part for the differences observed between decomposition in the presence and absence of oxygen.

Chlorination of PVC under mild conditions has been shown to produce little if any substitution of chlorine in the polymer molecules, but resulted in reduced rates of decomposition at processing temperatures.[19] This was attributed to addition of chlorine to unsaturated groupings. It has been suggested that the free chlorine atoms, required in the free-radical explanation of PVC decompositions, arise from dissociation of allylic carbon-chlorine bonds, which are known to be relatively weak;[1] but the high decomposition rates of peroxidised polymer suggest that allylic chlorine does not contribute much to the decomposition.[56]

5.3.4.3   Branching

The kinetics of vinyl chloride polymerisation indicate that branching must occur, but it has generally been regarded as difficult to estimate the amount.[12] Commercial polymers have been shown to contain between 5 and 10 per cent of long-chain branched molecules, while other estimates have indicated about 16 branches per molecule. Transfer reactions during polymerisation could lead to three different tertiary structures:

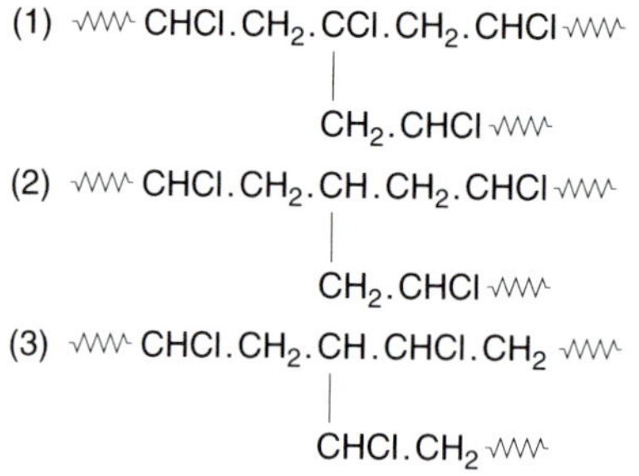

(1) ᴧᴧᴧ CHCl.CH$_2$.CCl.CH$_2$.CHCl ᴧᴧᴧ

CH$_2$.CHCl ᴧᴧᴧ

(2) ᴧᴧᴧ CHCl.CH$_2$.CH.CH$_2$.CHCl ᴧᴧᴧ

CH$_2$.CHCl ᴧᴧᴧ

(3) ᴧᴧᴧ CHCl.CH$_2$.CH.CHCl.CH$_2$ ᴧᴧᴧ

CHCl.CH$_2$ ᴧᴧᴧ

Dehydrochlorination of the first of these leads to a fourth structure:

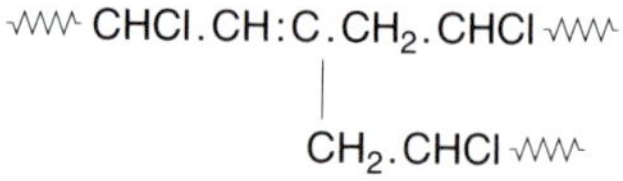

ᴧᴧᴧ CHCl.CH:C.CH$_2$.CHCl ᴧᴧᴧ

CH$_2$.CHCl ᴧᴧᴧ

Such determinations as have yielded positive results suggest, as would be expected, that labile chlorine is concentrated in high molecular weight molecules.[70] The amount of branching should depend on polymerisation temperature, and failures to correlate rates of dehydrochlorination with polymerisation temperatures,[28, 56] are rather odd, in view of the practical observation that, in a given series of similar polymers of varying molecular weight, stability as estimated by rate of colour development decreases with decrease in average molecular weight, i.e. with increase in polymerisation temperature. This does perhaps suggest that tertiary structures are not a major factor in the decomposition encountered in practical situations.

5.3.4.4.   Other structural irregularities

Although the structure of polyvinyl chloride is predominantly head-to-tail, it is not impossible that some head-to-head and tail-to-tail structures might occur to a small degree difficult to detect. Termination by combination would also lead to a tail-to-tail structure at the point where the combination of two growing chains had occurred. Removal of a hydrogen atom from a tertiary carbon atom in such a structure could result in a carbon-carbon double bond and a free chlorine atom, as required by radical-chain theories of decomposition.[39] It is suggested, however, that a concentration of one tail-to-tail structure in every one thousand normal linkages is unlikely to have a serious effect on the initiation of dehydrochlorination.[39]

There have been several proposals that oxidation of polyvinyl chloride could lead to peroxide and hydroperoxide groupings, and possibly other structures, and that these could be active centres for initiation of dehydrochlorination, both thermally and under the action of ultra-violet radiation. The evidence on this is confusing. Certainly, decomposition in the presence of oxygen is different to decomposition in its absence, and the formation of peroxidic groupings could be responsible. However, there are indications that ultra-violet degradation can differ from thermal decomposition in that reduction in molecular weight has been observed on outdoor exposure of clear PVC sheeting. This might well be expected to arise in intermediate peroxidation.

5.3.5   EFFECTS OF IMPURITIES

It is a common observation that vinyl chloride polymers of nominally identical type often exhibit markedly different heat-stabilities. Moreover, the reaction of such polymers to various stabilisers may also differ considerably, and rankings in order of stability may often vary according to the particular stabiliser employed. This seems to point almost exclusively to the idea that residual impurities are responsible. The stability of bulk-or mass-produced polymers is usually marginally superior to that of the best of suspension polymers; and this is usually attributed to the presence of small traces of residues from the suspension agents used in polymerisation. Likewise suspension polymers are generally more stable than emulsion polymers, certainly those isolated by spray-drying, which constitute the bulk of such products, and this is attributed to the relatively large concentrations of residual emulsifying agents, buffers, and other additives and their by-products remaining from the polymerisation system. With certain stabilisers, however, some emulsion polymers appear to react more favourably than other types, and this too is usually attributed to the particular nature of the emulsifying agent used.

Residual traces of initiators are also likely to affect thermal stability, but the experimental evidence offered in support of this belief[36] may have a different interpretation.[12]

Trace amounts (a few parts per million) of some metals, such as iron, appear to catalyse thermal decomposition considerably.[6, 18] A number of metal chlorides, such as ferric chloride, and those formed by the reaction of stabilisers with liberated hydrogen chloride, catalyse the decomposition. Zinc compounds in very small concentrations often exert a co-stabilising effect with other stabilisers, but in larger quantities can accelerate the thermal decomposition of some grades of commercial resin catastrophically. It has been suggested that the effect of cations is to catalyse decomposition of residual peroxides or peroxidic structures in some of the polymer molecules.[60]

5.3.6.   EFFECTS OF AIR AND OXYGEN

Effects of oxygen have already been alluded to on several occasions. The generalisation that degradation in the presence of oxygen results in less coloration than in its absence, in spite of the greater dehydrochlorination, [2, 6, 12, 13, 18, 25, 55] is not justified if stabilised compositions are included. When less coloration does occur, it is usually attributed to oxidation of unsaturated groups in polyene chains, ultimately resulting in shorter polyene sequences, and thus less colour. It is supposed that the oxidation produces hydroperoxides or peroxides, which decompose with resultant chain scission.[16, 52]

In practice the effects of oxygen can be important, since some processing, such as milling and calendering, is carried out in the open atmosphere, and even under the most favourable circumstances it is doubtful if air is ever completely excluded from the polymer in the melt state.

5.3.7   EFFECTS OF PHOTORADIATION

Studies of decomposition of unmodified vinyl chloride polymers under the influence of radiation, particularly ultra-violet radiation, are not only difficult but to some extent meaningless, since in application PVC is never submitted to such influences in the unstabilised state.

Furthermore, it has proved singularly difficult to design an apparatus which will reproduce reasonably accurately the effects of solar radiation, and most work has been carried out on commercial and experimental compositions exposed to a variety of actual atmospheric conditions, particularly at testing stations located in parts of the world where the amount of sunshine is high, and where the ultra-violet content of that sunshine is also high.

The effects of usual various additives are considerable, and there is usually a large difference in performance between clear and opaque compositions. There are indications that in clear compositions, at any rate, chain scission can occur without discoloration. Presumably this involves a peroxidation mechanism, as already referred to (Section 5.3.6). The generalisation that photodegradation follows the same pattern as thermal decomposition[7, 12, 41] does not always appear to be supported by practical observation.

It has been claimed[7] that only a small range of wavelengths in sunlight has a degrading effect on PVC, the maximum effect being at 310 μm, in the centre of the ultra-violet range.

There is evidence that the rate of decomposition of a PVC composition under the influence of sunlight is greater, the more the composition has been decomposed thermally during processing.

Introduction of vinyl acetate decreases ultra-violet stability, more or less in proportion to its its concentration.

5.3.8   MECHANICAL EFFECTS

Many polymers suffer chain scission during processing under conditions of high shear, as in milling, intensive mixing, and extrusion, and PVC has been said to behave thus. Macroradicals arising from such chain scission may react by disproportionation to give smaller molecules, or by addition to other polymer molecules to produce chain-branching, or by involvement in cross-linking reactions. It has been claimed that PVC tends to cross-link on thermal decomposition, but that milling produces a reduction in melt viscosity due to chain scission.[12, 71, 72] Radical acceptors may act as stabilisers against mechanical degradation, but other stabilisers such as lead stearate may be equally effective.[4] Discoloration of PVC during milling is reduced by small proportions (6 per cent) of maleic anhydride, and although this is attributed to reaction with double bonds, cross-linking and chain scission as a result of shear are also thought to occur.[73]

5.4   STABILISATION

If knowledge of the mechanisms of decomposition is far from complete or satisfactory, it is understandable that an explanation of the nature of stabilisation appears to be even more remote, particularly in view of the number and variety of materials which have some stabilising effect against thermal decomposition of polyvinyl chloride. It is therefore proposed to discuss the practical aspects of stabilisation, which have largely been developed in arbitrary fashion, before proceeding to an attempt to review such theories have been advanced.

5.4.1   GENERAL CONSIDERATIONS

Practical stabilisation of polyvinyl chloride has been reviewed frequently (e.g. refs. 9, 11,

61, 74–76, 96, 98), and trade literature is often very informative on this subject. The situation is confused, not only by the fact that very many different substances are offered and used as stabilisers, but also that a major proportion of commercial stabilisers are known only by their trade-names, or are at best only partially specified. As long ago as 1964, a by no means complete survey revealed over 1400 different trade materials, and something like 400 different types of compound offered as stabilisers for PVC. To describe or list all these, even if practicable, would be rather pointless. Quite apart from the magnitude of the subject, the situation is changing constantly, and anyone involved in the practical selection of stabilisers will find it necessary to consult the voluminous trade literature and seek the advice of the various suppliers. What will be attempted here is a review of the more important stabilisers or stabiliser types, which, it is hoped, will put the situation in perspective. It is believed that the materials mentioned will be found to be no less than adequate in the current commercial situation, and only very specialised demand should make it necessary to look beyond these confines for more exotic stabilisers.

Most of the currently available types of stabiliser have been available for many years, and developments in recent decades have been mainly by way of modifications to existing types. Perhaps this is not surprising in view of the fact that the 400 types referred to above represent only a fraction of the many more compounds that had been examined for potential stabilising effect and rejected by the mid 1960s.

5.4.1.1   Types of stabiliser

The vast majority of stabilisers for polyvinyl chloride can be loosely classified under the following broad headings:

(1) Inorganic metal salts, particularly of lead and barium.
(2) Metal soaps, or other salts of organic acids, particularly of barium, cadmium, lead, calcium, magnesium, zinc, lithium, manganese, strontium, aluminium, and sodium.
(3) Metal complexes, particularly of barium, cadmium, and zinc.
(4) Organotin compounds.
(5) Epoxy compounds.
(6) Organic phosphites, chelators, and antioxidants.
(7) Miscellaneous compounds.

Each of these will be considered in turn, though as will be seen later PVC compositions often contain combinations of stabilisers from more than one group.

5.4.2   INORGANIC METAL SALTS

Apart from a few odd compounds, such as barium silicate, barium polyphosphate, cadmium perborate, and sodium hydrogen phosphate/silicate, none of which has had wide acceptance, this group of stabilisers consists of lead compounds, particularly basic. Of this latter group, basic lead carbonate (otherwise known as 'white lead'), and tribasic lead sulphate (TBLS) dominated the market for a long period, and TBLS still accounts for the stabilisation of a fair proportion of PVC compositions. During processing metal compounds used as stabilisers react with liberated hydrogen chloride to form the corresponding salts. Most of these are water-soluble and so increase the water absorption of the PVC. Since lead chloride is insoluble in water, the use of lead compounds as stabilisers does not lead to increased water absorption in this way.

Basic lead carbonate was used in the early days of PVC, and is still regarded as an efficient cheap stabiliser where its deficiencies are not relevant. It is a dense white powder (s.g. 6.8) of composition approximating to $2PbCO_3.Pb(OH)_2$ (i.e. 85 per cent PbO); but commercial grades vary in the ratio of carbonate to hydroxyl groups, and this affects both the efficiency as a stabiliser and the basic white colour which the compound imparts. It can therefore be an embarrassment to change from one grade to another without thorough testing to establish identity of behaviour. To reduce the toxic hazard arising in the handling of the material it is usually employed in the form of a thick paste with plasticiser (e.g. 7 parts white lead: 1 part plasticiser). This stabiliser is employed in concentrations up to about 10 phr (parts per hundred of resin), but there is rarely any advantage in exceeding 5 phr. It suffers from a number of deficiencies in particular applications. First, even in the lower possible concentration range (about 2 phr), products of any thickness are substantially opaque, and even thin sections are at best translucent. This is not always a disadvantage, of course. Secondly, like all lead compounds, basic lead carbonate is toxic and therefore not generally usable in foodstuffs packaging and other applications where toxic hazards might arise. Thirdly, in contact with many sulphur compounds, lead sulphide is formed, leading to staining. Finally, the compound decomposes to yield gaseous carbon dioxide in the higher range of PVC processing temperatures. This deficiency would arise especially in the processing of unplasticised PVC, where high machine temperatures are employed and where considerable frictional heat is often generated owing to the relatively high melt viscosity of unplasticised PVC melts, leading to porosity in extrudates and mouldings. Even with plasticised compositions this problem can sometimes rise, for example, in high-speed extrusion of small sections, where high frictional heat is developed.

It is such cases which have led to the use of the more stable basic lead sulphates. Of these TBLS is the most widely known and used, but other varieties are also available, for example tetrabasic lead sulphate. Like its carbonate analogue, TBLS is a dense white powder (s.g. 7.1), of composition indicated by $3PbO.PbSO_4.H_2O$ (i.e. 90.1 per cent PbO), also available as a thick paste with plasticiser (e.g. 4 parts TBLS:1 part plasticiser). However, in unplasticised PVC compositions, even the small amount of plasticiser added in such a paste is sufficient to lower the softening tempeature to an often unacceptable degree, and in such cases the dry powder has to be employed. On the production scale this necessitates special provisions to ensue the good health of the process workers. Typically this would involve a self-contained air-conditioned chamber for handling, weighing, and mixing, within which workers would wear protective clothing and face masks fitted with 'long-breathing' air lines. Naturally the time any one worker can operate continuously under such conditions is limited, and this must be taken into account when estimating labour requirements and costs. Apart from its better stability, permitting higher processing temperatures, tribasic lead sulphate is very similar in effect to basic lead carbonate, and is used in similar concentrations. It is, however, slightly more expensive, but there can be commercial advantages to using it instead of the carbonate in order to avoid the disadvantage of purchasing and stocking two materials where one would suffice.

A third member of this little group of stabilisers is dibasic lead phosphite, another dense white powder (s.g. 6.94), of composition indicated by $2PbO.PbHPO_3.\frac{1}{2}H_2O$ (i.e. 90.2 per cent PbO), also available as a thick paste with plasticiser (e.g. 3 parts dibasic lead phosphite:1 part plasticiser). Its outstanding property is that it imparts particularly good resistance to photodegradation, and therefore finds some use in outdoor applications. It does, however, suffer from the same deficiency as basic lead carbonate in tending to 'gas' at relatively high processing temperatures, and is also more expensive than the basic carbonates and sulphates. In any case, in outdoor applications of unplasticised PVC, which constitute by far the greater proportion, tribasic lead sulphate appears to perform adequately in opaque compositions, and in clear compositions dibasic lead phosphite cannot be used because of its high pigmenting and opacifying power. It should be noted that finely divided lead phosphites decompose readily at high temperatures forming phosphorus/lead/oxygen mixtures that, once ignited, will continue to burn.

Most other compounds of lead used as stabilisers for PVC have lower pigmenting and opacifying powers than the basic carbonates, sulphates, and phosphites, and are not widely used. Various varieties of lead silicates and orthosilicates, sometimes in combination with silica gel, have been used where translucent effects have been required, and also to impart a dry feeling to PVC leathercloth and calendered sheeting.[74] Lead chloro-silicate has also been used. Of inorganic salts of other metals, sodium carbonate has found some use as a stabiliser in non-toxic compositions, but it is not very effective. Sodium hydrogen phosphate/silicate, barium silicate, and barium polyphosphate have also been offered as stabilisers, but their use is certainly not widespread. Somewhat surprisingly, calcium carbonate does not appear to be an effective stabiliser for PVC, in spite of occasional claims to the contrary.

5.4.3   METAL SOAPS

The stearates, laurates, napthenates, and ricinoleates of many metals are or have been used as stabilisers for PVC. Acetates, benzoates, caprates, caprylates, fumarates, heptoates, hexoates, maleates, myristates, octoates, palmitates, phthalates, salicylates, and versatates of a limited number of metals have also been offered as stabilisers. The use of most of these compounds is complicated and limited by the fact that they are efficient lubricants for PVC, so that the concentrations required for adequate stabilisation under severe processing conditions often cannot be employed since they would result in excessive lubrication and attendant problems such as 'plate-out'.

In addition to the lead compounds mentioned in Section 5.4.2, several lead compounds with organic acids are effective stabilisers for PVC. Lead stearate was once a fairly popular stabiliser for translucent compositions, but nowadays its main function is as a lubricant. Dibasic lead stearate is a rather better stabiliser than the normal salt, but is also mainly used as a lubricant, frequently in combination with the normal salt to achieve a precise level of lubrication. Other lead compounds of this type include dibasic lead phthalate,[77] which finds a limited usage as a stabiliser in plasticised PVC insulation for cables running at relatively high temperatures, in gramophone record compositions, and in foamed leathercloth.[74] Its use in records arises from the fact that, unlike most other salts of organic acids, it appears to be non-lubricating, though the same may be said of tetrabasic lead fumarate.[73] In PVC foams made using azodicarbonamide as a chemical blowing agent, dibasic lead phthalate has a 'kicking' action. Lead salicylate has been used in vinyl/asbestos floor tiles, where it has been said to neutralise any iron present by chelation.[74] Tribasic lead maleate is particularly effective where stabilisation against ultra-violet radiation is required,[75, 76] but, unlike dibasic lead phosphite, can be used in translucent compositions. None of the many other forms of lead salts appears to have more than a very limited usage, at least in the UK, though several are offered by many suppliers.

The most important stabilisers of this class are the soaps of barium, cadmium, zinc, calcium, and magnesium. Cadmium stearate is capable of producing plasticised PVC compositions of very low colour and high clarity, provided that processing conditions are not severe. However, at the concentrations which can be used (i.e. not much more than 1 phr, owing to incompatibility: see above), the stabilising effect is not very prolonged, and discoloration occurs rapidly once the time limit has been passed. On other hand, the use of barium stearate as sole stabiliser leads to slightly yellow base colours, but the initial colour is maintained for relatively long processing times, and then deteriorates fairly slowly. In other words, cadmium stearate is said to be a good 'short-term stabiliser' whereas barium stearate is a poor 'initial stabiliser' but a good 'long-term' stabiliser. Combining both stearates, in an attempt to get the best of both worlds, gives results that are better than additive, a typical example of the phenomenon known by the much-abused names 'synergism' or 'synergistic effect'. Co-precipitating the two stearates yields an even better result, and barium/cadmium stearates prepared in this way have been available commercially for many years, as also have a corresponding series of laurates. In combinations of this kind the ratio of one metal to the other may be varied over quite a wide range, the optimum ratio depending very much on the nature of the particular commercial resin and other ingredients in the composition, being determined by trial and error. The original barium/cadmium stabilisers were in fact combinations of barium ricinoleate and cadmium octoate, but when these were later combined with epoxydised oils and antioxidant organic phosphites, particularly in calendered sheeting, exudation and tackiness resulted from interaction between the ricinoleate and the epoxy compound. As a consequence the salts were displaced by barium/cadmium laurates.[74] Barium stearate is occasional encountered as a stabiliser in the absence of other metal compounds, or in combination with small amounts of zinc, but this is usually only in special applications, such as flooring. Cadmium compounds are now recognised to be toxic, and the metal tends to be retained by body tissue. Consequently they are now specifically banned from a wide range or products.[108, 109] Improved cadmium-free stabilisers have been and are being developed, and barium compounds are now commonly used in combination with compounds of other metals such as zinc, calcium, and lead.[99, 100, 103] Small amounts of zinc compounds enhance the stabilising action of barium/cadmium and other metal compounds.

Of other metal soaps the most commonly encountered are calcium stearate, the most well-known, and probably the least effective, non-toxic stabiliser for PVC, and magnesium stearate, also non-toxic. The two compounds are often used in combination together and with small amounts of zinc soaps, forming a series analogous to but much less effective than the barium/cadmium and barium/cadmium/zinc compounds. Lithium and manganese stearates are also non-toxic, but since the latter yields a basic brown colour its use is naturally restricted. As a matter of fact sodium and potassium stearates can also be reasonably effective stabiliser/lubricants, but are rarely encountered in industrial practice. Although, as indicated previously, many other metal soaps (e.g. strontium and aluminium stearates) and combinations of soaps are offered as stabilisers, their actual commercial usage is very limited.

Over the past decade or so there has been some success in developing more effective calcium, barium, magnesium, and zinc soap combinations in efforts to replace cadmium compounds on toxicity grounds.[99, 100, 103] One example is a calcium/zinc alkyl maleate.

5.4.4   METAL COMPLEXES

Most of the metal soaps are effective lubricants for and have a limited compatibility with PVC. This is manifest by migration to the surface of the soaps themselves, or acids formed from them by reaction with hydrogen chloride, particularly under high shear conditions, or it may even appear as exudation on standing, especially in light. For a given composition and set of conditions there is usually a maximum concentration of any particular metal soap above which this separation becomes evident and objectionable. In addition to separation of the metal soap itself, the substance frequently carries with it pigment or filler particles, depositing them on the surface of processing equipment, where they spoil the surface of the PVC, or deposit to be picked up by material processed subsequently. This phenomenon, known as 'plate-out', can be particularly troublesome in calendering, but its occurrence is by no means limited to that process. The maximum tolerable concentration of a metal soap to avoid plate-out usually decreases with increasing shear and temperature, and consequently became very apparent as the speeds and temperature of calendar operation were increased several years ago. This development thus soon revealed the deficiencies of the barium/cadmium soap stabiliser systems in common use, and it became necessary to find other stabilisers which do not behave in this way. As a result, barium/cadmium and barium/cadmium/zinc complexes such as phenates and alkylated phenates were introduced, and have probably been the most commonly used stabilisers.

Whereas the meal soaps are generally powdery or wax-like solids, the complexes are usually liquid. The constitution of commercial products is rarely disclosed, but they are probably mainly phenates or substituted phenates, alone or in combination with salts of short-chain aliphatic organic acids such as octoates, or benzoates, which are much less lubricating than the salts of long-chain acids, such as laurates and stearates. Commercial stabilisers range in price from the relatively cheap, at about twice the price of white lead, weight for weight, to about eight times that value, the majority of general purpose complexes costing about three times as much as white lead. As might be expected, stabilising performance improves with increase in price, and the most expensive materials are capable of producing clear colourless compositions equivalent to those obtainable with many of the more expensive organotin compounds.

The concentrations of barium/cadmium and barium/cadmium/zinc complexes employed for PVC stabilisation are normally in the region of 2–3 phr; but they are usually used in combination with other stabilisers, particularly with epoxy compounds and to a lesser extent organic phosphites.

While the barium/cadmium and barium/cadmium/zinc complexes have been the most popular stabilisers of this type, many other complexes and complex mixtures are available. Thus, there seems to be some advantage in including aluminium and calcium in some complex mixtures, particularly for the stabilisation of plastisols.

As with metal soaps there is a series of calcium/magnesium/zinc complex compounds analogous to the barium/cadmium/zinc series, possessing the advantage of non-toxicity but the disadvantage of inferior powers of stabilisation; though improved combinations of calcium, barium, magnesium, strontium, and zinc complexes have been introduced with the object of eliminating cadmium while retaining adequate stability.[99, 100] Little has been published about their precise nature.

5.4.5   ORGANOTIN COMPOUNDS

Although the best and most expensive of the barium/cadmium/zinc complexes can yield PVC compositions of high clarity substantially free of colour, organotin compounds as a class are generally superior, and are selected where maximum clarity and freedom from discoloration are required and cost is not too important. Even the cheapest of these compounds is appreciably dearer than the dearest of the barium/cadmium/zinc type.

Tin soaps and complex compounds analogous to those of other metals described in Sections 5.4.3 and 5.4.4 are available, but their use is not widespread, and here were are concerned with alkyl and aryl substituted tin compounds.[78] Roughly speaking there are essentially four classes of organotin stabiliser. The simplest comprises butyltin compounds, the best known of which are dibutyltin dilaurate (DBTL) and dibutyltin maleate (DBTM). The latter can be polymeric and is available in a variety of forms of different molecular weight. Other varieties of this type of stabiliser include dibutyltin laurate/maleate, dibutyltin dinonyl maleate, tributyltin laurate, and dibutyltin diacetate. As a class these compounds are excellent stabilisers for PVC, used alone at concentrations of about 2 phr or in combination with other stabilisers. Their use sometimes involves problems of formulating with regard to lubricants, because some seem to cause sticking of PVC compositions to hot metal processing equipment.

The second type of organotin stabiliser comprises dibutyltin compounds in which various kinds of sulphur-containing groups are incorporated. These are commonly known as 'thio-organotin compounds'. As a class they are superior to the non-sulphur-containing butyltin compounds, both in stabilising power and the level of clarity which can be achieved by their use. They do, however, suffer from the fact that they usually possess rather unpleasant smells which are imparted to a greater or lesser degree to the PVC compositions in which they are incorporated. The most common butyltin sulphur-containing stabilisers are mercaptans, mercapto esters, mercaptoamides, derivatives of mercaptobenzothiazole, xanthates, and dithiocarbamates, exemplified in dibutyltin dilauryl mercaptan, dibutyltin thiomaleate, dibutyltin dinonyl thioglycollate, and dibutyltin dioctyl thioglycollate.

The butyltin compounds as a class are generally toxic, and are thus barred from many interesting applicants, such as packaging of food-stuffs. A number of monobutyltin compounds derived from butyl thiostannic acid have been developed[79] and are claimed to be non-toxic,[80] but they do not appear to have become generally available. Octyltin compounds analogous to the dibutyltin compounds previously discussed were developed specifically to overcome the toxicity problem, and here is much evidence to suggest that these compounds are of a very low level of toxicity.[81] The most common octyltin stabilisers are the dioctyltin laurate, maleate, dinonyl thioglycollate, and dioctyl thioglycollate. Their efficiency as stabilisers is comparable with the butyltin compounds, but their costs are higher.

Other octyltin compounds, that have been approved for food packaging materials by the US Food and Drugs Administration, include di-*n*-octyl cis-butenedioate polymer and di-*n*-octyltin-S,S'-bis(iso-octyl/mercaptoethanoate).[107] Dimethyltin-S,S'-bis(iso-octyl mercaptoethanoate) was introduced as a stabiliser for PVC potable water piping.[107]

In the mid-1970s a new class of organotin stabilisers, the so-called 'estertins', was intro-

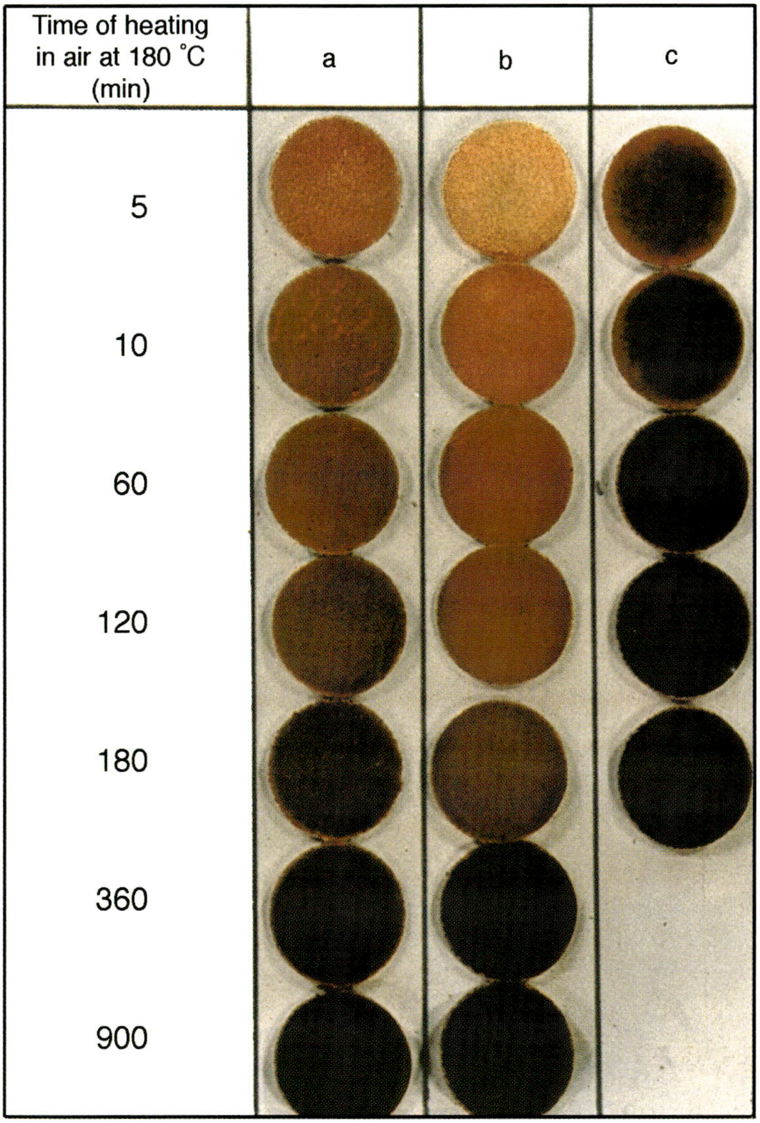

Time of heating in air at 180 °C (min)	a	b	c
5			
10			
60			
120			
180			
360			
900			

Plate 5.1 Thermal degradation of PVC resins: (a) suspension homopolymer; (b) spray-dried emulsion homopolymer; (c) suspension copolymer (10% vinyl acetate).

Time of pressing at 180℃ (min)	a	b	c	d
15				
30				
45				
60				
75				
90				
120				

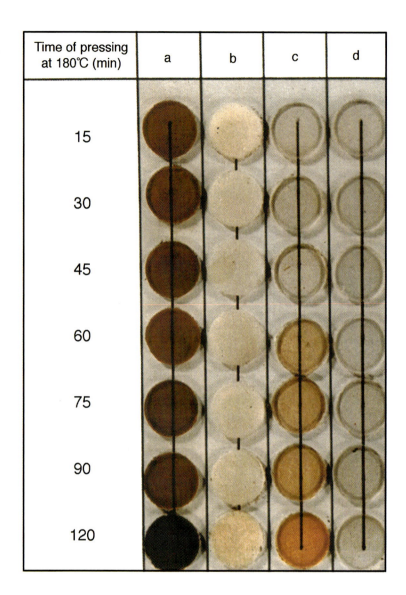

Plate 5.2 Stabilisation of PVC (all at 2 phr): (a) calcium stearate; (b) basic lead carbonate; (c) barium/cadmium complex; (d) sulphur-containing organotin compound.

duced. These are esters of dialkyltin dipropionic acid or trialkytin propionic acid.[98, 103, 105] They appear to have similar activity to the older alkyltin compounds.

5.4.6 EXPOXY COMPOUNDS

The exopy group $\overset{\displaystyle O}{\overset{\displaystyle \diagup\diagdown}{-CH-CH-}}$ seems to have a generally stabilising effect on PVC compositions, possibly owing to its ready reaction with hydrogen cloride to form chlorhydrins:

$$ \overset{\displaystyle O}{\overset{\displaystyle \diagup\diagdown}{-CH-CH-}} + HCl \longrightarrow -CHOH.CHCl- $$

Epoxy compounds are not sufficiently effective for use as sole stabilisers, but they are frequently employed with other compounds to impart a degree of 'long-term stability', and barium/cadmium stabilisers are rarely employed without them. Provided that other factors do not interfere, the stabilising effect of epoxy compounds generally increases with oxirane content up to about 9 per cent.[82]

Many liquid epoxy compounds are reasonably effective plasticisers for PVC and thus contribute to the plasticiser content in flexible compositions, occasionally even finding use as sole plasticiser. In unplasticised compositions this is often an embarrassment, even at the low concentrations of 2–5 phr commonly employed for stabilisation, leading to reductions in softening temperatures; and non-plasticising solid epoxy compounds are then preferable. Very many epoxy compounds have been recommended as stabilisers for PVC, but in the main they fall into five ill-defined groups:

(1) Glycidyl ethers, such as those of resorcinol, di-isobutylphenol, and tetraphenyolmethane; and 1-phenoxy-2,3-epoxypropene.
(2) Esters, either glycidyl esters such as the oleate and laurate, or esters of epoxyacids, such as butyl, capryl, and octyl epoxystearate, butyl epoxyoleate, octyl epoxyphthalate, and d-isodecyl expoxytetrahydrophthalate.
(3) Epoxydised oils, such as soya bean, linseed, and castor oils.
(4) Stearin derivatives, such as epoxy butyl stearin, epoxy acetoxy stearin and epoxy isobutyl acetoxy stearin.
(5) Epoxy resins.

Of these epoxydised soya bean oils, esters of epoxyacids, and epoxy resins are the most commonly encountered, the first two mainly in plasticised compositions, and the third in unplasticised compositions where it is desired to maintain the softening temperatures as high as possible. It has been suggested that epoxy resin and dibutyltin dilaurate exhibit marked synergism. On the other hand it has been suggested that in respect of outdoor weathering epoxy stabilisers are more effective in combination with barium/cadmium stabilisers than with organotins.[104]

5.4.7   ORGANIC PHOSPHITES, CHELATORS, AND ANTIOXIDANTS

The confusion and aura of mystery arising from the lack of understanding of the mechanisms of degradation and stabilisation is particularly manifest in the above materials. This confusion is not helped by the fact that commercial materials are much more often than not specified simply as 'phosphite', 'chelator', or 'antioxidant'. These materials are not stabilisers in their own right, but augment the activity of other stabilisers, particularly those based on barium, cadmium, and zinc, and usually in combination with epoxy compounds. The organic phosphites are generally regarded as acting as antioxidants, but it has been suggested that they also act as chelators for metal ions which might accelerate degradation. Of those phosphites which have been specifically named, triphenyl, tritolyl, tris-nonylphenyl, diphenyl iso-octyl, diphenyl isodecyl, di-isodecyl phenyl, tri-isodecyl, and tri-octyl phosphites are offered commercially. The concentrations of organic phosphites employed in PVC are usually quite low, generally being in the region of 0.25–0.5 phr, and certainly not above 1 phr.

Other types of chelators and antioxidants are rarely characterised specifically, but antioxidants of the phenolic type, particularly 2,4,6-trisubstituted phenols and diphenylolpropane derivatives are probably quite commonly used. This type of additive is useful in reducing degradation of plasticisers,[83] particularly in compositions designed to operate at relatively high temperatures.

5.4.8   MISCELLANEOUS COMPOUNDS

A number of organic compounds outside the scope of the preceding sections are often used as stabilisers, particularly in unplasticised PVC, where the non-toxic ones can be particularly valuable. Urea and thiourea, and most of their substituted derivatives, have some stabilising effect, particularly in unplasticised compositions. The most common and probably the most effective of these is diphenylthiourea, though monophenylurea and monophenylthiourea are occasionally encountered. On the Continent α-phenylindole seems to be preferred, and it does seem to have advantages over diphenylthiourea in clear compositions. Other organic compounds used as stabilisers on a limited scale include esters (particularly the ethyl ester) of β-aminocrotonic acids, and aminophenyl sulphones.

Most of these compounds are usually employed in conjunction with other stabilisers, generally at concentrations of about 0.5-1 phr.

Nearly 20 years ago antimony mercaptides, mainly antimony trimercaptide, were introduced,[106] but in spite of the fact that they are claimed to be similar to, but more effective at low concentrations than liquid organotin stabilisers, they have made little progress so far. They appear to have marked synergism with calcium stearate.[103]

5.4.9   SELECTION OF STABILISERS

From the foregoing sections it will be apparent that selection of stabilisers in commercial production practice can be complicated. What usually happens is that, within a company concerned with formulating, a body of knowledge and experience is built up and continually added to, so that it is rarely necessary to start from scratch.

In the selection of types the problem need not be severe, as was suggested in Sub-section

5.4.1. In the current state of the market, if the composition has to meet stringent requirements of non-toxicity, the choice is limited to a few accepted materials, and the most commonly used type of system would be a combination of epoxydised soya bean oil (3–5 phr) with a calcium-magnesium/zinc complex (1.5–3 phr). Improvements in the performance of such systems appear to have eliminated any need to boost their effect with indole or thiourea derivatives.

An alternative but more expensive system is one based on octyltin stabilisers, where legislation permits their use and where it is agreed that they are acceptable. Such a system permits a high degree of clarity and good stability. Octyltins may also be used in combination with other stabilisers, including calcium/magnesium/zinc, epoxy, and organic types. In the specific case of unplasticised PVC pipe and fittings for potable water it is usually possible to use tribasic lead sulphate, provided that the concentration is kept sufficiently low for the amount extracted to be within safe limits. This usually means a concentration of about 2 phr, which of course requires particular care in the design of processing equipment and in the actual processing itself.

Where clarity is required but toxicity is not important, sulphur-containing butyltin stabilisers give best results but are expensive. At somewhat lower cost and with practically as good results with the best grades, barium/cadmium or barium/cadmium/zinc complexes are selected, usually at about 2 phr concentration, and in combination with 2–5 phr of epoxydised oil in plasticised compositions or with about 1 phr of epoxy resin in unplasticised compositions. About 0.5 phr of an organic phosphite might also be included for best results.

In opaque compositions about 5 phr of tribasic lead sulphate is probably as good a choice as any with a marginal commercial advantage in using basic lead carbonate in plasticised compositions. There is a trend to use barium/cadmium stabilisers even in opaque compositions and to obtain the opacity by means of a pigment.

More detailed specific recommendations are considered later in conjunction with specific processing and application requirements (Chapter 8).

Most manufacturers of stabilisers now offer what are termed 'package' systems, each of which consists of a combination of two or more individual stabilising substances formulated to meet specific situations and often designed to meet a specific need of a particular customer.

Selection between competitive materials will have to rely on advice from suppliers of raw materials, especially polymers, or will involve more or less extensive programmes of test. Experimental procedures suitable for such testing are described later (Chapter 18). Plate 5.2. illustrates some typical results of tests on a few different stabiliser systems.

5.4.10   MECHANISMS OF STABILISATION

Although most if not all stabilisers for vinyl chloride polymers react with hydrogen chloride, this is certainly not always the only or even the main function. Indeed, it is also a fact that mere ability to react and neutralise hydrogen chloride is not sufficient to confer on a substance the attribute of stabilising PVC. The fact that PVC will continue to decompose if the hydrogen chloride is removed suggests, of course, that some other activity could be useful in stabilisation. The situation is further complicated by the fact that the metal chlorides form by reaction of some stabilisers can apparently act as catalysts for further decomposition.[7, 84, 85]

From a consideration of possible mechanisms (Section 5.3) one can generalise and speculate that a substance could contribute effective stabilising activity on vinyl chloride polymers by the following types of action.

(1) By reaction with liberated hydrogen chloride, especially if the resultant product is neutral to the decomposition reaction.
(2) By reaction with polymer molecules at points of deviation from the idealised structure so as to reduce the number of positions where decomposition might be more readily initiated.
(3) By interference with the propagation reaction.
(4) By reaction with oxygen to form products without activating effect on the decomposition.
(5) By reaction with impurities, e.g. metals, also to form products without activating effect on the decomposition.
(6) In terms of reducing colour development, by reacting with double bonds or other chromophoric structures.

It seems likely that all of the above are involved at times, those relevant in any particular case being dependent on the nature of the stabiliser system used. It has been shown by radioactive labelling of different parts of some metal soap and organotin stabilisers that some part of these stabilisers is retained by polymer after heating to normal decomposition temperatures, though the bonding appears to be weak.[84, 86] It also appears to be associated with the carboxylate parts of the stabiliser molecules rather than the metals.[86]

At present, however, although a number of theories have been advanced, there is little confirmation or co-ordination between them, and it is probably not worth enlarging on the subject here. The interested reader is referred to the reference listed in Gedde's review[12] and elsewhere.[2, 84–86, 103, 104]

5.4.11   STABILISATION AGAINST PHOTODEGRADATION

Although stabilisation against photodegradation may be aided by chemical means, optical filtration of the harmful radiation seems to be the most effective method.[7] While the effect may be at a maximum with a wavelength of 310 nm,[7, 87] some degradation is observed at wavelengths above 455 nm, and screening of radiation of wavelengths below this value is required for good stability.[88] From the many ultra-violet absorbing chemical available certain 2-hydroxybenzophenone derivates, phenyl salicylate and similar compounds, and 2-hydroxphenylbenztriazol derivatives appear to be most effective for PVC.[7] Many pigments, particularly if organic, and in combination with rutile (titanium dioxide) are effective ultra-violet stabilisers.[88] The value of dibasic lead phosphite may also be due, at least in part, to its opacity.

The decomposition of PVC is often complicated by ingredients other than the polymer. Thus, plasticisers may decompose or be lost by volatilisation. Chlorinated paraffin extenders are usually very detrimental to stability to sunlight, but here dibasic lead phosphite has a remarkably good effect, even when compared with white lead. Barium/cadmium stabilisers have been thought to yield better light-stability than organotin compounds, and epoxy compounds have been shown to have a light-stabilising effect,[89] but these effects could be due to the elimination or avoidance of certain unstable groups in the polymer, rather than direct stabilisation against ultra-violet radiation.

REFERENCES

1. R.R. STROMBERG, S. STRAUS and B.G. ACHHAMMER, *J. Polym. Sci.,* **35**, 355 (1959)
2. C.F. BERSCH, M.R. HARVEY and B.G. ACHHAMMER, *J. Res. natn Bur. Stand.,* **60**, 481 (1958)
3. P. BRADT and F.L. MOHLER, *J. Res. natn. Bur. Stand.,* **55**, 325 (1955)
4. I. OUCHI, *J. Polym. Sci.,* A, 3, 2685 (1965)
5. E.C.A. HORNER, *Chemy Ind.,* **12**, 308 (1960)
6. B. BAUM, *S.P.E. Jl,* **17**, 71 (1961)
7. W. JASCHING, *Kunststoffe,* **52**, 8, 458 (1962)
8. E.LO. SCALZO, *Materie plast.,* **28**, 682 (1962)
9. F. CHEVASSUS and R. DE BROUTELLES (Trans. by C.J.R. Eichhorn and E.E. Sarmiento), *The Stabilization of Polyvinyl Chloride.* Arnold, London (1962) (Original French Edition, Amphora, 1957)
10. S.L. MADORSKY, *Thermal Degradation of Organic Polymers,* Interscience, New York (1964)
11. G.YA. GORDON, *Stabilisation of Synthetic High Polymers,* Israel Program for Scientific Translations (1964)
12. W.C. GEDDES, *RAPRA tech. Rev.,* 31, May (1966)
13. D. DRUESEDOW and C.F. GIBBS, *Natn Bur. Stand.,* Circular 525, 69 (1953)
14. C.B. HAYNES, *Natn Bur. Stand.,* Circular 525, 107 (1953)
15. D.K. TAYLOR, *Trans. J. Plast. Inst.,* **28**, 170 (1960)
16. A. RIECHE, A. GRIMM and H. MUCKE, *Kunststoffe,* **52**, 265 (1962)
17. W.I. BENGOUGH and H.M. SHARPE, *Makromolek. Chem.,* **66**, 31 (1963)
18. E.J. ARLMAN, *J. Polym. Sci.,* **12**, 547 (1954)
19. B. BAUM and L.H. WARTMAN, *J. Polym. Sci.,* **28**, 537 (1958)
20. G. TALAMINI and G. PEZZIN, *Makromolek. Chem.,* **39**, 26 (1960)
21. A.S. DANYUSHEVSKII, N.YA. PARLASHKEVITCH, Z.N. FROLOVA and I.S. SHENTSIS, *Soviet Plast.,* **2**, 63 (1961)
22. D. BRAUN and M. THALLMAIER, *Kunststoffe,* **56**, 80 (1966)
23. A. CITTADINI and R. PAOLILLO, *Chimica Ind., Milano,* **41**, 10, 980 (1959)
24. C. CORSO, *Chimica Ind., Milano,* **43**, 8 (1961)
25. D. DOLEZEL and J. STEPEK, *Chemicky prumysl,* **10**, 381 (1960)
26. M. LISY and S. VARDA, *Chemické zvesti,* **14**, 14 (1960)
27. P. BERTICAT and C. VALLET, *Compt. rend.,* **261**, 2102 (1965)
28. A. GUYOT, P. ROUX and P. QUANG THO, *J. appl. Polym. Sci.,* **9**, 1823 (1965)
29. M. LISY, *Chemické zvesti,* **19**, 84 (1965)
30. N.V. MICKHAILOV, L.G. TOKAREVA and U.S. KLIMENKOV, *Kolloid. Zh.,* **18**, 578 (1956)
31. Z.V. POPOVA and D.M. YANOVSKII, *J. appl. Chem. USSR,* **33**, 182 (1960)
32. A. GUYOT and J.P. BENEVISE, *J. appl. Polym. Sci.,* **6**, 98 (1962)
33. J. URBANSKI, *Polim., Tworzywa wielkoczast.,* **8**, 331 (1963)
34. ASTM D 793–49
35. J. STEPEK, Z. VYMAZEL and B. DOLEZEL, *Mod. Plast.,* **40**, 146 (1963)
36. G. TALAMINI, G. CINQUE and G. PALMA, *Mat. plat. elast.,* **30**, 317 (1964)
37. G.YA. GORDON, *Soviet Plast.,* **4**, 9 (1965)
38. Z. VYMAZEL, B. DOLEZEL and J. STEPEK, *Kunststoffe,* **56**, 86 (1966)
39. W.C. GEDDES, *RAPRA Res. Rep.,* 154 (1966)
40. R.F. BOYER, *J. phys. Coll. chem.,* **51**, 80 (1947)
41. V.W. FOX, J.G. HENDRICKS and H.J. RATTI, *Ind. Engng Chem. analyst. Edn,* **41**, 1774 (1949)
42. A. CHARLESBY and M.G. ORMEROD, *Vth Int. Symp. Free Radicals,* Uppsala, 111 (1961)
43. G.J. ATCHISON, *J. appl. Polym. Sci.,* **7**, 3, 1471 (1963)
44. J.P. ROTH, P. REMPP and J. PARROD, *Polym. Sci.,* **C**, **4**, 1347 (1963)
45. W.I. BENGOUGH and I.K. VARMA, *Eur. Poly, Jnl,* **2**, 61 (1966)
46. D. BRAUN and M. THALLMAIER, *Makromolek. Chem.,* **99**, 59 (1966)

47. G.C. Marks, J.L. Benton and C.M. Thomas S.C.I. Conference, *Advances in Polymer Science and Technology*, London, Paper 11 (Sept. 1966)
48. J.B. Decoste, *S.P.E. Jl*, **21**, 764 (1965)
49. P.V. McKinney, *J. appl. Polym. Sci.*, **9**, 3359 (1965)
50. E.A. Collins and C.A. Krier, *J. appl. Polym. Sci.*, **10**, 1573 (1966)
51. E.W. J. Michell and D.G. Pearson, *J. appl. Chem.*, **17**, 171 (1967)
52. D. Druesedow and C.F. Gibbs, *Mod. Plast.*, **30**, 123 (1953)
53. N. Grassie, *Trans. Faraday Soc.*, **48**, 379 (1952); **49**, 835 (1953)
54. N. Grassie, *Chemy Ind.*, **6**, 161 (1954)
55. W.I. Bengough and H.M. Sharpe, *Makromolek. Chem.*, **66**, 31 and 45 (1963)
56. A. Crosato-Arnaldi, G. Palma, E. Peggion and G. Talamini, *J. appl. Polym. Sci.*, **8**, 747 (1964)
57. F. Sondheimer, D.A. Ben-Afraim and R. Wolovsky, *J. Am. chem. Soc.*, **83**, 1675 (1960)
58. P.J. Flory, *J. Am. chem. Soc.*, **61**, 1518 (1939)
59. A.A. Berlin, Z.V. Popova and D.M. Yanovskii, *S.P.E. Trans.*, **3**, 27 (1963)
60. D.E. Winkler, *J. Polym. Sci.*, **35**, 3 (1959)
61. G.M. Dyson, J.A. Horrocks and A.M. Fernley, *Plastics*, **26**, 288, 124 (1961)
62. J. Novak, *Kunstoffe*, **52**, 269 (1962)
63. D.H.R. Barton and P.F. Onyon, *Trans. Faraday Soc.*, **45**, 725 (1949)
64. D.H.R. Barton and K.E. Howlett, *J. chem. Soc.*, 155 and 165 (1949)
65. M. Asahina and M. Onozuka, *J. Polym. Sci.*, A, **2**, 3505 (1964)
66. J.J.P. Staudinger, *Plast. Prog., Lond.*, **9** (1953)
67. H.V. Smith, *Rubb. J. int. Plast.*, **138**, 966 (1960)
68. A. Potocki, A. Balas and B. Dudek, *Polim., Tworzywa wielkoczast.*, **8**, 16 (1963)
69. G. Chiltz, P. Goldfinger, G. Huybrechts, G. Martens and G. Verbeke, *Chem. Rev.*, **63**, 355 (1963)
70. W.I. Bengough and M. Onozuka, *Polymer*, **6**, 625 (1965)
71. B.A. Kargin, T.L. Sogolova, G.L. Slonimskii and E.V. Reztsova, *Zh. fiz. Khim.*, **30**, 1903 (1956); *RABRM Translation* No. 568
72. A.A. Berlin, G.S. Petrov and Y.F. Prosvirkina, *Zh. fix. Khim.*, **32**, 2565 (1958); *Khim. Nauka Prom.*, **2**, 522 (1957); *RABRM Translation* No. 615
73. N.K. Baramboim and G.K. Pytov, *Nauch, Trudy Mosk. Teckhnol. Inst. Legk. Prom.*, **18**, 48 (1960); *RAPRA Translation* No. 983
74. J.A. Rhys, 'PVC Statilisation', Chapter 3 of *Advances in PVC Compounding and Processing* (M. Kaufman, ed.), Maclaren, London (1962)
75. W.S. Penn, *Rubb. Plast. Wkly*, July to August (1961)
76. W.S. Penn, *PVC Technology*, Maclaren, London (1966)
77. *Brit. Pat.* No. 687 425
78. V. Yngve, *U.S. Pat.* 2 191 463
79. C. Dorfelt, *German Pat.* 1 078 772; *U.S. Pat.* 3 021 302
80. H.V. Smith, *Br. Plast.*, **37**, 8, 445 (1964)
81. J.M. Barnes and H.B. Stoner, *Br. J. ind. Med.*, **15**, 15 (1958)
82. H.C. Murfitt, *Br. Plast.*, **33**, 12, 578 (1960)
83. *Resin Rev.* (pub. Rohm & Haas), **18**, 4, 17 (1968)
84. H.F. Frye and R.W. Horst, *J. Polym. Sci.*, **45**, 1 (1960)
85. G.M. Dyson, J.A. Horrocks and A.M. Fernley, *Plastics*, **26**, 288, 124 (1961)
86. H.F. Frye, R.W. Horst and M.A. Paliobagis, *J. Polym. Sci.*, A, **2**, 1765, 1785 and 1801 (1964)
87. R.C.Hirt, N.Z. Searle and R.G. Schmitt, *S.P.E Trans.*, **1**, 1, 21 (1961)
88. J.B. Decoste and R.H. Hansen, *S.P.E. Jl*, **18**, 4, 431 (1962)
89. A. Merz, *Kunststoffe-Plast.*, **6**, 2, 169 (1959)
90. J.H. Nelson, 'PVC Processing', *PRI International Conference*, Royal Holloway College, Egham Hill, Surrey, England (6–7 April 1978), Paper 2
91. D. Braun, *Degradation and Stabilisation of Polymers* (G. Geuskens., ed.) Applied Science Publishers Ltd., London (1975) Chapter 2
92. A.A. Yossin and M.W. Sabaa, *J. Polym. Sci., Polym. Chem. Ed.*, **18**, 2513, 2523 (1980)
93. W.H. Starnes and D. Edelston, *Macromolecules*, **12**, 797 (1979)
94. W.H. Starnes, *Developments in Polymer Degradation*, – **3** (N. Grassie, ed.), Applied Science Publishers Ltd. (1981), p. 135

95. T. Hjertberg and E.M. Sorvik, *Degradation and Stabilisation of PVC* (E.D. Owen, ed.) Elsevier Applied Science Publishers (1984) Chapter 2

96. W.V. Titow, *PVC Technology*, Applied Science Publishers Ltd. (1984) Chapter 9

97. T.C. Jennings and C.W. Fletcher, *Encyclopedia of PVC* (L.I. Bass and C.A. Heiberger, eds.) Marcel Dekker, New York (1988) Vol. 2, Chapter 2

98. W.V. Titow, *PVC Plastics*, Elsevier Applied Science (1990) Chapter 4

99. P.J. Donnelly, 'PVC Processing', *PRI International Conference*, Royal Holloway College, Egham Hill, Surrey, England (6–7 April 1978), Paper 1

100. Ciba-Geigy, *Plastics & Rubber Weekly* (9 Jan. 1993)

101. R.M. Lum, *J. Appl. Polym. Sci.*, **20**, 1635 (1976)

102. B.B. Troitskii, V.A. Sozorov, F.F. Minchuk and L.S. Troitskaha, *Eur. Polym. J.* **11**(3), 277 (1975)

103. W.V. Titow, *Developments in PVC Production and Processing – 1* (A. Whelan and J.L. Craft, eds.) Applied Science Publishers Ltd. (1977) Chapter 4

104. E. Szaks and R.E. Lally, *Polym. Engng. Sci.*, **15**(4), 277 (1975)

105. D. Lanigan, Krauss-Maffei 5th International Extrusion Symposium, Linz/Asten (8–11 Sept. 1976)

106. D. Dieckmann, *SPE 34th ANTEC Proceedings*, 507 (1976)

107. P. Smith and L. Smith, *Chem. in Britain*, **11**(6) 208 (1975)

108. European Communities Council Directive 91/338/EEC (18 June 1991); Offl. J. Eur. Com., No. L186/59–63 (12 July 1991)

109. U.K. Dept. of the Environment, *Environmental Protection (Controls on Injurious Substances) (No. 2) Regulations 1993*; SI No. 1643 (31 July 1993)

6
Plasticisation

6.1 INTRODUCTION

At one time plasticisers were commonly used to reduce the softening and thus the processing temperatures required for PVC, but while this device is still used occasionally (e.g. in injection moulding of some rigid PVC products), plasticisation of PVC is almost always done in order to achieve flexibility.

As with degradation and stabilisation, there is extensive literature on the subject of plasticisation in general[1-5, 57, 58] and that of PVC in particular.[1, 6, 59] While the broad theories advanced, whether from a thermodynamical[7-10] or an empirical[11, 12] viewpoint, are generally well presented and easy to understand, elaborations of theory and application to practical plasticisation have often been confused, and thus require considerable care when used to resolve actual problems of formulation.

Plasticisation involves two distinct processes or functions. The processor is not only concerned about the way in which plasticisers modify the properties of PVC products, but also with the rates at which and the mechanisms by which plasticisers are induced to penetrate the polymer matrix there to exert their influence on properties.

In developing a PVC formulation for a given application, the first consideration will usually be the flexibility required, and this, taken in conjunction with the dimensions of the finished product, will determine within fairly narrow limits the required concentration of any particular plasticiser. The precise plasticisers, or combination of plasticisers, will be determined by other property and commercial considerations which will be discussed later.

6.2 GENERAL CONSIDERATIONS

The effects of plasticisers on properties are, of course, of prime concern in considering their use in particular manufactured products. In addition to the main effect of producing flexibility practically every other property of PVC is changed by the introduction of plasticisers, and attempts have been made to relate these to each other.[13] Qualitatively, it is found that as plasticiser concentration increases, there are corresponding *reductions* in modulus, tensile strength, hardness, density, T_g, softening temperatures, brittle temperatures (however measured), and volume resistivity, while at the same time there are *increases* in elongation at break, toughness, softness, dielectric constant, and power factor. The relationships between changes in these properties and plasticisers content depend on the nature of the particular plasticiser. Electrical properties are very dependent on degree of purity, and care must be exercised when drawing theoretical conclusions from values of these, as small amounts of impurity can produce effects which completely overshadow those due to basic plasticiser structure.

In addition to the above rather obvious properties, the requirements of an end-product often demand attention to others dependent on the plasticiser. Thus, effects on weathering behaviour would clearly need to be considered for outdoor applications. Changes in flammability may occur and need to be taken into account. Loss by slow volatilisation, depending on the nature of the plasticiser and the ambient conditions, can lead to stiffening, general deterioration of mechanical properties, and cracking. Loss by extraction by liquids can be equally serious in packaging and piping applications. Migration to adjacent materials can lead to crazing or deterioration in electrical insulation properties of the latter. Several other specific effects can occur, and are considered in more detail later.

Before any of the aforementioned effects on the properties of PVC can be achieved the molecules of a plasticiser have to be induced to assume the appropriate locations in the polymer matrix, and once there to stay there. The thermodynamics of these two functions appear to be closely related. Plastisication may involve diffusion into particles in a more physical sense as well as diffusion on a molecular scale.

6.3   THEORY OF PLASTICISATION

6.3.1   GENERAL MECHANISM

Several theories have been advanced to explain the mechanism of plasticisation,[57, 59] but they tend to be inadequate to explain all aspects of plasticiser action, and there is considerable overlap between them. As shown earlier (Section 4.2) polyvinyl chloride has largely a head-to-tail arrangement of vinyl chloride units, with variable amounts of chain-branching, and only a small amount of chrystallinity. The highly electro-negative nature of chlorine leads to dipoles along the polymer molecules, and these result in high concentrations of secondary valency forces, thus resulting in reduction of flexibility in individual molecules and consequently account for the rigidity of unplasticised PVC. Chlorine atoms are sufficiently bulky to separate the polymer chains, and van der Waals forces, which are responsible for cohesion in polyethylene and similar polymers, are comparatively ineffective. Reduction of the dipole bonding therefore reduces the restrictions on deformation of the molecules, so that increased flexibility and flow result. Plasticisers for PVC have molecules with polar or polarisable groups which have been envisaged as bonding with the polymer dipoles, and non-polar parts which act as shields between polymer dipoles, thus resulting in less cohesion overall, with consequent increased freedom of molecular movement.

6.3.2   THEORIES OF PLASTICISER ACTION

After more than four decades during which extensive studies have been carried out, a single satisfactory comprehensive theory of plasticiser action in PVC has not appeared, and recent work[59] has cast doubt on some of the views that have been tacitly accepted over the past thirty years or so.

6.3.2.1 *Lubricity theory*

As its name implies, the lubricity theory postulates that plasticisers act as lubricants between polymer molecules, enabling the latter to move more readily with respect to each other.[60–62] It is assumed that there is no bonding between polymer molecules and at most only weak bonding between plasticiser and polymer molecules. By itself the lubricity theory is of little value, but it has some applicability at higher temperatures where plasticiser action can be related to coefficients of friction.[63]

6.3.2.2. *Gel theory*

This theory postulates that the rigidity of a polymer arises from a three-dimensional honeycomb structure resulting from loose attachments at points along the polymer molecules.[62] A plasticiser acts by breaking these attachments and masking centres of force. Some plasticiser molecules remain unattached to polymer molecules except indirectly through other plasticiser molecules, but are effective in swelling the gel structure, and so permitting easier movement between polymer molecules and hence increasing flexibility.

6.3.2.3 *Free volume theory*

Free volume theory is based on a consideration of glass transition temperatures (T_g). Defined as the difference between specific volume at the temperature under consideration and the specific volume at absolute zero, free volume is difficult to determine with certainty because some substances exhibit unexpected behaviour near 0K. Free volume increases with increase in the number of end-groups or side-chains present. Plasticiser molecules are smaller than polymer molecules, hence their introduction into a polymer matrix raises the proportion of end-groups and lowers T_g.

   The theory has been developed mathematically to yield an equation relating change in glass transition temperature to the concentration and physical properties of added plasticiser,[57, 64] but, although predictions that follow from it agree qualitatively with experimental data, it is less successful numerically. Thus Zhurkov's Rule that equal molar concentrations of different plasticisers have the same effectiveness, in accordance with the free volume equation, is not borne out by measurements of mechanical properties of PVC containing different plasticisers over a range of concentrations. Also the phenomenon of 'anti-plasticisation', where small proportions of plasticisers lower T_g but increase tensile strength and modulus, does not fit the equation, partly because of its dependence on the glass transition temperature of unmodified polymer.

6.3.2.4 *Solvation–desolvation equilibrium*

Doolittle[7] proposed a 'mechanistic' theory which embraced aspects of lubricity, gel, and free-volume theories. It postulated that plasticiser molecules separate and become attached to points of attachment along the polymer molecules, but are continually being replaced by other molecules. It was implied that no stoichemical relationship exists between polymer and plasticiser. With PVC it is almost implicit that some such relationship should exist.

   From torque rheometer experiments Hartmann[16] concluded that mole fractions of plasticiser to vinyl chloride units of 0.08 and 0.15 of any phthalate have special significance in completing 'solution'. Barshtein and Kotlyarevskii [65] assumed that PVC has a helical repeat unit $C_{28}H_{42}Cl_{14}$ of 'molecular weight' 875, and derived the following equation for the amount of plasticiser required to block attachments between the polymer molecules:

$$PHR = 100M/875 = 0.114M$$

where M is the molecular weight of the plasticiser. From this equation 45 phr of dioctyl phthalate would be required, equivalent to a 'mole fraction' of about 0.07, i.e. of the same order as Hartmann's 0.08.

6.3.2.5   Generalised structure theories

In these the polymer is regarded as having regions of order and disorder, the former constituting micelles that may be large enough to be crystallites, with lower free volume than amorphous regions. Small amounts of plasticiser increase free volume, permitting some increase in crystallite regions and resulting in increased rigidity. This is in accordance with Horsley's studies[66] using X-ray diffraction, which showed an increase in order as the proportion of dioctyl phthalate was increased from zero up to 10–15 per cent, beyond which the order became reduced. The changes in order paralleled changes in tensile strength, modulus, impact strength, and refractive index at T_g, whereas the value of T_g dropped fairly rapidly and continuously as plasticiser was introduced, from 60 to 20°C. This is the phenomenon often referred to as 'antiplasticisation' referred to in Section 6.3.2.3. It was also shown that quenching from the 'melt' inhibited antiplasticisation. It has been suggested that when larger proportions of plasticiser are added crystallites may still form, i.e. crystallinity may increase, but amorphous regions become so swollen that the whole mass becomes softer and more flexible, the regions of order remaining contributing to the strength of PVC. However this seems to be at variance with Horsley's observations using X-ray diffraction.

Moorshead[12] adopted the suggestion that the dominant force of cohesion in PVC is dipole–dipole attraction, and found that compounds that plasticise PVC are those that have suitable dipole structures to produce similar cohesion to the polymer. Plasticiser molecules are supposed to be held in place by polar and polarisable groups attached to the polymer dipoles. Non-polar and non-polarisable groups in the plasticiser molecules are thus held between the polymer molecules, separating them and reducing overall cohesion. The concentration of dipoles is revealed in dielectric constant or permittivity, which suggest that a good plasticiser for PVC should have similar permittivity to the polymer. According to Moorshead all true plasticisers for PVC have permittivities between 4 and 8, equivalent to one polar group per 6 to 14 carbon atoms. On the other hand unplasticised PVC appears to have a permittivity of between 2 and 3. It is interesting to note that the permittivity of a plasticised PVC composition is higher than that predicted from summation of the volume fractions of polymer and plasticiser,[57] and Dannis[67] has used this fact to follow the rate of absorption of plasticiser into unplasticised PVC.

Recent work by Howick and colleagues[59] using NMR and molecular modelling techniques concluded that polarity of a plasticiser is important but not the determining factor. Interaction between polar centres of plasticiser and polymer was found to be probably electrostatic and not hydrogen bonding. Relaxation rate data revealed two phases, a polymer/plasticiser phase and a rigid PVC phase, and it was suggested that plasticiser does not penetrate amorphous polymer surrounding crystalline regions.

6.3.2.6   Interaction parameters

No discussion of theories of plasticisation would be complete without reference to some of the interaction parameters that have been proposed, though it has to be admitted that, because of the large amount of experimental data on the plasticisation capabilities of a wide range of compounds now available, they are of only limited potential value. The Hildebrand 'solubility parameter'[68] has been used to estimate compatibility between resin and plasticiser based on the solvating power of the latter. A solubility parameter is the square root of a 'cohesive energy density', which is the energy of vaporisation per molar volume and is thus a measure of the strengths of the attractive forces between the molecules of a substance. Obviously it cannot be determined directly for a polymer, especially for one like PVC which degrades at temperatures well below vaporisation, but solubility parameters can be estimated by Small's additive method.[43] Use of solubility parameters to predict plasticising behaviour is based on the supposition that the closer the correspondence between the parameters of polymer and plasticiser the greater the activity of the latter, but this is only very approximately borne out by experimental data.

A more successful parameter, at least for non-polymeric plasticisers,[58] is the 'polarity parameter',[70] which is proportional to the molecular mass of the plasticisers and the ratio of the number of non-aromatic, non-carboxylic acid, carbon atoms to the number of polar groups in the molecule.

Agnagnostopoulus and co-workers[17] related 'solid–gel transition' temperatures to Flory-Huggins 'interaction parameters', but, as suggested later, 'solid-gel' transitions are dependent on time as well as temperature, and only appear to be critically dependent on the latter because of the extreme dependence of transition rate on temperature. Bigg's 'activity parameters' are also based on Flory-Huggins interaction parameters.[71]

6.3.2.7   *Conclusions*

The existence of so many different approaches attempting to explain and predict plasticiser activity, and the less than complete correlation between the various parameters used to assess compatibility,[76, 77] reflect the facts that the subject is very complex and that to date no completely satisfactory and comprehensive theory has appeared. From the practical point of view this is now little consequence, but it is intellectually unsatisfactory. Certainly the simple picture of plasticised PVC consisting of a fairly uniform matrix of polymer molecules held apart to degrees depending on the proportion and nature of the plasticiser, the molecules of the latter being held in place by dipole bonding with polar regions in the polymer molecules, cannot be sustained. Boo and Shaw[72, 73] have postulated a two-phase system to explain the broad modulus/temperature response of PVC plasticised with dialkyl phthalates.

Howick[59] found that relaxation rate behaviour showed two phases; a PVC/plasticiser phase, and a phase yielding similar data to pure polymer. As mentioned in Section 6.3.2.5, it was also suggested that plasticiser concentration varies throughout the amorphous regions, and that plasticiser does not enter amorphous polymer surrounding crystalline regions. Although polarity of plasticiser molecules was confirmed as important it was shown to be not the determining factor. No evidence for hydrogen bonding was found, and it was concluded that interaction between polar centres of plasticiser and polymer is probably electrostatic.

Somewhat dubious support for the proposition that plasticised PVC is a multi-phase system comes from the 'folk-lore' sometimes held by those in the PVC industry decades ago that removal of plasticiser from a plasticised compound by volatilisation or extraction exhibited two stages, a proportion being readily removable, the remainder proving more stubbornly held in the polymer/plasticiser matrix. As far as the author is aware this suggestion has never been investigated systematically.

6.3.3   THE PROCESS OF PLASTICISATION (*See* Section 9.3.2 and Plate 9.1)

While low molecular weight solvents can enter a polyvinyl chloride molecular matrix at room temperatures, it is generally thought that thermal energy is required to reduce the cohesion arising from dipole attraction to a sufficient degree to permit plasticiser molecules to penetrate between the polymer molecules.[11] Indeed, many workers have postulated the existence of 'critical solution temperatures' for specific polymer/plasticiser pairs, below which plasticiser molecules do not penetrate into the polymer matrix,[14–19] and these have been related to depressed melting points in solution.[20] However, strong theoretical objections can be raised to the idea of critical temperatures in plasticisation processes,[21] and studies on polymer/plasticiser mixtures at constant temperatures below 'clear point' temperatures tend to support the idea that only classical diffusion and solution are involved,[22] the observation of 'pseudo-critical' changes being due to the extreme temperature dependence of the phenomena.

McKinney[74] followed the interaction between butyl benzyl phthalate and a porous particle suspension resin using differential thermal analysis. Mixtures of polymer and plasticiser stored at temperatures above ambient were found to develop the DTA characteristics of dry blends at rates that increased with storage temperature, the rate being very temperature dependent. Extrapolating results obtained over the range 60–100°C indicated that a time of the order of four years would be required at 25°C. Smith and Viney[22] maintained suspensions of polymer and plasticiser at various temperatures below 'clear points',[28] and determined 'swelling rates' and 'times to clear point' of individual polymer particles under the microscope. The results demonstrated a reasonable linearity of dependence of log (swelling rate) on reciprocal K, and on log (plasticiser viscosity). Plots of individual swelling on increasing temperature at 2 K/min revealed a marked inflexion around the PVC glass transition temperature. Plots of log (reciprocal time to clear point) against reciprocal K were also linear, but because the time scale involved in relation to the time scale of the project it was not possible to continue results below 100°C. Rates of interaction were found to increase with decrease in molecular weight of the polymer, and the log 1/t vs log 1/K plots appeared to extrapolate to the same point near the glass transition temperature of PVC, though the significance of this should not be stressed, because of the insufficiency of the data. It may be of interest to note that extrapolation of the results to temperatures below 100°C indicated 'gelation times' of the order of five days at 92°C and 500 days at 81°C. These are much longer than the periods indicated by McKinney's results. The situation is complicated by the possibility that polymer particles may have a skin of some form which resists penetration by plasticiser (see Section 4.4). Some commercial resins are claimed to be free of any such skin and consequently to absorb plasticiser at lower temperatures than other commercial polymers.

Classical theory of diffusion in isotropic materials is based on the hypothesis that the rate of transfer of a penetrating substance across unit area is proportional to the concentration gradient at right angles to the section,[23] thus:

$$F = - D\delta c/\delta x$$

where F = rate of transfer per unit area

c = concentration of diffusing substance

x = space co-ordinate normal to section

D = diffusion coefficient

For diffusion though a plane sheet of thickness l, with concentrations c_1 and c_2 on either side, we have:

$$(c-c_1)/(c_2-c_1) = x/l$$

This gives the concentration c at distance x from the surface, i.e. the concentration gradient is linear and the rate of diffusion constant, hence:

$$F = - D.dc/dx = D(c_1-c_2)/l$$

for a hollow sphere the equations become somewhat more complicated but can be reduced to the expression:

$$F = 4\pi D(c_1-c_2)ab/(b-a)$$

where a and b are the inner and outer radii, respectively. For a solid sphere $a = 0$ and the expression reduces to zero, which is not much help. Derivation of more useful expressions is complicated by the fact that, while the concentration of plasticiser external to a particle may remain constant throughout a large part of the plasticisation process, the concentration at the centre presumably builds up progressively from zero to an equilibrium value. Moreover, the shapes of polymer particles can be highly complex and their sizes are often variable over a wide range; and although it is not impossible to determine the surface areas of complex particles, the other geometric dimensions of many particles are practically impossible to determine or express mathematically.

Work on the dissolution of polymethyl methacrylate[24, 25] suggests that there might be other complicating factors. In this work a plot of the log of the velocity of penetration against the reciprocal of the temperature reveals two breaks, one of which is related to a glass transition temperature, T_g. The other is related to a 'gel temperature', below which dissolution occurs by the development of cracks in the polymer and breakdown, and above which substantial swelling occurs. The first penetrating front of solvent molecules is postulated as penetrating readily through micro-molecular channels. In the region of the glass transition the 'spongy hole' system collapses, so that solvent entering above the glass transition temperature has to create 'holes', giving rise to an increase in activation energy of dissolution in the elastic state. In such dissolution it is envisaged that solvent penetrates faster than the polymer molecules can disentangle and pass into solution. The penetrating front of solvent creates a swollen surface layer which is in the same elastic state as the dissolving polymer itself. This layer is limited on one side by unpenetrated polymer and on the other by a fluid layer which serves as a disentanglement zone where polymer molecules can become free to float off into solvent. It seems quite possible that a similar picture might be applicable to the penetration of polymer particles by plasticisers. It must be confessed, however, that in spite of the immense amount of work already expended on this superficially simple process, much remains to be done before an accurate analysis of the process can be made.

In the production of plasticised powders or plastisols, interaction between plasticiser and polymer under conditions of little or no shear is clearly most significant. In conversion process such as compounding, dry blend extrusion, or dry bend injection moulding, however, relative intensive shearing of the material is invariably involved. While some work under such conditions has been done, using the Brabender Plastograph or laboratory-scale intensive internal mixers of the Banbury type,[15, 26, 27] detailed theoretical interpretation has so far not been possible, partly because the nature of the individual polymer particles appears to play a very important part in determining what shear conditions will be developed in particular circumstances, and partly because measurement of shear in such systems is necessarily on a gross scale and cannot identify localised conditions within the polymer/plasticiser mixture.

Rates of interaction between polymer and plasticiser are dependent not only on temperature and shear, but also on the precise nature of the polymer and the plasticiser, but again, owing to the obscure state of the general theory of the interaction only limited theoretical guidance can be offered. Little has been published about the effects of variations in polymer parameters, but as a general rule, for a given plasticiser and type of polymer, interaction becomes more rapid with decrease in molecular weight of the polymer and with the introduction of vinyl acetate units into the polymer structure. Emulsion polymers also appear to be attacked more readily than suspension polymers of the same molecular weight. These effects are illustrated in Table 6.1 by the 'critical solution temperatures' which were, in fact, 'clear points' obtained by heating stirred mixtures of polymer (1 g) in plasticiser (20 ml), at a rate of 2°C/min.[28] 'Pseudo-melting points' determined by means of a Brabender Plastograph[20] are also indicated for comparison.

It is highly likely that plasticiser viscosity plays an important part in polymer/plasticiser interactions. Certainly it would be expected that absorption by entry into the pores of porous resin particles would be more rapid the lower the viscosity. This could well be why temperatures of around 100°C are usually necessary to produce plasticised dry blends. In compounding under conditions of high shear, however, the situation could be complicated by the possible effect of plasticiser viscosity on the shear stresses developed.

Most investigations into the effects of variations on polymer/plasticiser interactions have been concerned with variations in the nature of the plasticiser. Thus 'pseudo-melting points',[18, 20] fusion temperatures,[29] 'solid–gel transition temperatures',[30] 'pregel temperatures',[31] and 'gelling temperatures'[10, 32] have been determined for a variety of plasticisers, and related to the Flory-Huggins interaction parameter x, dielectric constants, molar volumes, and to ratios of number of carbon atoms to ester groups in plasticiser molecules (A_p/P_o), as shown in Table 6.2.

6.3.4.   PROPERTIES OF PLASTICISED PVC

While the rates of interaction between polymer and plasticiser can be important, the prime concern is usually with the properties (including cost) of fully processed material.

6.3.4.1   Compatibility

Continued compatibility under conditions of service is naturally essential. If a plasticiser is not completely compatible it may be possible to compound it to form an apparently homogeneous product, but on standing at atmospheric temperatures it will migrate to the surface,

Table 6.1 VARIATION OF 'CLEAR POINTS'[28] AND 'PSEUDO-MELTING POINTS'[20] WITH NATURE OF PLASTICISER, AND WITH MOLECULAR WEIGHT AND TYPE OF POLYMER

Polymer	ISO Viscosity No.	Plasticiser	Clear point (°C)	Pseudo-m.p. (°C)
EP suspension	125	TXP	105	—
EP suspension	96	DBP	—	81
EP suspension	115	DBP	—	95
EP suspension	125	DBP	108	—
EP suspension	185 (estd,)	DBP	—	103
Emulsion (spray-dried)	46	DOP	72	—
Suspension	46	DOP	92	—
Emulsion (spray-dried)	85	DOP	108	—
Suspension	85	DOP	108	—
EP suspension	96	DOP	—	110
EP suspension	105	DOP	113	—
EP suspension	115	DOP	—	117
Emulsion (spray-dried)	125	DOP	110	—
EP suspension	125	DOP	118	—
Suspension	125	DOP	118	—
Emulsion (coagulated)	140	DOP	107	—
Emulsion (spray-dried)	145	DOP	118	—
EP suspension	185 (estd.)	DOP	—	128
EP suspension	125	DNP	126	—
EP suspension	96	DIDP	—	121
EP suspension	115	DIDP	—	131 (estd.)
EP suspension	185 (estd.)	DIDP	—	135
EP suspension	96	DOA	—	129
EP suspension	115	DOA	—	140 (estd.)
EP suspension	185	DOA	—	150
EP suspension	125	DAS	140	—
EP suspension	96	DOS	—	145
EP suspension	115	DOS	—	152
EP suspension	185	DOS	—	165
EP suspension	125	PPS	160(?)	—

Note: The term 'EP' is retained in this Table to distinguish porous particle polymers from others. The term was still current at the time these results were obtained.

there to form a separate liquid layer. The rate at which this occurs and the magnitude of the exuded layer will depend on the degree of incompatibility and on the nature and concentration of plasticiser. Incompatibility can sometimes appear to be less than it is, owing to the slow rate of migration of plasticiser molecules once compounded into the plastic mass, because of their relatively large size. In many cases incompatibility effects are accelerated or induced by ultra-violet light.

Compatibility and incompatibility have been related to the parameters already discussed in connection with the mechanics of polymer/plasticiser interaction, and in broad terms it can be suggested that the more likely a plasticiser is to attack the polymer and become integrated into the polymer matrix the more likely it is to stay there. While incompatibility is difficult to measure and express in mathematical terms, it can be demonstrated relatively easily. Specimens stored under normal conditions will generally develop a more or less sticky or slipper film of liquid within a few days or weeks if incompatibility is present. A

Table 6.2

Reference	t_m (°C) (18)	χ (18)	t_{sm} (°C) (20)	Di-electric constant (20, 33)	A_p/P_o (30)	SGTT (°C) (30)	Pregel (°C) (31)	Fusion (°C) (29)	Gel temp. (°C) (10) (32)	
Dibutyl phthalate	92	0.02	90	6.4	4	86	62	63	55	68
Dihexyl phthalate	107	0.04			6	104		107		
Dioctyl phthalate	117	0.06	110	5.18				110		
Di-iso-octyl phthalate							77	110		
Di-'alphanyl' phthalate*							75			
Dinonyl phthalate							99	124	95	122
Di-isodecyl phthalate			123	4.5	10	139	102	130		
Ditridecyl phthalate†					13	151	133			
Butyl benzyl phthalate			95	6.5				76		
Dibutyl adipate					6	98				
Dihexyl adipate	124	0.3								
Dioctyl adipate			130	4.1	11.5	142		133	95	120
Dibutyl sebacate					8	117	77			
Dioctyl sebacate	151	0.53	144	3.4	12	150	131	150	100	
Butyl acetyl ricinoleate	156	0.73								
Trioctyl phosphate					8	121	94		85	110
Tritolyl phosphate							70			
Trixylyl phosphate							73	73	5	95

* Alphanol is a mixture of C^7–C^9 alcohols.
† Industrial tridecanol is a mixture of isomeric C_{13} alcohols; cf. Section 6.3.5.1.

more rapid demonstration can be achieved by preparing a sheet specimen about 1.25 to 1.8 mm thick and bending it over to form a circular fold of about 5 mm external diameter. Drops of liquid will usually appear within 24 h if incompatible.

In addition to the parameters discussed earlier (Sections 6.3.2.6 and 6.3.2.7), compatibility can also be studied by processing in a torque rheometer,[78] and by examination of dynamic mechanical properties of plasticised compositions.[79] In the former compatibility can be related to fusion temperature and time, and to torque at the fusion peak. In the latter elastic moduli are determined at relatively low temperature (e.g. 140°C) for identical compositions moulded at different temperatures, when minima are observed that correlate with solid–gel transition temperatures.

In practice it is found that three broad classes of plasticising materials exist. The most important plasticisers are those which are completely compatible over the whole range of practical compositions and conditions. These are known as primary plasticisers. Secondary plasticisers, on the other hand, are compatible only over a limited range of concentrations, although this range can usually be extended by the additional incorporation of primary plasticiser. Sometimes secondary plasticisers are not diffentiated from and are classed as extenders.[58] Plasticiser extenders, or more simply, just extenders, are generally incompatible alone, but can become compatible over a limited range of concentrations by the incorporation of primary plasticiser. Table 6.3 indicates a few of the more important materials in each class. Further classifications have usually been based on chemical structure dependent on the presence or otherwise of polar groups, cyclic structures, and aliphatic structures, as, for example, in the scheme[12] of Table 6.4. Chlorinated paraffins are placed in Class 3 struc-

Table 6.3

Primary plasticisers	Secondary plasticisers	Extenders
Dibutyl phthalate (DBP)	Dioctyl adipate (DOA)	Chlorinated paraffins
Dioctyl (i.e. 2-ethylhexyl) phthalate (DOP)	Di-iso-ictyl adipate (DIOA)	Mineral oil extracts
Di-iso-octyl phthalate (DIOP)	Di-'alphanyl' adipate (DAA)	
Di-'alphanyl' phthalate (DAP)	Dioctyl sebacate (DOS)	
Diononyl phthalate (DNP)	Di-iso-octyl sebacate (DIOS)	
Di-isodecyl phthalate (DIDP)	Di-'alphanyl' sebacate (DAS)	
Ditridecyl phthalate (DTDP)	Di-iso-octyl azelate (DIOZ)	
Tritolyl phosphate (TTP)	Polypropylene adipate (PPA)	
Trixylyl phosphate (TXP)	Polypropylene sebacate (PPS)	
Trioctyl trimellitate		

Table 6.4

Class	Groups			Type	Examples
	polar	cyclic	aliphatic		
1	√	√	—	Primary	TTP, TXP
2	√	√	√	Primary	DBP, DOP, DIOP
3	√	—	√	Secondary	DOA, DIOS, PPA
4	—	√	√	Extender	Alkylate oils
5	—	√	—	Extender	Aromatic hydrocarbons
6	—	—	√	Lubricant (q.v.)	Paraffin wax

turally, but their dipole concentrations are so high that they generally behave as extenders rather than as secondary plasticisers. This is in line with their high dielectric constants, in the region of 12.

6.3.4.2　*Efficiency*
It has already been stated that the main use of plasticisers in PVC is to achieve flexibility. For a given flexibility, different concentrations of different plasticisers will in general be required. In other words, the plasticisers may be said to vary in efficiency; the smaller the concentration of a particular plasticiser required to yield a particular flexibility, the greater its efficiency.

Since efficiency is mainly of interest commercially because of its bearing on costs (usually volume costs), it would be preferable for it to take into account the relative densities of the compounded materials, which are related to both the relative densities and the concentrations of the plasticisers concerned.

Flexibility is not an easily measured property, and in practice some other loosely dependent property has to be matched. Properties that have been used include British Standard Softness, moduli, tensile strengths, brittle temperatures, and lowering of transition temperatures. It has been suggested[34] that flexibility is dependent on modulus at very low extensions, i.e. of about 1 or 2 per cent, but the practical difficulties of measurement appear to have delayed acceptance of this as a means of assessment.

Table 6.5 presents some values for 1–2 per cent moduli for varying concentrations of dioctyl phthalate.

Table 6.5 VARIATION OF 1–2% MODULUS WITH
CONCENTRATION OF DOP

Concentration (phr)	1–2% modulus (kg/cm²)
0	150
6.5	200
10	190
20	100
30	11.5
40	5
50	1.7
60	1
70	0.6
80	0.5

Assuming that the transition temperature decreases linearly with plasticiser concentration, then:

$$t_m = t_o - \alpha W/M$$

where t_o and t_m are the transition temperatures of pure polymer and plasticised polymer, respectively; W is the weight per cent of plasticiser of molecular weight M[1, 35] and α may be regarded as 'the efficiency' of the plasticiser. However, whatever means is used to evaluate a plasticiser or formulate a plasticised composition, the fact must not be forgotten that the main objective in the vast majority of practical situations is to obtain a particular degree of flexibility in a finished product under conditions of use. Where flexibility itself is not important the most common justification for using a plasticiser is to achieve a greater toughness than is obtainable with unplasticised polymer.

The value of much of the reported work on comparison of different plasticisers is nullified by the fact that comparisons have often been carried out at constant concentration, authors frequently seeming oblivious to the fact that, once it has been established that different plasticisers have different efficiencies, there is little point in comparing them at constant concentrations; and it is far more meaningful to compare them at such concentrations as yield the same flexibility, or other property which might be of importance in a particular set of circumstances. Valuable information appears in the trade literature of the polymer and plasticiser suppliers. Some of these provide most comprehensive series of curves relating properties to concentrations for important plasticisers[36] and Figures 6.1 and 6.2 are based mainly on this information.

A number of apparent discrepancies occur in the curves, and others relating softness and other properties to plasticiser content. This is largely due to relatively poor accuracy of test methods, but sometimes arises from the methods of specifying concentration. Most commonly concentrations of plasticisers and other additives to PVC are expressed in 'parts per hundred of resin' (PHR or phr), and published data often take little or no account of the effects of other additives. This method, however, is most useful in day-to-day formulation practice, since concentrations of individual ingredients can be varied simply, without the need for calculation involved in expressing concentrations as percentages of the whole. For more fundamental purposes volume or molar concentrations would be more valuable.

It will be noticed that the curves of softness against concentration tend towards parallelism and near linearity. This in itself is somewhat surprising at first sight; even more so is

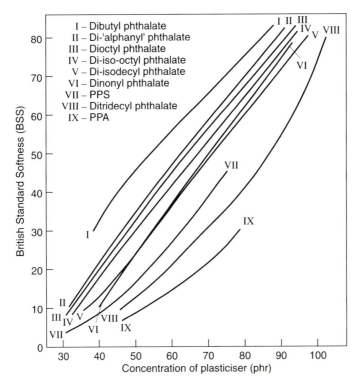

I – Dibutyl phthalate
II – Di-'alphanyl' phthalate
III – Dioctyl phthalate
IV – Di-iso-octyl phthalate
V – Di-isodecyl phthalate
VI – Dinonyl phthalate
VII – PPS
VIII – Ditridecyl phthalate
IX – PPA

Figure 6.1. Variation of British Standard Softness with concentration for a number of phthalate and polyester plasticisers[36]

the fact that the curves do not appear to extrapolate to the origin. This is presumably due to so-called 'antiplasticisation' that occurs over a range of relatively low plasticiser contents, up to about 15 phr, apparently due to crystallisation.[66] This leads to increase in tensile strength and modulus, but decrease in impact strength, while glass transition temperature falls steadily with increase in plasticiser content from zero upwards. However, arising from the fact that the curves have similar slopes, it is apparent that the differences in efficiency between plasticisers is less important at high plasticiser concentrations than at low. Mathematical definition of efficiency in terms of concentration required to produce a specific softness is, for the same reason, rather meaningless. Even the qualitative comparison of efficiency of plasticisers is sometimes complicated by the surprising fact that occasionally experimental curves cross. Even the rough generalisation that in the alkyl phthalate series efficiency decreases as the size of the alkyl group increases is seriously transgressed when various branched-chain alkyl groups are included, as well as linear groups.

6.3.4.3　Tensile properties
Plasticisation of PVC modifies not only flexibility and softness, but also all other mechanical properties. Thus tensile strength and moduli decrease and elongation at break increases with increase in plasticiser content, the precise relationship depending on the particular plasticiser, as illustrated in Figures 6.3, 6.4, and 6.5, which were compiled from similar sources

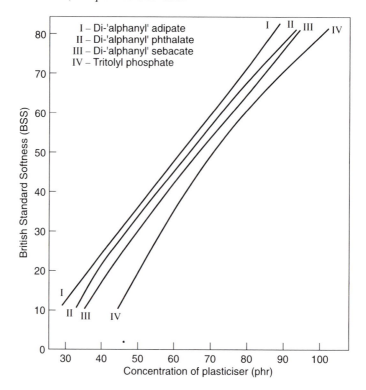

Figure 6.2. Variation of British Standard Softness witth concentration for some 'alphanyl' esters
and tritolyl phosphate[36]

to those of the previous two figures. For any given plasticiser the actual values for each property are also dependent on the polymer used, particularly on its average molecular weight.

Stress relaxation and variation of modulus with temperature of PVC plasticised with dioctyl phthalate (di-2-ethylhexyl phthalate) have been studied and reported in some detail.[37]

Since it is generally more meaningful to compare the performance of plasticisers at similar flexibility or some other dependent property, it would perhaps be better to plot variation in properties against flexibility. As previously indicated, this property is difficult to quantify, but plots of its nature have been made against BS Softness as a compromise.[38] Some of these suggest that in a series of similar plasticisers, for example alkyl phthalates, the relationship between tensile strength and softness is the same. Markedly different plasticisers, such as aryl phosphates, however, show an appreciably different relationship[38] (Figure 6.6). Extension and correlation of published data are complicated by the fact that different compositions, and conditions and methods of test have been employed by different workers, and while there is generally qualitative agreement, precise comparisons cannot often be made.

There is evidence that plasticised PVC undergoes a progressive marked stiffening and hardening with time at room temperature, and that this is reversed by reheating to processing temperatures. It has been suggested that the stiffening is due to slow crystallisation, though it may be that some form of bonding between polymer chains without major mor-

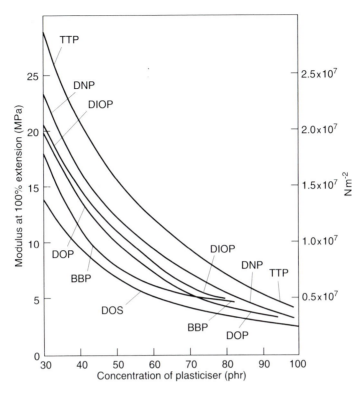

Figure 6.3. Variation of modulus at 100 per cent extension with concentration for various plasticisers

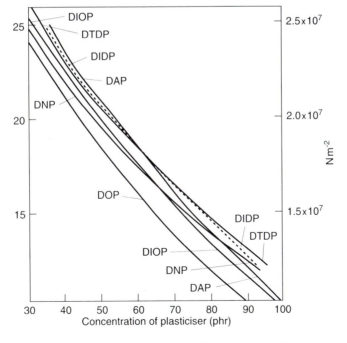

Figure 6.4. Variation of Ultimate Tensile Strength with concentration for various plasticers[36]

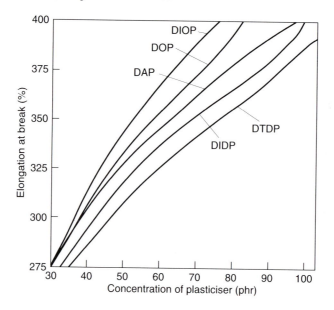

Figure 6.5. Variation of elongation at break with concentration for various plasticisers[36]

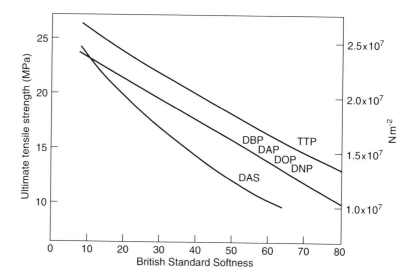

Figure 6.6. Variation of Ultimate Tensile Strength with British Standard Softness for
various plasticisers[38]

phological changes involving crystallite formation is occurring. Howick's studies[59] suggest
that such bonding is unlikely to be dipole bonding as has previously been proposed.[76] The
phenomenon does deserve consideration during formulation and production, if only to ensure that by the time the customer receives his product it has more or less reached its ultimate state and that this corresponds to his requirements. It is for this reason that the British
Standard Softness Test prescribes an interval of seven days between preparation and testing

of a specimen of compound, which is of course inconvenient in manufacturing. For this reason broad-based NMR is sometimes used to estimate plasticiser content at the time of production.

6.3.4.4   Mechanical properties at low temperatures

As the temperature of a plasticised PVC composition is reduced, its modulus increases, i.e. it becomes progressively stiffer and more brittle. Various methods of test of this behaviour have been devised, some aimed at measuring the stiffening and some aimed at determining the onset of brittleness. There is generally little correlation between the results obtained by various different methods of test, and commonly little correlation between any of them and performance in practical commercial situations. Nevertheless it can be stated qualitatively that for any selected flexibility or softness at room temperature the rates of stiffening and onset of brittleness with reduction in temperature can vary appreciably from one plasticiser to another. Stiffening and onset of brittleness with reduction in temperature are likely to be less marked with plasticisers whose viscosities have relatively low temperature-dependence than with those having markedly temperature-dependent viscosities. As a general rule the alkyl aliphatic esters as secondary plasticisers yield reduced rates of stiffening and lower brittle temperatures than phthalates, which, in turn are 'better' than aryl phosphates[38] (Figures 6.7 and 6.8). Of the phthalates the 'linear' variety, that is those having mainly linear alkyl groups, exhibit superior low temperature performance to those having branched-chain alkyl groups.

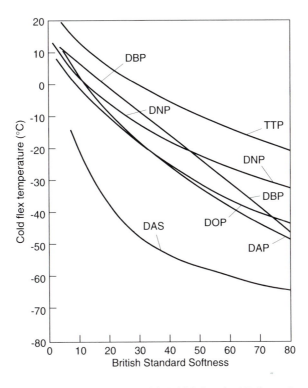

Figure 6.7. Variation of Cold Flex Temperature with British Standard Softness for various plasticisers[38]

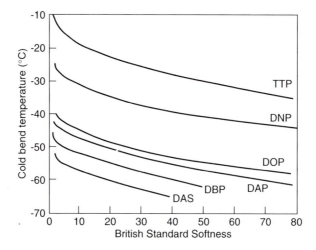

Figure 6.8. Variation of Cold Bend Temperature with British Standard Softness for various plasticisers[38]

6.3.4.5 Behaviour at elevated temperatures

On increasing the temperature of a plasticised PVC it becomes progressively softer and more flexible, and this can limit use in some applications at elevated temperatures. Softening temperatures, however, are often of little guidance, as witness the use of fairly soft grades for insulation of 105°C electric cables, where the support for the insulation is sufficient to prevent distortion.

A more serious problem in applications at elevated temperatures arises from the slow volatilisation and decomposition of plasticisers. The first of these processes is easily controlled by the use of less volatile plasticisers. In the phthalate series, for example, as a general rule the larger the alkyl groups the lower the volatility, so that requirements for increased permanence at elevated temperatures are usually met by replacing the conventional 'general purpose' di-iso-octyl or di-actyl phthalates by dinonyl, di-isodecyl, or ditridecyl phthalates, according to particular needs. In practice one is usually aiming to meet the requirements of a specification, which is more often than not for electric cable insulation or sheathing. Since, as a general rule, the higher phthalates are both less efficient and more costly than the octyl and lower varieties, the amounts of the former are usually kept near to the minimum necessary to meet the specification requirements. Trimellitic esters such as tri-2-ethylhexyl trimellitate are also suitable in products where low volatility is required.[75] For demanding applications polymeric plasticisers are sometimes necessary.[39-41]

Reducing physical volatility, however, is not all that may be involved in reducing loss of plasticiser at elevated temperatures, and chemical decomposition with subsequent loss of reaction products may also be involved. This can sometimes be counteracted by incorporation of stabilisers, usually antioxidants.[39] Some commercial plasticisers are supplied with antioxidant already incorporated, though sometimes additional antioxidant may be desirable.[75]

Table 6.6 illustrates the relative volatility of a number of different common plasticisers.[36] The figures quoted are weight losses expressed as percentages of plasticiser present in compound containing an original 50 phr. The determinations were made by a method similar to BS 2782 Part 10F:1956, involving heating for 24 h at 373 K (100°C) in circulating air.

In Table 6.7 the weight losses after 24 h at 358 K (85°C) have been determined on com-

Table 6.6

Plasticiser	% weight loss
Dibutyl phthalate (DBP)	30
Di-isobutyl phthalate (DIBP)	30
Di-n-heptyl phthalate (DnHP)	3.3
Di-'alphanyl' phthalate (DAP)	2.1
Dioctyl phthalate (DOP)	2.4
Di-iso-octyl phthalate (DIOP)	2.1
Dinonyl phthalate (DNP)	2.1
Di-isodecyl phthalate (DIDP)	1.3
Ditridecyl phthalate (DTDP)	0.6
Di-'alphanyl' adipate (DAA)	5.4
Dioctyl adipate (DOA)	5.7
Triethylene glycol dicaprylate	10.7

Table 6.7

	Concentration	Weight loss (24 h at 85°C)	
	phr	% on compound	% on plasticiser
Dibutyl phthalate	54	7.2	21.6
Di-isobutyl phthalate	62.6	9.3	25.4
Di-'alphanyl' phthalate	58.7	0.35	0.99
Dioctyl phthalate	62.8	0.35	0.95
Di-iso-octyl phthalate	64	0.4	1.07
Dinonyl phthalate	69.5	0.35	0.9
Di-isodecyl phthalate	66.5	0.3	0.79
Ditridecyl phthalate	74.5	0.3	0.74
Di-iso-octyl adipate	52.7	0.55	1.68
Dibutyl sebacate	45	3.1	10.1
Di-iso-octyl azelate	56.3	0.35	1.02
Tritolyl phosphate	71	0.3	0.76
Trixylyl phosphate	75	0.25	0.61
Epoxydised soya bean oil	66.5	0.2	0.53
Butyl epoxystearate	55	1.2	3.6
Iso-octyl epoxystearate	56.3	0.6	1.75

positions containing the quantity of each plasticiser required to yield a tensile modulus at 23°C of 7580 kN/m^2 (7.58 MPa) at an elongation of 100 per cent.[47]

Because of the variety of test methods and conditions employed it is often difficult to compare a required pair of plasticisers from published data. It has been suggested that loss by volatility is proportional to the vapour pressure of a plasticiser at the temperature under consideration,[41] implying that the strength of polymer/plasticiser interactions is unimportant from this point of view. If this is so it should be possible to asses the performance of a plasticiser by consideration of its vapour pressure data, and thus avoid the troublesome business of heat loss testing of compounded specimens. Table 6.8 gives some published vapour pressure data,[41] and also boiling points of a number of plasticisers.

As early as 1947 it was pointed out that under most service conditions the rate of loss is most likely to be controlled by gas-phase diffusion, since the activation energy for discussion is less than the latent heat of vaporisation and diffusion constants in the gas phase

Table 6.8

	Temperature (°C) at a vapour pressure of		Boiling point °C at torr[36]
	0.5 μm[41] Hg	5 μm[41] Hg	
Dibutyl phthalate	29	75	182/5
Di-isobutyl phthalate	—	—	168/5
Dicapryl phthalate	66	121	—
Di-*n*-hexyl phthalate	54	120	—
Di-*n*-heptyl phthalate	—	—	230/6
Di-*n*-octyl phthalate	82	132	—
Dioctyl phthalate	68	120	230/4
Di-iso-octyl phthalate	72	121	230/4
n-Octyl *n*-decyl phthalate	80	138	
Dinonyl phthalate	72	123	246/5
Di-isodecyl phthalate	92	141	250/4
Ditridecyl phthalate	—	—	285/3.5
Dioctyl adipate	58	105	232/10
n-Octyl *n*-decyl adipate	72	128	—
Dinonyl adipate	61	110	—
Dioctyl azelate	76	125	—
Dioctyl sebacate	84	131	—
Trioctyl phosphate	57	101	—
Tritolyl phosphate	78	136	—
Epoxydised soya bean oil	over 150	over 150	—

are relatively insensitive to temperature; the large effect of thickness was also noted (Table 6.9.[43]

The possibility of chemical decomposition of plasticisers on heating plasticised PVC has already been referred to. This can result not only in weight loss but also in accelerated discoloration.[40] In alkyl ester plasticisers the structure of the alkyl groups markedly affects the stability. Free radical oxidisation has been claimed as the main cause of decomposition; this is relatively slow in normal straight-chain hydrocarbons but generally increases in rate with introduction of branching.[44] Phthalates in general oxidise more slowly than sebacates, rates of decomposition as assessed by colour development increasing in the order dinonyl-, dioctyl-, dialphanyl-, di-isodecyl-, di-iso-octyl phthalate, in agreement with the order expected from their degrees of chain branching. Many different antioxidants reduce decomposition rates effectively, including amines, phenols, amine-aldehyde and ketone condensates, phenothiazines, phenol-aldehyde and ketone condensates, and bis(hydroxyphenyl) sulphides, though all but the last two cause severe discoloration of PVC.[39] Table 6.10 illus-

Table 6.9

	Vapour pressure at 100°C (μm)	Relative vapour pressure	Relative loss rate
Dibutyl phthalate	48	1	1
Dihexyl phthalate	3.7	0.077	0.24–0.40
Dioctyl phthalate	1.1	0.023	0.12–0.30
Dibutyl sebacate	12	0.25	0.25–0.52
Butyl acetyl ricinoleate	1.0	0.021	0.37–0.74
Tritolyl phosphate	0.63	0.013	0.05–0.10

Table 6.10 EFFECTS OF SELECTED ANTIOXIDANTS ON DECOMPOSITION
OF PLASTICISED PVC

Antioxidant (concentration in brackets in columns 2 and 3)	Time to discolour at 150°C in air (min)	Weight loss at 100°C (%/day)
None	140	0.75
2,2-Bis(4-hydroxyphenyl)propane (diphenylolpropane)	27.0 (0.1%)	0.4 (0.2%)
2-methyl-3-hydroxy-4-t-butylphenyl sulphide	250 (0.1%)	—
Bis-3-methyl-4-hydroxy-5-t-butylphenyl sulphide	—	0.25 (0.2%)
Diphenylamine-acetone condensate plus disalicylidene-ethylenediamine(1:1)	—	0.4 (0.02%)

trates the effect of added antioxidants on the decomposition of di-sio-octyl phthalate as assessed by the time taken to discolour to a standard level and by loss in weight.[40]

The use of antioxidants of the types tabulated is valuable in preventing discoloration during processing and in assisting the retention of mechanical properties in high temperature applications.[39]

6.3.4.6. Windscreen fogging

'Fogging' of windscreens can arise by condensation of volatiles coming from fumes entering the ventilation system and from material within a vehicle, where high temperatures can arise, particularly if the vehicle stands in direct sunlight for any length of time. Plasticised PVC can contribute to this fogging through volatilisation of emulsifiers, stabilisers and plasticisers, or their decomposition products.

Manufacturers now recommend specific polymers and other components for compositions where the potential for this problem exists. Generally the lower the volatility of a plasticiser the lower the risk of fogging, though it is necessary to keep degradation during processing to a minimum in order to avoid the production of volatile material in sufficient quantity to cause the problem. Trimellitate esters and linear alkyl phthalates are commonly used for 'non-fogging' applications.[58]

6.3.4.7 Loss of plasticiser by extraction

In a number of actual and potential applications contact between plasticised PVC and liquid is involved, either continuously, as for example in conveying and packaging, or intermittently, as in products which are periodically washed or sprayed. In gross cases, stiffening and cracking of the PVC can occur, but even in less severe cases liquids can permeate into and through the PVC and can be seriously contaminated by extracted plasticiser. Even distilled water appears to extract small amounts of most plasticisers, but aqueous soap and detergent solutions are more severe. Organic liquids and liquid/solid mixtures as a rule extract greater quantities of plasticiser than aqueous materials, but there is considerable variation in effect between different extractants and between different plasticisers. Often specific behaviour between a particular extractant and plasticiser cannot be predicted, and special tests may have to be made if the appropriate information cannot be found in the journals or trade literature. Among the more important likely extractants are lubricating fuel, and edible oils.

Comparative extraction figures for a number of plasticisers are set out in Table 6.11.[36, 42] In each case the figures have been calculated on the basis of a value of unity for the plasticiser with the lowest extraction figure.

It is difficult to express extraction of plasticiser in absolute terms, and the figures in Table 6.11 are therefore only related in the vertical columns, with no relationship horizontally. The dimensions of the PVC article or specimen, concentration of plasticiser, the nature of contact with and any motion of the extractant, and temperature, all affect the rate and absolute amounts extracted. To give an idea of the relative effects of different extractants Table 6.12 sets out the weight percentages of a few plasticisers extracted by a number of different solvents, using standard test specimens 76 mm × 5 mm × 1.25 mm immersed at 23°C.[41]

Quite apart from any deterioration in the PVC article, contamination of an extractant by plasticiser can be important, particularly, of course, where the extractant is a foodstuff or beverage, or some other material which is to be introduced into the human body, for example in blood transfusion apparatus and feeding tubes. Blood extracts di-2-ethylhexyl phthalate when held in bags made from PVC containing that plasticiser, but the blood is preserved longer than when held in glass containers. The human digestive tract also extracts this plasticiser from feeding tubes extruded from PVC compounded with it. Toxicity is considered in more detail later (Chapter 17).

Appreciable extraction of a completely non-toxic plasticiser might still be objectionable, however, since a tainting of flavour is, in any case, likely to result, even if the mere idea of having plasticiser in foodstuffs is accepted. At the other end of the scale the phosphates are sufficiently toxic to be barred from consideration for use in any applications where toxicity is relevant.

Table 6.11 RELATIVE AMOUNTS OF PLASTICISER EXTRACTED

Reference	Water (42)	Soapy water (36)	(43)	Detergent (42)	Mineral oil (36)	(42)	Olive oil (36)	Petrol (36)
Dibutyl phthalate	10	16	20	6	8.5	4.2	9.4	3.6
Di-isobutyl phthalate	10	13.9	26	12	2.7	6	3	2.4
Di-*n*-heptyl phthalate	—	14.1	—	—	9	—	13	4.4
Di-'alphanyl' phthalate	2	12.4	8	4	6	7.4	9.4	4
Dioctyl phthalate	2	12.4	6	3.6	5.4	6	11	3.1
Di-iso-octyl phthalate	1	12.4	8	3.3	6	5.2	10	3.4
Dinonyl phthalate	1	11.9	2	1	3.7	7.6	13	4.3
Di-isodecyl phthalate	3	5.6	3	2.7	5.5	28	16.7	4.4
Ditridecyl phthalate	4	1	1	4	25	80	20.3	5.1
Di-'alphanyl' adipate	—	17	—	—	26.2	—	29	3.9
Dioctyl adipate	—	17.4	—	—	25	—	25	3.2
Di-iso-octyl adipate	2	—	10	4.7	—	33	—	—
Di-iso-octyl azelate	3	—	1	5.3	—	52	—	—
Dibutyl sebacate	4	—	13	5.7	—	36	—	—
Pentaerythritol ester	—	3.2	—	—	1.7	—	4.7	4.2
Tritolyl phosphate	2	—	6	2	—	1.2	—	—
Trixylyl phosphate	2	—	6	1.7	—	1	—	—
Epoxydised soya bean oil	2	—	7	5	—	2.8	—	—
Butyl epoxystearate	4	—	12	6	—	27	—	—
Iso-octyl epoxystearate	6	—	7	3.3	—	76	—	—
Polyesters	—	3–15	—	—	1–2	—	1–3.5	1–2.5

Table 6.12 % EXTRACTION BY SOLVENTS FROM COMPOSITIONS OF EQUAL MODULUS
(7.7 MN/m^2)

Extractant	Water	1% Soap	1% Detergent	Mineral oil	Iso-octane	Di-iso-butylene	Iso-octane (70)+ toluene (30)
Time of immersion (h)	240	240	240	240	24	24	24
Dibutyl phthalate	0.5	1	0.9	1.05	1.85	2.1	14.5
Di-iso-octyl phthalate	0.05	0.4	0.5	1.3	20.2	23.4	16.4
Di-isodecyl phthalate	0.15	0.15	0.4	7.05	28.5	29.3	20.8
Ditridecyl phthalate	0.2	0.05	0.6	20	33	33.8	32.6
Di-iso-octyl azelate	0.15	0.05	0.8	13	26.3	27	19
Dibutyl sebacate	0.2	0.65	0.85	8.95	5.2	7	15.3
Trixylyl phosphate	0.1	0.3	0.25	0.25	0.7	1.7	18.9
Epoxydised soya bean oil	0.1	0.35	0.75	0.7	1.3	3	12.1

6.3.4.8   *Other plasticiser migration problems*

In a number of applications PVC is used in situations adjacent to other plastics materials and migration of plasticiser to the surface or into the matrix of such materials may lead to some form of deterioration. Thus migration of plasticiser from PVC cable sheathing to poly-ethylene insulation of conductors can result in serious reduction in resistivity and increase in power factor of the insulant. Another example is the use of PVC insulated or sheathed cables in conjunction with acrylic or polystyrene lighting fittings, where plasticiser can cause crazing of the associated product. Other PVC products where similar problems can arise are in refrigerator gaskets, and where plasticiser in PVC components could cause crazing of some car finishes.

Clearly the effect is dependent on interaction between the plasticiser and the second ma-terial, and not just on migration. In this kind of situation 'non-migratory' polymeric plasti-cisers (such as polypropylene adipates or sebacates) perform best, but good results are often obtained with other high molecular weight plasticisers, such as epoxydised soya bean oil and esters of pentaerythritol.

6.3.4.9   *Electrical properties*

As a general rule the electrical resistivity of PVC decreases progressively with the intro-duction of plasticiser. Quoted values can be misleading, however, because the resistivity of the unplasticised composition is very dependent on the nature of the resin, and the nature and concentration of other additives, especially pigments, and because the effects of plasti-cisers themselves are very dependent on the purity, and not only on their chemical structure. The figures[36] in Table 6.13 are typical of those published in trade literature.

6.3.5   INDIVIDUAL PLASTICISERS

The number of substances offered commercially as plasticisers for PVC is large; the num-ber of substances suggested is much larger. Unlike stabilisers, most of the common com-mercial plasticisers are well characterised chemically, though a few are offered under code names without disclosure of chemical structure. Comprehensive compilations of plasti-

Table 6.13 VOLUME RESISTIVITY (Ω cm $\times 10^{12}$) VS
PLASTICISER CONTENT (phr)

phr	40	60	80
Di-'alphanyl' phthalate	800	38	5.1
Di-iso-octyl phthalate	940	39	6.2
Dinonyl phthalate	3000	90	7.6
Di-isodecyl phthalate	600	31	3.4
Ditridecyl phthalate	480	13	4.7

cisers and their physical properties have been published,[57, 58, 80, 81] though it should be noted that minor variations in properties may be found between different samples of nominally the same plasticiser, due to the fact that the alcohol feedstocks are frequently composed of mixtures of closely related alcohols rather than a single pure compound.

6.3.5.1 Esters of aromatic acids

Phthalates (Table 6.14) constitute the major group of plasticisers in the UK and on a worldwide scale, accounting for around 70 per cent of the plasticiser market for many years.[45, 57, 75]

Higher available phthalates include dicapryl, dilauryl, and di-2-butoxy-ethyl phthalates; mixed esters, such as butyl isodecyl, 'alphanyl' capryl, 'alphanyl' iso-octyl, and 'alphanyl' decyl phthalates; and the phthalate of di-2-methoxyethanol. At least one ester of isophthalic acid, the dioctyl ester (DOIP), is also offered as a plasticiser. Some of the above names are a little misleading. Thus[39] 'octyl' normally refers to the 2-ethylhexyl group; 'alphanyl' is derived from 'alphanol', which is a mixture of straight-chain and branched heptanols, octanols, and nonanols, with a small proportion of cyclic alcohols; 'iso-octyl' is derived from 'iso-octanol', which is a mixture of 3,5-, 4,5-, and 3,4-dimethylhexanols, with 3-, 4-, 5-, and 6- methylheptanols and other isomers; 'nonyl' to the 3, 5, 5-trimethylhexyl group; 'isodecyl' is derived from an industrial isodecanol which is a mixture mainly of dimethyloctanols and trimethylheptanols;[45] and 'tridecyl' is derived from industrial tridecanol, which is a mixture of isomeric branched-chain primary alcohols having thirteen carbon atoms per

Table 6.14 PROPERTIES OF PHTHALATE PLASTICISERS[36,42]

	Molecular weight	Relative density		Viscosity (cP)			Refractive index		Boiling point
		20/ 20°C	25/ 25°C	0°C	25°C	60°C	n_D^{20}	n_D^{25}	°C/torr
Dibutyl	278	1.048	1.045	61.5	16	5.1	1.494	1.492	310:182/5
Di-isobutyl	278	1.041	1.041	180	29	7.2	1.491	1.488	312:168/5
Di-n-heptyl	362	0.987	—	—	30	—	—	1.483	230/7
Di-'alphanyl'	380	0.994	0.996	210	47	11.0	1.490	1.488	—
Dioctyl	390	0.986	0.985	340	54	10.5	1.488	1.484	386:230/4
Di-iso-octyl	390	0.986	0.984	320	54	13.0	1.488	1.485	231/4
Dinonyl	418	0.972	0.971	490	75	15.0	1.485	1.483	246/5
Di-isodecyl	447	0.969	0.966	470	81	17.5	1.485	1.483	251/4
Ditridecyl	530	0.949	0.950	2112	197	29	1.485	1.484	285/3.5
Butyl benzyl	312	—	1.115	—	50	—	1.540	—	240/15

alkyl group, Di-2-ethylhexyl phthalate is almost unique among the common plasticisers in being almost pure single isomer. This is the reason for its preferment in a number of critical applications, such as those in medicine and surgery.

Esters of another aromatic acid, trimellitic (1,2,4-benzene tricarboxylic) acid, such as di-octyl trimellitate, are of particular interest where low volatility at high temperatures is required. As with phthalates, straight-chain alkyl trimellitates have better low temperature properties than branched-chain varieties.[75]

6.3.5.2   *Esters of aliphatic acids*
Many different aliphatic acids have been used to produce plasticisers of this group, which are mainly of the secondary type. The most common acids are adipic, sebacic, azelaic, and caprylic, though esters of other acids such as citric, trimellitic, ricinoleic, and pelargonic are also available, usually for specific purposes. The alkyl groups in the esters tend to be the same as those used for phthalates, presumably because of the commercial availability and relatively low cost of the appropriate alcohols. In addition, however, some esters of glycols and other polyhydric alcohols are offered.

The adipates, sebacates, and azelates are mainly employed for 'low temperature/ applications, where they defer the onset of stiffening as temperature is reduced. Their use for this purpose is somewhat expensive, and nowadays similar esters of nylon waste acids are available, which perform as well but are cheaper. As a class they tend to be more volatile than the phthalates, so that formulation to meet both high and low temperature requirements in one and the same compound can be difficult.

Acetyl tributyl citrate and a few other esters of this type have been produced specifically for applications where non-toxicity is important, though acetyl butyl citrate itself has now been shown to be more toxic than di-2-ethylhexyl phthalate. for cling film d-2-ethylhexyl adipate is the preferred plasticiser in the UK.

Other available plasticisers of this type include acetyl trioctyl citrate, and iso-octyl glutarate and succinate.

6.3.5.3   *Polymeric plasticisers*
The number of different truly polymeric plasticisers is small, though the term has sometimes been applied to other high molecular weight plasticisers, such as epoxydised oils, which are not polymeric. This type of plasticiser has been developed as non-migratory and non-volatile, for packaging and high temperature applications respectively. The only plasticisers of this type of real commercial importance are the adipates and sebacates of propylene glycol (PPA and PPS respectively), sometimes chain-stopped with lauric acid to limit molecular weight during their production by polycondensation. the corresponding ethylene glycol polyesters are not suitable because they tend to form crystalline wax-like solids at room temperatures. As a class these materials give the desired low migration and volatility performance, depending on molecular weight, but they are correspondingly inefficient, expensive, and difficult to homogenise into compounded PVC.

6.3.5.4   *Organic phosphates*
Because of their generally poor colour and toxicity, use of this type of plasticiser nowadays tends to be limited to compositions which need to be as non-flammable as possible. They accounted for only 8 per cent of the plasticiser consumption in the UK during 1967.[45] The

Table 6.15 PROPERTIES OF ALIPHATIC ESTER PLASTICISERS[36, 42]

	Molecular weight	Relative density		Viscosity (cP)			Refractive index	Boiling point
		20/25°C	25/25°C	0°C	25°C	60°C	n_D^{25}	°C/torr
Di-'alphanyl' adipate	360	0.933	—	—	10.3*	—	1.4469 (20°C)	—
Dioctyl adipate	371	0.927	0.923	—	10.2*	—	1.453	232/10
Di-iso-octyl adipate	371	—	0.928	41.5	14.7	5.7	1.447	—
Di-isodecyl adipate	426	0.915	—	—	22*	—	1.452	242/4
Dioctyl azelate	413	0.918	—	—	—	—	—	237/5
Di-iso-octyl azelate	413	—	0.919	58	19	7	1.450	—
Dibutyl sebacate	314	—	0.936	21	8.6	4.1	1.443	—
Di-'alphanyl' sebacate	416	—	0.916	—	19	—	—	—
Dioctyl sebacate	426	—	0.915	—	19	—	1.451	—
Di-iso-octyl sebacate	426	0.912	0.912	—	23	—	1.450 (20°C)	—
Triethylene glycol dicaprylate	426	—	0.965	—	16.2	—	1.446 (20°C)	220/0.2
Diethylene glycol dipelargonate	418	0.966	—	—	—	—	—	229/5

* Estimated.

Table 6.16 TYPICAL PROPERTIES OF POLYESTER PLASTICISERS

	Molecular weight	Relative density	Viscosity at 25°C(P)
Polypropylene adipate/laurate	2200	1.08	23
Polypropylene adipate/laurate	3300	1.10	80
Polypropylene adipate	6000	1.15	2000
Polypropylene sebacate	8000	1.06	2200

Table 6.17 PROPERTIES OF COMMERCIAL PHOSPHATE PLASTICISERS

	Relative density	Viscosity (cP)			Refractive index	Boiling point
	25/25°C	0°C	25°C	60°C	n_D^{20}	°C/torr
Trioctyl phosphate (TOP)	0.926	—	11.5†	—	—	220/5
Tritolyl phosphate (TTP)	1.165	800	72	10.3	1.557	420–430
Trixylyl phosphate (TXP)	1.138	1450	100	13.0	1.553	270/3
Cresyl* phenyl phosphate (CPP)	1.202	—	38.4	—	—	

* More properly 'tolyl'. † Estimated.

only compounds of commercial importance are tritolyl phosphate (TTP) and trixylyl phosphate (TXP), though other phosphates, such as the phenyl, tri-'alphanyl', trioctyl, and tri-iso-octyl esters are available.

6.3.5.5   *Epoxy plasticisers*
This group of plasticisers (Table 6.18) comprises various types of expoxydised oil, and esters of epoxy acids. Of the first of these, epoxydised soya bean and linseed oils are by far

Table 16.8 PROPERTIES OF EPOXY PLASTICISERS

	Relative density	Viscosity (cP)			Refractive index
	25/25°C	0°C	25°C	60°C	n_D^{20}
Epoxydised soya bean oil	0.993	2240	380	63	1.474
Butyl epoxystearate	0.905	50	17.5	6.7	1.452
Iso-octyl epoxystearate	0.897	74	23.5	8.1	1.454

the most important, although other oils are used to some extent, e.g. tall oil, a complex mixture obtained from pine-wood pulp. These materials vary from grade to grade in epoxy content and therefore in stabilising properties, but they are probably used more for stabilisation than plasticisation. As a rule these materials are regarded as non-toxic, but there are prescribed limits on concentration and degree of epoxydisation in food contact applications.

The main epoxy acid used is epoxystearic, and its butyl and iso-octyl esters, in particular, are used as stabilising plasticisers.

Although sometimes used as sole plasticisers, epoxy plasticisers are not usually regarded as primary because they tend to sweat out under some adverse conditions.[47]

6.3.5.6   *Plasticiser extenders*
During 1967 some 14 per cent of the plasticiser usage in the UK was in the form of extenders, mainly chlorinated paraffins (Table 6.19).[45] Bt 1981 this was down to around 10 per cent, in spite of improvements in colour and heat-stability obtained by advances in chlorination techniques and the use of straightchain hydrocarbon feedstocks.[75] As indicated previously plasticiser extenders are not compatible with PVC in the absence of sufficient quantities of primary, or possible secondary plasticisers. The main justification for their use is to reduce cost, although the chlorinated materials are sometimes valuable in reducing inflammability. The materials in this class are ill-defined chemically, but are generally mineral oil extracts or chlorinated hydrocarbons such as paraffins and paraffin waxes. The former type are usually very cheap but often have somewhat objectionable odours, and are restricted to the cheapest of compositions, for example cheap hose and carpet backing. The chlorinated materials are generally more expensive and less efficient, and are also of higher density than most plasticisers, so that their value should be very carefully examined before deciding to include them in formulations.

Upper compatibility limits for a particular extender depends on the nature and concentration of other plasticisers present, and also appear to be somewhat increased by the inclusion of filler. At one time it was believed that a definite minimum ratio of a particular primary plasticiser to a particular extender had to be maintained to ensure compatibility, but it now appears that with at least some extenders the minimum ratio varies with total plasticiser plus extender content and itself passes through a minimum (Figure 6.9 and 6.10). With chlorinated compounds the chlorine content has a pronounced effect on compatibility, as is perhaps not unexpected. Compatibility appears to approach that of a primary plasticiser as the chlorine content increases towards that of polyvinyl chloride (theoretical 56.8 per cent). Compatibility increases with decrease in molecular weight of the resin,[48] with the introduction and increase in concentration of vinyl acetate into the resin, and also with increase in content of filler.[48] In some compositions such as heavily filled polymer or vinyl

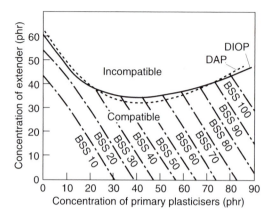

Figure 6.9. Compatibility of a chlorinated paraffin extender (Cereclor S52) in PVC plasticised with DIOP or DOP[49]

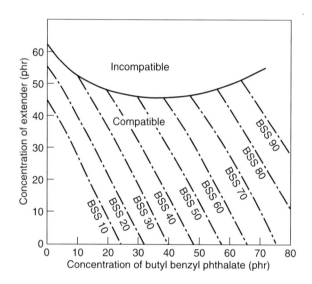

Figure 6.10. Compatibility of a chlorinated paraffin extender (Cereclor S52) in PVC plasticised with BBP[49]

acetate copolymer, compatitibity of some chlorinated hydrocarbons becomes so high that they then qualify for classifications as primary plasticisers. The main extenders of this type are chlorinated liquid paraffins or paraffin waxes of chlorine content ranging from 30 to over 70 per cent (Table 6.19), the interesting range for PVC plasticisation being from about 40 to 55 per cent. At one time they were invariably pale yellow or straw coloured viscous liquids, and were not suitable for colourless clear or pale-coloured compositions. With improvements in feedstock and production techniques introduced a few decades ago they can now be substantially colourless and have much improved heat-stability and compatibility.

Electrical properties are such as to permit the use of these extenders in insulation wherever plasticised PVC would be appropriate.

On exposure outdoors extenders generally tend to sweat out from PVC and form objec-

Table 6.19 SOME TYPICAL PROPERTIES OF CHLORINATED
PARAFFIN EXTENDERS

Molecular weight	Chlorine content	Density (25°C)	Viscosity at 25°C(P)
530	42	1.6	25
400	51	1.25	17

tionable sticky surface layers, and for this reason they should not normally be used in applications outdoors without prior exhaustive trials.

6.3.6   SELECTION OF PLASTICISERS

To the uninitiated, selection of plasticisers may appear to be most complicated and lengthy, but in practice the situation is usually much simpler. In commercial practice, just as with stabilisers, over a period of time there is built up a body of knowledge and experience on the basis of which plasticisers can usually be selected to meet new compound requirements, so that it is not necessary to go back to first principles and examine a large number of commercial offers. Of course offers of a new commercial material may involve detailed evaluation, even if it is nominally chemically identical to a known plas-ticiser, and even more so if it is chemically unknown as a plasticiser. A new application, with new performance requirements, may also necessitate a detailed investigation, but these situations are now comparatively rare.

For the novice, and for that matter the expert, a relatively simple approach can resolve most plasticiser formulation requirements.

Provided that the required technical performance can be secured by using a number of alternative plasticisers or plasticiser combinations, the decision in most cases will rest with that offering the lowest volume cost. This will be dependent mainly on the price, the efficiency, and the density of each plasticiser.

At any point of time there will therefore usually be only a few materials available with combinations of properties and prices which justify their selection for the majority of 'general purpose' applications. Thus, at the time of the first edition of this book (1972) the first choice in the UK would have rested between di-iso-octyl phthalate (DIOP) and di-alphanyl phthalate (DAP), which were both slightly cheaper than the di-octyl (di-2-ethylhexyl) ester (DOP). However, changes in the availability of feedstock alcohols affect the availability and costs of all alkyl phthalates, and some of the mixed alcohol feedstocks are no longer freely available, so that the corresponding esters disappear from the market. The situation should therefore be reviewed as changes in availability and costs occur. Of course this is true, not only of the various 'octyl' phthalates, but also where an application requires special properties obtainable from a number of different plasticisers. So for most 'general purpose' applications the most readily available 'octyl' phthalate would be selected. Dicapryl phthalate (DCP) and di-n-heptyl phthalate might also be considered as alternatives. Although still used to some extent, dibutyl phthalate is really too volatile under normal mild conditions for use even in non-demanding applications. Where colour and clarity are not highly critical part of the primary plasticiser may be replaced by extender.

In applications where even the octyl phthalates are too volatile, it becomes necessary to substitute them partially or completely by the higher phthalates, di-isodecyl phthalate (DIDP) and ditridecyl phthalate (DTDP). The concentrations of these are usually kept as low as possible commensurate with achieving the other required properties, because they are relatively inefficient and expensive.

At the other end of the scale, if retention of flexibility at low temperatures is required, linear phthalates will provide some improvement over the branched-chain variety, or part of the primary plasticiser may be replaced by an adipate, azelate, sebacate, or similar material. Di-sio-octyl azelate appears currently to be the cheapest of the chemically well-characterised compounds of this type, but even cheaper and more or less equivalent ill-defined esters are available, and would probably be first choice. All plasticisers of this type are more expensive than the octyl phthalates, and consequently, the concentrations of these, too, are normally kept to the minimum required to achieve the desired low temperature performance.

Where contact with some extractant is involved it may be necessary to consult the literature to discover how different plasticisers react to the particular extractant, or even carry out actual trials. For contact with oils and fatty substances the polyester plasticisers offer excellent permanence, but they are inefficient, expensive, and difficult to compound, requiring extended compounding cycles. As a compromise at somewhat lower cost and inferior though still good performance, expoxydised soya bean oils can be used.

For low inflammability tritolyl and trixylyl phosphats and the chlorinated extenders offer excellent performance, though their use is usually limited to compositions which do not have to possess the full potential of PVC clarity and freedom from colour.

For electrical applications the same kind of considerations apply, but it is additionally necessary to take account of the electrical properties if any plasticiser whose use is contemplated, with particular note of the fact that adventitious impurities can have grossly exaggerated effects, and careful selection between competitive offers of nominally the same materials should be made.

In some applications where PVC is used in combination with other materials, good bonding is required and adhesives may be used to attain adequate bond strength. A typical example is the manufacture of footwear with PVC soles and leather uppers. In this application, although linear phthalates provide better low temperature properties than branched-chain phthalates, the later give better adhesion. Alkyl phenyl/cresyl sulphonates give even better adhesion, but have inferior low temperature stress-cracking properties.[82]

REFERENCES

1. I. Mellan, *The Behaviour of Plasticisers*, Pergamon, Oxford (1961)
2. I. Mellan, *Industrial Plasticisers*, Pergamon, Oxford (1963)
3. D.N.S. Buttrey, *Plasticisers*, Macmillan, London (1957)
4. P.F. Bruin, (ed.), *Plasticiser Technology*, Reinhold, New York (1965)
5. 'Plasticization and Plasticizer Processes', *Adv. Chem. Sr.*, **48**, Am. Chem. Soc. (1965)
6. W.S. Penn, *Rubb. Plast. Wkly,* **141**, 11–17 (1969)
7. A.K. Doolittle, *J. Polym. Sci.*, **2**, 2, 121 (1947)
8. W.R. Moore, *An Introduction to Polymer Chemistry*, Chapter 4, University of London Press (1963)
9. K. Ueberreiter, *Adv. Chem. Ser.*, **48**, 35, Am. Chem. Soc. (1965)
10. J.S. Luna, *Ion. Madr.*, **22**, 247, 67 (1962) 248; 156 (1962) 249, 226 (1962); *RAPRA Translation* No. 1093

11. L.E. NIELSON, R. BUCHHAHL and R. LEVREAULT, *J. appl. Phys.,* **21**, 6, 607 (1950)
12. T.C. MOORSHEAD, Chapter 2 in *Advances in PVC Compounding and Processing* (M. Kaufman, ed.), Maclaren, London (1962)
13. J. DELMONTE, *Plastics in Engineering*, Penton, Croydon, Surrey (1946)
14. J.F. EHLERS and K.R. GOLDSTEIN, *Kolloidzeitschrift*, **118**, 137 (1950)
15. A. HARTMANN and F. GLANDER, *Kolloidzeitschrift*, **137**, 79 (1954)
16. A. HARTMANN, *Kolloidzeitschrift*, **142**, 123 (1955)
17. C.E. ANAGNOSTOPOULOS, A.Y. CORAN and H.R. GAMRATH, *J. appl. Polym. Sci.,* **4**, 11, 181 (1960)
18. K. THINIUS, *Chemie, Physik und Technologie der Weichmacher*, VEB Verlag, Berlin, 33 (1960)
19. F. BARGELLINI, G. BABRIS and F. CHOZZINI, *Poliplastic,* **11**, 70, 5 (1963)
20. F. BARGELLINI, *Materie plast.,* **30**, 4, 378 (1964)
21. P.A. SMALL, unpublished (1961)
22. G. MATTHEWS, D.H. SMITH and J.A. VINEY, 'Rate and temperature relationships for plasticiser solvation of pvc particles', PVC Plasticiser Group Meeting, Univ. of Technology, Loughborough (8 April, 1981)
23. J. CRANK and G.S. PARK (eds), Chapter 1 in *Diffusion in Polymers*, Academic Press, New York (1968)
24. F. ASMUSSEN and G. RAPTIS, *Diplomarbeit*, Federal University, Berlin (1965)
25. K. UEBERREITER, Chapter 7 in *Diffusion in Polymers*, (J. Crank and G.S. Parks, eds) Academic Press, New York (1968)
26. P. SCHMIDT, *Kunststoffe,* **41**, 23 (1951); **42**, 142 (1952)
27. H.S. BERGEN and J.R. DARBY, *Ind. Engng Chem., ind. Edn,* **43**, 2404 (1951)
28. G.A.R. MATTHEWS, E.C.A. HORNER and W.W. VINCENT, unpublished work (1956–1957)
29. L.A. MCKENNA, *Mod. Plast.,* **35**, 10, 142 (1958)
30. G.J. VAN VEERSEN and A.J. MEULENBERG, *Plastica,* **17**, 440 (1964)
31. E.H. POAKES, *Br. Plast.,* **39**, 92 (1966)
32. W.J. FRISSELL, *Mod. Plast.,* **38**, 232 (1961)
33. F. BARGELLINI, *Materie Plast.,* **28**, 4, 372 (1962)
34. J.H. GISOLF, *Plastica,* **15**, 10, 498 (1962)
35. R.F. BOYER, *TAPPI,* **34**, 8, 357 (1951)
36. *Plasticisers for PVC*, BP Chemical (U.K.) Limited (1968)
37. R.B. TAYLOR and A.V. TOBOLSKY, *J. appl. Polym. Sci.,* **8**, 4, 1563 (1964)
38. *'Corvic' vinyl chloride polymers and copolymers*, I.C.I. Ltd. (1962)
39. H.C. MURFITT, *Br. Plast.,* **33**, 578 (1960)
40. I. PHILLIPS, *Br. Plast.,* **37**, 261 (1964) and **37**, 325 (1964)
41. W.J. FRISSELL, *Ind. Engng Chem., ind. Edn,* **48**, 6, 1096 (1956)
42. *Plasticisers for PVC*, A. Boake, Roberts & Co. Ltd. (1963)
43. P.A. SMALL, *J. Soc. chem. Ind., Lond.,* **66**, 17 (1947)
44. D.C. ATKINS, H.R. BAKER, C.M. MURPHY and W.A. ZISMAN, *Ind. Engng Chem., ind., Edn,* **39**, 491 (1947); C.M. Murphy and H. Rayner, *ibid*, **44**, 1607 (1952)
45. T.C, MOORSHEAD, *Rubb. Plast. Age,* **49**, 240 (1968)
46. *Plasticisers in PVC,* The Geigy Company Ltd. (1960)
47. *Resin Rev.,* **18**, 4, 16 (1968)
48. *'Cereclor' S52 in vinyl compositions*, I.C.I. Ltd., Mond Division
49. *'Cereclor' S52 Technical Supplement 4*, I.C.I. Ltd., Mond Division
50. *Brit. Pat.* 960 506
51. *'Cereclor' S52 Technical Supplement 6*, I.C.I. Ltd, Mond Division
52. D.G. BOLLER, *Officiel Matier Plast.,* **14**, 5, 449 (1967)
53. K.M. BELL and B.W. MCADAM, *Kunststoffe,* **57**, 7, 526 (1967)
54. D.G. BOLLER, *Kunststoffe,* **57**, 8, 614 (1967)
55. K.M. BELL and J.L. FOWLER, *Kunststoffe,* **57**, 9, 719 (1967)
56. P.F. WOERNER, *Mod. Plast.,* **45**, 3, 153 (1967)
57. J.K. SEARS and J.R. DARBY, *The Technology of Plasticisers*, John Wiley, New York (1982)
58. D.F. CADOGAN and C.J. HOWICK, *Ulmann's Encyclopedia of Industrial Chemistry*, VCH Publishers, New York, vol. A20, 439–458 (1992)
59. C.J. HOWICK, *PVC '93 – The Future*. Inst. of Materials Conf., Brighton, England (27/28 April 1993) Paper 38
60. F.W. CLARK, *Chem. Ind.,* **10**, 225 (1941)

61. R. HOUWINK, *Fundamental of Synthetic Polymer Technology in its Chemical and Physical Aspect*, Elsevier, New York (1949)
62. R. HOUWINK, *Proc. XIIth Int. Congr. Pure Appl. Chem.*, London, 1947, **5**, 575 (1953)
63. K. NAKAMURA, *J. Polym. Sci.*, **13**, 137 (1975)
64. G. KANIG, *Kolloid Z.*, **190**(1), 1 (1963)
65. R.S. BARSHTEIN and G.A. KOTLYAREVSKII, *Sov. Plast.*, **7**, 18 (1966)
66. R.A. HORSLEY, *Plastics Progress* (P. Morgan, ed.), Iliffe, London, 77 (1957)
67. M.L. DANNIS, *J. Appl. Phys.*, **21**, 505 & 510 (1950)
68. J. HILDEBRAND and R. SCOTT, *The Solubility of Non-Electrolytes*, Reinhold, New York, 3rd. Edn. (1949)
69. J.A. BRUSON, *Plastics Materials*, Newnes-Butterworths, London, 3rd Edn, (1975) chapter 5, esp. pp 75–86
70. G.J. VAN VEERSON and A.J. MEULENBERG, *SPE Tech. Pap.*, **18**, 314 (1972)
71. D.C.H. BIGG, *J. Appl. Polym. Sci.*, **19**, 3119 (1975)
72. H-K. BOO and M.T. SHAW, *J. Vinyl Tech.*, **9**(4), 168 (1987)
73. H-K. BOO and M.T. SHAW, *J. Vinyl Tech.*, Proceedings ANTEC '89', SPE Conference, 753 (1989)
74. P.V. MCKINNEY, *J. Appl. Polym. Sci.*, **11**, 193 (1967)
75. D. L. BUSZARD, *PVC Processing II*, PRI Intl. Conf., Brighton, England, (26–28 April 1983) Paper 22
76. G. MATTHEWS, *Vinyl and Allied Polymers*, Vol. 2 (1972)
77. L.F. RAMOS-DE VALLE and M. GILBERT, *Plast. Rubb. Proc. Appn.*, **13**, 151 (1990)
78. L.F. RAMOS-DE VALLE and M. GILBERT, *Plast. Rubb. Proc. Appn.*, **13**, 157 (1990)
79. L.F. RAMOS-DE VALLE and M. GILBERT, *Plast. Rubb. Proc. Appn.*, **15**, 207 (1991)
80. L.G. KRAUSKOPF, *Handbook of PVC Formulating* (E.J. Wickson, ed.), John Wiley & Sons, Inc., New York (1993) Chapter 5
81. J.T. LUTZ, *Handbook of PVC Formulating* (E.J. Wickson, ed.), John Wiley & Sons, Inc., New York, (1993) Chapters 6 & 7
82. J. BALDWIN, Private communication

7
Other aspects of formulation

7.1 INTRODUCTION

In previous chapters stabilisers and plasticisers for PVC have been considered separately. This chapter deals with other additives, such as lubricants, fillers, processing aids, impact modifiers, and colorants. This does not imply that these materials are necessarily less important, but in general the choice of different materials is relatively limited and can consequently be dealt with in less space.

7.2 LUBRICANTS

As the term implies, the functions of lubricants in PVC compositions are regarded as reducing adhesion between the hot plastic material and metal processing equipment, and in easing the flow or polymer molecules relative to each other. Substances that fulfil the former function are known as 'external lubricants', and the latter as 'internal lubricants'. The predominant activity of a lubricant seems to be closely related to its compatibility with the polymer, external lubricants being generally incompatible and internal lubricants compatible. Compatibility can be assessed by haze tests in transparent compositions, and by determining changes in T_g, the greater the compatibility the greater the reduction in T_g, but practical evaluation of processing behaviour is essential. Lubrication can be very critical if maximum outputs and best possible quality are to be achieved. With unplasticised PVC in particular, optimum formulations are often specific to particular machines and processing conditions, and need to be adjusted if performance is to be matched in machines of only slightly different design. Naturally this makes formulation extremely difficult and troublesome, particularly as important differences often only show up in full-scale production, so that development of new compositions can be very expensive in materials and time.

An important function of lubricants is to control 'melting' rate, particularly in extrusion of unplasticised dry blends.[21] The behaviour of lubricants, particularly in unplasticised PVC compositions, has been studied experimentaliy in some detail by extrusion experiments and using torque rheometers.[31]

Like other additives substances added for lubrication purposes may have other effects on processing behaviour and final properties. Some highly compatible lubricants are claimed to have synergistic stabilising effects with some stabilisers. Compounds containing hydroxyl groups, e.g. long-chain partial esters of polyfunctional alcohols, appear to be particularly effective in this respect.[22]

7.2.1   EXTERNAL LUBRICANTS

In by far the major part of PVC processing, lubrication is required to prevent excessive sticking. Processing at the usual 'melt' temperatures of 160°C to 200°C and even above leads in the absence of lubricant to sticking to the hot processing equipment, usually metal. Quite apart from the consequent loss of material, interference with removal of the PVC from the equipment and general messiness, the adhering polymer will invariably become over-heated and decompose excessively, leading to deterioration of processed material and corrosion of the equipment.

On the other hand, complete absence of adhesion between plastic material and equipment eliminates shear required for thermal plasticisation and homogenisation.

For these reasons external lubricants are usually included in PVC compositions to reduce and control adhesion to the required level. Little fundamental guidance can be offered for the section of lubricants and it is usually necessary to rely on past experience and trial and error experiments, very often on extended production scale. It must be noted, however, that lubrication can be very critical in terms of processing and product performance, and that small-scale laboratory experiments are often of little help in development of compositions for production use. It has been suggested that for unplasticised PVC, lubricants should have low solubility in the polymer but should be strongly polar, so as to have affinity for both the PVC and the metal surfaces of the processing equipment.[1]

An objectionable side effect sometimes displayed by external lubricants arises from their low compatibility with the PVC composition. As the concentration of a lubricant is increased it will often reach the point where it is exuded to the surface during processing, carrying with it pigment or filler articles and forming a deposit on the surfaces of the equipment. Because it was first encountered in calendering where pigment from one batch was deposited on the calender bowls and contaminated subsequent material, the phenomenon is generally known as 'plate-out'. Its tendency to occur increases with increasing shear rates, and so it became evident as calendering speeds increased with increasing production demands. It can nevertheless, also occur in other processes, such as extrusion, where it becomes evident in the form of a deposit on die lips. This effect often limits the permissible concentration of a lubricant and may even preclude it as the sole lubricant if the limiting concentration is inadequate to provide sufficient lubrication.

For plasticised PVC the most commonly employed lubricants are stearic acid and its salts with calcium, lead (both normal and dibasic salts), cadmium, and barium; myristic acid; paraffin wax and various other waxes; partially oxydised polyethylene; low molecular weight polyethylene; and certain esters, such as ethyl palmitate. Concentrations employed are usually between 0.5 and 1 phr, but 0.25–0.6 phr for stearic acid. Selection of lubricants for unplasticised PVC can be difficult and is considered in more detail later.

7.2.2   INTERNAL LUBRICANTS

The term internal lubricant is sometimes applied to lubricants of any kind which are included in a composition rather than being applied separately to processing equipment. In the present context the name is applied to materials which lubricate the flow of a PVC composition within the melt. It is not pretended, however, that lubricants can be definitely categorised as 'external' or 'internal', since it is likely that most lubricants, if not all, act in both ways, the classification depending on the predominant behaviour. Moreover the formulation of lubricants is very arbitrary, and it is often not clear just how a particular substance is performing.

Internal lubricants are rarely necessary in plasticised PVC, but they are almost essential nowadays for unplasticised PVC compositions of high quality and performance.

In addition to the substances listed as external lubricants, which may or may not have internal lubrication effects, the following are used: wax derivatives, such as montan wax ester derivatives; amide waxes, such as stearamide; glyceryl esters, such as those of stearic, oleic, and erucic acids, particularly glyceryl monostearate (GMS) and glyceryl monoricinoleate; long-chain alcohols, such as stearyl and cetyl alcohols; and long-chain esters, such as cetyl palmitate. Concentrations vary widely and combinations of two, three, and even four different lubricants are frequently employed in unplasticised PVC to get the desired balance of behaviour. Total concentrations rarely exceed 4 phr.

7.3 FILLERS

Fillers are employed in PVC, as in many other plastics materials, for a variety of reasons, including not only cost savings but also technical improvements.

7.3.1 UNPLASTICISED PVC

In unplasticised PVC the use of fillers is essentially limited to the increase in toughness by the incorporation of small proportions of synthetic calcium carbonates of fine particle size, any cost saving being incidental and minimal. A typical material suitable for this purpose is a precipitated calcium carbonate of ultimate particle size less than 100 nm, e.g. about 75 nm.[2] Powdered substances like this agglomerate rather severely and are normally coated with a few per cent of lubricant such as stearic acid or calcium stearate to reduce this. This treatment is also said to improve wetting by polymer during processing. About 10 phr of such a filler, thoroughly blended into unplasticised PVC, can increase impact strength over six-fold, the actual improvement depending on other ingredients of the composition, the nature of the specimen, and the temperature.[3] Ground chalks with lubricant coatings are also used, in window frame profiles, for example. These act as gentle scourers of die and calibrator surfaces, as well as contributing to pigmentation at low cost.

7.3.2 FILLERS IN PLASTICISED PVC

7.3.2.1 *Introduction*
As a general rule, unlike the effect of fine particle fillers in unplasticised compositions, fillers of any kind in plasticised PVC generally lead to a deterioration in most mechanical properties.

The fillers most commonly employed in plasticised PVC are the precipitated forms of calcium carbonate referred to previously, various comminuted forms of natural calcium carbonate, and related minerals such as calcium magnesium carbonate (dolomite). Other fillers used include china clay, calcined clays, and other clays, barytes, talc, alumina, and kieselguhr. Substances containing zinc, such as lithopone and zinc oxide, should be avoided since they catalyse thermal degradation of PVC.

The natural forms of calcium carbonate are obtained from a variety of mineral deposits, ranging from soft chalks or whitings to the harder limestones and marbles, and thus vary considerably in purity. They also vary in particle size and shape, depending on the nature of the grinding process used to produce them. Like the precipitated calcium carbonates, and for the same reasons, the finer mineral forms are frequently coated with a few per cent of lubricant. As a general rule the cost of the natural calcium carbonates and similar minerals increases with increasing purity and decreasing particle size, the dearest being comparable in price with the synthetic forms. Even allowing for the relatively high density of these fillers (in the region of 2.6–2.9), the most expensive grades of these fillers offer appreciable cost savings in the range of concentrations possible in plasticised compositions for extrusion, moulding and calendering. The concentrations used in such compositions range up to about 60 phr, but higher concentrations are sometimes used where the application permits, and very much higher concentrations are common in flooring compositions.

7.3.2.2   Visual effects
The most obvious difference observed between different fillers is that the very fine synthetic materials scarcely affect the surface gloss obtainable with unfilled compositions, even at the 60 phr level. As the particle size increases mattness becomes apparent in the surfaces of extruded, moulded, or calendered products.

In addition to the mattness or gloss, other visual effects can be important. Thus, the base colour of unpigmented PVC can vary considerably according to the chemical nature, whiteness, and purity of the filler, from quite a good bright white to dull grey, or even yellow, green, and purple tints. Other things being equal, with a 'white' filler whiteness increases as particle size decreases. Heat stability as assessed by colour development also varies widely, mainly dependent on the purity of the filler. Less immediately obvious, but nevertheless important, is the behaviour on bending of coloured plasticised PVC. With the coarser fillers bending produces whiteness where the bending is most intense; and some of this is retained on recovering the original shape. With the fine synthetic calcium carbonate fillers this defect is absent or almost so.[2] Similarly, scratching the surface of pigmented plasticised PVC containing white mineral fillers produces unsightly white marks which are virtually non-existent if a fine synthetic calcium carbonate filler is used, even at the high loadings common in vinyl floor tiles.[2]

Most fillers cause translucence in an otherwise transparent composition at quite low concentrations and give near opacity at quite moderate concentrations, but some grades of alumina or hydrated alumina can often be incorporated at quite high concentrations without serious loss of clarity. This has the advantage of reducing surface tack and blocking, which is otherwise quite difficult to achieve in soft clear material.

7.3.2.3   Effects on mechanical properties
At a given loading of a calcium carbonate filler, tensile strength and elongation at break appear to increase with decrease in particle size,[4, 5] but no evidence appears to be available for other types of filler. The particle size of any grade will usually vary over a wide range, the lower end of a mineral filler rarely reaching down to 1 μm, the upper end attaining around 55 μm. Average particle sizes vary from about 1.5 to about 25 μm.

On the basis of studies of the effect of one particular form of calcium carbonate, it has been suggested that the presence of filler does not basically change the nature of mechanical behaviour of plasticised PVC, but that small proportions (about 10 per cent by volume) act almost like additional plasticiser in that moduli are lowered and extensions at break are increased.

Most of the data presented on the effects of fillers on the physical properties have shown the variation with properties with filler content, the concentration of plasticiser and other ingredients being maintained constant. This is somewhat unrealistic, since the introduction of fillers reduces flexibility and softness which are usually properties that one wishes to keep more or less constant for a given application.[2, 4–7] There is some evidence that plasticiser absorbed by filler can become bound and fail to contribute to plasticisation of the polymer. It is even suggested that, as a consequence, the order of mixing of ingredients of a given filled plasticised composition can affect the softness of the fully processed material. This is on the basis that addition of the filler to the mix before the plasticiser can result in absorption of the latter by filler particles, thus reducing the quantity available for plasticisation, whereas thorough mixing of plasticiser with the polymer before introduction of the filler leaves the maximum possible proportion of plasticiser available for plasticisation, thus yielding a softer and more flexible product. For these reasons the oil absorption of a filler can be important, and is often quoted in technical sales literature. Typical values for calcium carbonate types of filler run from 20 to 60 g of plasticiser per 100 g of filler. Evidence has been presented to indicate that oil absorption is approximately inversely proportional to filler bulk-density,[5] but it seems doubtful if this can be universally true since the fine structure of the particles of a filler would also be expected to affect absorption.

Fairly detailed comparisons of the effects of different types of fillers have been made,[4, 5, 6] but since mechanical properties such as tensile strength and elongation at break appear to be dependent on the specific surface and fine particle content,[4] care must be exercised in applying the results to particular commercially available fillers.

In general softness decreases with increase in concentration of filler, although there appears to be a limited range of concentrations (zero to about 5 per cent) for stearate-coated natural whiting, where the reverse is true.[6] For a number of fillers (stearate-coated natural whiting, precipitated calcium carbonate, china clay, coated precipitated calcium carbonate, calcined china clay, and carbon black), the relationship between BS Softness and concentration of filler is substantially linear, the gradient increasing in the order given.[6] With asbestos the relationship was not linear, the gradient decreasing with increasing concentration. Thus with a basic composition of equal parts of polymer and tritolyl phosphate, the values for BS Softness listed in Table 7.1 have been quoted.

Comparable results have been quoted for a few fillers in compositions plasticised with a wide range of concentrations of DAP with chlorinated paraffin extender.[8] Such results are easiest presented in the form of triangular diagrams,[8] but these are often difficult to interpret rapidly, and the data of Table 7.2 have been extracted for illustration.

For the reasons discussed previously, it would perhaps be more meaningful to compare compositions at equal softness, and an example of such a comparison is given in Table 7.3, where a series of formulations designed to give an equal BS Softness of 50 is set out.

The concentrations indicated in Tables 7.1, 7.2, and 7.3 cover the range most commonly employed in all grades of plasticised PVC other than in vinyl flooring, where the concentrations of filler can be much higher. Nevertheless, the same general principles apply.

Table 7.1   VARIATION OF BS SOFTNESS NOS. WITH CONCENTRATION FOR A NUMBER OF FILLERS

Concentration (weight %)	0	10	20	30	40	50
Coated natural whiting	88	86	76	66	54	42
Precipitated calcium carbonate	88	78	68	56	45	34
China clay	88	78	67	55	43	33
Coated precipitated calcium carbonate	88	75	64	52	41	32
Calcined china clay	88	73	60	46	34	23
Carbon black	88	66	45	28	13	—
Asbestos	88	65	48	36	30	37

Table 7.2   VARIATION OF BS SOFTNESS NOS. WITH CONCENTRATION OF FILLERS

Polymer/plasticiser ratio	60/40				55/45			
Filler concentration (%)	0	10	20	30	0	10	20	30
Surface treated calcium carbonate	52	47	37	28	69	63	54	50
China clay	52	47	37	24	69	58	50	39

Table 7.3   VARIATIONS OF PLASTICISER AND FILLER CONCENTRATIONS AT CONSTANT SOFTNESS ((BSS 50) PLASTICISER = 2:1 MIXTURE OF DAP AND CHLORINATED PARAFFIN)

	A		B		C		D		E		F	
	%	Parts	%	Parts	%	Parts	%	Parts	%	Parts	%	Parts
Polymer	61	100	54	100	50	100	46	100	43	100	40	100
Plasticiser	39	64	36	67	35	70	34	74	32	74	30	75
Calcium carbonate (surface treated)	0	0	10	19	15	30	20	44	25	58	30	75
Polymer	61	100	54	100	49	100	45	100	41	100	37	100
Plasticiser	39	64	36	67	36	73	35	78	34	83	33	89
Clay	0	0	10	19	15	31	20	44	25	61	30	81

As already indicated, tensile properties are dependent, with some fillers at least, on the particle size. Effects of variation of concentration of filler on tensile properties vary from one filler to another. While it is generally considered that tensile strength steadily decreases with increase in filler content, there is evidence[6] that some materials, such as carbon black, finely divided china clay, and calcined china clay, used at concentrations up to about 20–30 per cent by weight of total composition, can yield compounds with higher tensile strengths than the unfilled compound. Variations in elongation at break do not follow the same pattern, but there is still considerable difference in effect from one filler to another. Tables 7.4 and 7.5 set out some values estimated from the literature.[6] The compositions were all based on equal concentrations of polymer and plasticiser (TTP).

Apart from the fact that carbon black appears to yield the highest tensile strengths and correspondingly the lowest elongations at break, these figures seem to show little correlation. The behaviour of carbon black is interesting, but is of relatively little commercial significance, since, although it is widely used as a pigment, this is generally at concentrations of only about 2 per cent. Much higher concentrations are used in a few special applications where low electrical resistivity is required.

Table 7.4 VARIATIONS OF TENSILE STRENGTH (MN/M²) WITH FILLER CONCENTRATION

% Filler	0	10	20	30	40	50
Carbon black	11.5	13.5	13.8	11.2	—	—
China clay	11.5	13.2	12.2	9.3	7.3	5.3
Calcined china clay	11.5	11.8	11.2	7.7	5.6	5.2
Coated whiting	11.5	11.2	10.4	8.7	6.8	4.5
Precipitated calcium carbonate	11.5	10.4	9.3	8.3	6.8	5.3
Asbestos	11.5	9.3	8.0	6.8	5.9	5.3

Table 7.5 VARIATIONS OF ELONGATION AT BREAK (%) WITH FILLER CONCENTRATION

% Filler	0	10	20	30	40	50
Coated whiting	400	400	400	390	370	310
Precipitated calcium carbonate	400	390	370	350	330	300
China clay	400	385	355	330	290	250
Calcium silicate	400	380	320	180	85	120
Asbestos	400	360	300	250	180	45
Carbon black	400	330	260	180	75	—

Table 7.6 MODULI (MN/M²) AT 100% ELONGATION OF COMPOSITIONS OF EQUAL SOFTNESS OF TABLE 7.3

	A	B	C	D	E	F
% Filler	0	10	15	20	25	30
Surface treated calcium carbonate	7.5	6.89	6.84	5.52	5.52	5.58
Clay	7.5	6.84	5.8	5.65	5.52	4.83

Table 7.6 sets out published[6, 7] moduli at 100 per cent elongation for the compositions A–F of BS Softness 50 listed in Table 7.3, but again little of theoretical value can be deduced from these figures.

It might be argued that the lack of correlation between moduli and BS Softness casts doubt on the value of the latter as indications of flexibility of PVC compositions, but it must be remembered that moduli for materials of this type are rather arbitrary and do not always have any immediately obvious fundamental significance. As is pointed out in Chapter 6 (Section 6.3.4.2) it has been suggested that in flexing strains of about 1–2 per cent only are involved, and that, since stress/strain curves for the compositions under consideration are never linear, measurement of moduli at this low order of strain would offer more useful numerical guidance to flexibility.[9] Practical difficulties have limited acceptance of this as a general means of evaluation.

7.3.2. *Effects on behaviour at low temperatures*
Inclusion of filler leads to deterioration in low temperature flexibility of plasticised PVC, although again the effects vary quite widely from one filler to another. Of materials for which data are available carbon black seems to have the most severe effect and fine calcium carbonates the least. As indicated previously (Section 3.3.4.4), test methods for properties

at sub-zero temperatures do not correlate well with each other or with service experience, but the order of magnitude may be assessed from the values for cold flex temperatures (BS 2571)[6] quoted in Table 7.7. These are based on the same unfilled composition as that of Tables 7.1, 7.4, and 7.5.

Table 7.7   EFFECTS OF FILLER ON COLD FLEX TEMPERATURES (°C) OF PLASTICISED PVC
(BS SOFTNESS 88)

Concentration (weight %)	0	10	20	30	40	50
Whiting	−21	−21	−20	−19	−19	−18
China clay	−21	−21	−19	−18	−17	−15
Cork dust	−21	−20	−18	−16	−12	—
Carbon black	−21	−19	−17	−13	−10	—
Calcium silicate	−21	−18	−17	−17	−17	—

Requirements for retention of flexibility and avoidance of cracking at low temperatures in plasticised PVC applications are usually for intermittent periods only, for example in electric and telephone cables, where such performance is normally only required for repair work during cold weather. As before, then, it is more meaningful to compare compositions having similar flexibility at room temperature. This has been attempted in Table 7.8, based on data from trade literature[8] on two fillers. The figures are related to the same compositions as listed in Table 7.3.

Table 7.8   COLD FLEX TEMPERATURES (°C) (BS 2571) OF COMPOSITIONS OF EQUAL
SOFTNESS OF TABLE 7.3

	A	B	C	D	E	F
% Filler	0	10	15	20	25	30
Surface treated calcium carbonate	−27	−30	−32	−34	−35	−35
Clay	−27	−27	−29	−30	−32	−35

If the data of Table 7.8 are meaningful, the significance is that if plasticiser contents are increased to compensate for hardening due to introduction of filler, cold flex behaviour is not only maintained but even improved relative to the corresponding unfilled composition. This suggests that the effect of varying plasticiser content is more significant than that of varying filler content.

7.3.2.5   *Effects on behaviour at elevated temperatures*

Mechanical properties which are affected by increase in temperature, change in like manner whether a plasticised PVC composition is unfilled or filled, but little information is available on the precise behaviour of filled compositions.

At one time asbestos was used in floor tiles, particularly those for use in hot environments such as kitchens, presumably because of a real or imagined effect on mechanical properties at elevated temperature, but, because of the health hazard associated with its use, has been replaced by other fillers such as talc.

Important effects do arise in processing. As indicated previously (Section 7.3.2.2), fillers can affect heat stability and hence colour development during processing. Deleterious effects appear to be generally due to impurities, particularly ions of metals such as iron and copper. Somewhat surprisingly calcium carbonate does not appear to be a stabiliser for PVC. Indeed, its presence can lead to problems where hydrochloric acid resulting from decomposition reacts to form calcium chloride, greatly increasing the tendency to absorb water.

In the concentrations normally used for extrusion, moulding, calendering, etc., the melt flow behaviour of filled PVC compositions is rarely sufficiently affected to raise any serious processing problems. At high concentrations, however, compositions become increasingly stiff and 'dry' during processing, and mixing can be difficult. This frequently requires substitution of a homopolymer of normal average molecular weight (e.g. ISO Viscosity No. 125) by one of lower molecular weight, or by a copolymer, so as to achieve adequate wetting of the filler particles by plasticised polymer. Limited data on the effect of fillers on melt flow behaviour of plasticised PVC are available.[7] These are summarised in Table 7.9, where the effect of various fillers on melt flow behaviour is expressed in terms of the pressures required to extrude at a constant rate through a standard die, using a Macklow-Smith plastometer.

Table 7.9   MACKLOW-SMITH FLOW PRESSURES (MN/M²) FOR VARIOUS FILLERS

Concentration (weight %)	0	10	20	30	40	50	60	70
Precipitated calcium carbonate	4.22	—	4.22	4.22	4.22	4.22	4.22	4.22
China clay	4.22	—	—	4.5	4.8	5.25	7.6	>13.8
Woodflour	4.22	4.7	4.9	5.25	5.51	—	—	—
Calcium silicate	4.22	4.35	4.83	5.51	7.6	11.73	22.75	—
Carbon black	4.22	4.9	5.65	6.83	10.34	—	—	—

7.3.2.6   *Effects on electrical properties*

Apart from carbon black, the fillers used in PVC all appear to increase volume resistivity although the effect varies not only with the basic nature of each filler but also with purity, as might be expected. Some calcined china clays or kaolinites, particularly those of metamorphic origin,[6] can increase values of volume resistivity very markedly even at low concentrations of the order of 6 per cent, and are, therefore, very valuable in the more demanding electrical insulation applications, where they are often used in combination with larger proportions of the better quality calcium carbonates. Table 7.10 sets out some data taken from the published literature showing the variation in volume resistivity of a PVC compound with variation in concentration of a few fillers. The effects of carbon black are quite remarkable, and at least indicate that it is a suitable material to use as a black pigment in insulation grades.

Table 7.10   VARIATION IN VOLUME RESISTIVITY (Ω CM $\times 10^{11}$) WITH FILLER CONTENT[6]

Concentration (weight %)	0	10	20	30	35	40
Calcined china clay	71	250	530	820	1000	1100
Whiting	71	84	86	90	93	100
Carbon black	71	1300	1800	480	1	—

Other electrical properties, such as power factor, permittivity, and breakdown voltage, appear to be far more dependent on the other ingredients of PVC compositions and can be largely neglected from consideration of fillers.

7.3.2.7   *Effects on reactions to chemical environment*
With the exception of effects of fillers on water absorption and permeability,[5-7] little appears to have been published about the effects of fillers in PVC on its behaviour towards liquids, vapours, and gases. Apart from possible absorptive behaviour, the majority of fillers used are obviously inert to organic chemicals. Except at the upper end of the range of concentrations used in very heavily filled vinyl flooring, it seems likely that most filler particles would be surrounded by plasticised PVC matrix, so that the latter would then be the main point for attack. Nevertheless, filler particles would contribute to the overall permeability of a layer of plasticised PVC, and it would be possible to theorise about this, but insufficient data are available to make this worthwhile.

Water absorption[5-7] and permeability to water vapour[5] are generally increased by the introduction of fillers, but the effects are usually small at low concentrations, again possibly due to imbedding of the filler particles below the surface of the polymer matrix. Published literature[6,7] (Table 7.11) indicates a most peculiar behaviour with calcium silicate, the relationship between absorption and filler content exhibiting both a maximum and minimum. Data for carbon black indicate a very small effect, but this is at variance with practical problems which have arisen on the commercial scale. The data on which Table 7.11 is based were obtained by measuring the increase in weight of specimens 51 mm in diameter and 1.27 mm thick after immersion in distilled water at 50°C for 48 h.

Table 7.11   EFFECTS OF FILLERS ON WATER ABSORPTION (GRAMMES ABSORBED BY SPECIMEN)

Concentration (weight %)	0	10	20	30	40	50	60
Carbon black	0.02	0.03	0.04	0.04	0.03	—	—
Calcined china clay	0.02	0.02	0.02	0.02	0.02	0.02	—
Precipitated calcium carbonate	0.02	0.02	0.03	0.03	0.03	0.03	0.03
China clay	0.02	0.06	0.09	0.13	0.15	0.18	0.21
Woodflour	0.02	0.09	0.16	0.23	0.30	0.37	—
Calcium silicate	0.02	0.25	0.38	0.25	0.22	0.21	0.23

It would be expected that introduction of filler would introduce a susceptibility to chemicals with which it normally reacts; thus, calcium carbonate would obviously be expected to render a composition liable to attack by acids. It is of course possible that a measure of protection of the filler particles would be provided by the polymer/plasticiser matrix, which would tend to break down as the concentration reached such a high level that an appreciable proportion penetrated and contributed to the external surface. No detailed data on this appear to be available.

7.3.2.8   *Effects of fillers on behaviour in fires*
Although it is true that unplasticised PVC, and even PVC containing quite high concentrations of plasticiser, will not ignite in the way that acrylics, polystyrene, and polyolefins will, the widespread idea that PVC is inherently fire-resistant[23] can be misleading. Quite

apart from the effects of varying formulation on behaviour in fires even a simple unplasticised PVC will degrade in a fire and may contribute to the resultant hazards of fumes and smoke.

It would be expected that high concentrations of inorganic mineral fillers would reduce the flammability of an otherwise flammable material such as a PVC composition plasticised with phthalates, adipates, or sebacates, because they are effectively reducing the concentration of flammable material, and thus reducing availability of fuel to sustain the flame. This is true for inert fillers but unfortunately the most common, filler, calcium carbonate, reacts with hydrogen chloride formed by degradation of the polymer and decreases resistance to burning.

The relatively low flammability of PVC is attributable to its chlorine content.[24, 25] It has been claimed[24, 26] that incorporation of 'basic' fillers, such as calcium carbonate, which react with hydrogen chloride evolved during degradation, has little effect on flammability, but this is not true of all forms of calcium carbonate. The efficiency with which calcium carbonate reacts with the evolved hydrogen chloride, limiting its escape into the atmosphere surrounding the degrading mass, depends on the particle size of the filler.[27] The smaller the particle size the greater the efficiency of reaction, and very fine particle calcium carbonate can be used to retain all the evolved hydrogen chloride.[28] For example, in unplasticised PVC about 100 phr of calcium carbonate of 0.1 μm particle size results in essentially complete retention of the hydrogen chloride, but this is accompanied by an increase in flammability, as indicated by a fall in Oxygen Index from 48 to 29 per cent oxygen.[27, 29] If ignited and the source of ignition is removed, PVC plasticised with 80 phr of DOP continues to burn, whereas with 60 phr it will not. However, the incorporation of about 70 phr of a fine particle calcium carbonate into the latter composition converts it into one that will continue to sustain a flame after ignition and removal of the igniting source. With a calcium carbonate of particle size 1 μm the effect of hydrogen chloride retention seems to be balanced by what might be called an 'inert' filler effect and possibly evolution of carbon dioxide, so that burning behaviour is not changed appreciably. Coarser calcium carbonates reduce flammability.[27] The precise mechanisms involved are complex, and have not been completely resolved, but it seems that the action of chlorine in the PVC molecule in reducing flammability is chemical and not merely due to physical effects of hydrogen chloride in gases evolved by degradation.[30]

A conclusion from these observations is that it must not be assumed that a PVC product is necessarily of low flammability. This topic is discussed more fully in Chapter 17. As suggested earlier the inclusion of inert fillers would be expected to reduce flammability, and might, therefore, be regarded as being 'fire retardants'. Thus, incorporating silica into PVC in increasing concentrations results in a fairly steady increase in Oxygen Index. Other fillers reduce flammability by chemical reaction and are used essentially for their fire-retardant properties. Thus antimony trioxide is effective at concentrations as low as 5 per cent, which is fortunate since it is rather expensive, and it is only its inorganic nature which renders it eligible to be classed as a filler at all. Incidentally, it appears to be effective only in the presence of compounds containing chlorine, in this case the polymer. Antimony trioxide is losing favour because it is toxic and is also suspected of being carcinogenic,[48] and is tending to be replaced by other flame retardants such as alumina. Hydrates are particularly effective because, in addition to any chemical influence[31] on the course of combustion, heat is removed from the system by separation and volatilisation of water.[32, 33]

Other compounds that perhaps should more correctly be termed 'fire retardants' include magnesium hydroxide[36], molybdenum trioxide[36], zinc borate[35] and phosphate.[36]

7.3.2.9   Selection of fillers

Where the use of fillers is primarily to reduce cost, the cheapest available will naturally be used where possible. Some of the cheaper minerals have such low costs that transport over large distances contributes a major part to the overall cost of their use. With this type of material therefore there is a tendency to use such minerals as are readily available locally, so that there is variation in usage from place to place.

Where quality is not particularly important, ground limestones and similar minerals are sometimes employed. If higher purity, with better colour, better electrical insulation properties, or more uniform mattness are required, the more expensive ground calcium carbonates may be selected. If high gloss is desired lubricant-coated synthetic calcium carbonates or suitable grades of ground mineral calcium carbonate or dolomite may be used.

For maximum electrical resistivity a proportion of calcined clay may be used, although there is some disagreement about the value of this material; for minimum flammability, antimony trioxide, especially its trihydrate was preferred, but as explained above, it is now being replaced by other substances or one of the fire-retardant combinations now available.[32–36]

For improvement of impact resistance of unplasticised PVC, 10–15 phr of a lubricant-coated synthetic calcium carbonate is appropriate.

7.4   IMPACT MODIFIERS

As indicated previously (Section 7.3.1) the impact resistance of unplasticised PVC can be increased considerably by the inclusion of around 10 per cent of synthetic calcium carbonate, and here the filler ought, perhaps, to be considered as an 'impact modifier' or 'toughening agent'.

Apart from the synthetic calcium carbonates, impact modifiers are generally of low T_g, and are designed to form a discrete rubbery phase. As with some other rigid polymers, notably polystyrene, rubbers (particularly of the synthetic variety) at concentrations up to about 15 per cent increase the toughness of unplasticised PVC, but additional additives may be desirable to prevent loss of impact strength on ageing, particularly if exposed outdoors.

The most effective impact modifiers appear to be certain rubbery acrylic copolymers, which also effect marked improvements in the region of 5–15 per cent concentration, as well as having excellent stability. This type of modifier can also be used in transparent compositions without markedly affecting clarity.

Other materials used to increase the toughness of unplasticised PVC include acrylonitrile/butadiene/styrene (ABS) and methyl methacrylate/butadiene/styrene (MBS) terpolymers, ethylene/vinyl acetate (EVA) copolymers, chlorinated polyethylenes, butyl and octyl acrylate polymers,[40] and blends of nitrile rubbers with polyolefin graft copolymers.[34, 37]

Table 7.12 gives some published[10] values for improvement in impact strength of a typical unplasticised PVC, as measured by the notched Izod impact test (ASTM D–256), and by a 'bottle-drop' impact test, obtained by incorporation of an acrylic copolymer impact modifier and ABS. The bottle-drop tests were carried out by dropping capped four-ounce

Table 7.12   EFFECT OF IMPACT MODIFIERS ON TOUGHNESS OF UNPLASTICISED PVC

Concentration (%)	0	5	10	15	20
Notched Izod impact strength (ft lb/in) (Nm/m)					
Acrylic modifier	0.5(0.27)	2(1.1)	14(7.5)	21(11.3)	22(11.9)
Mean bottle failure height (ft) (m)					
Acrylic modifier		7(2.13)	7.15(2.18)	10(3.05)	—
ABS	—	6(1.83)	7(2.13)	8(2.44)	—

bottles filled with water on to a steel plate over a range of heights. The mean height to pro-duce failure was recorded.

Most impact modifiers affect other mechanical properties as well as toughness, generally in the directions expected for increased rubberiness or plasticisation, but the effects can be quite small. Thus hardness, tensile strength and modulus, flexural strength and modulus, light transmission, and heat distortion temperature are all reduced to some extent, while elongation at break is increased. Typical values of some of these properties for an unmodi-fied composition and for one containing 20 per cent of an acrylic modifier are listed in Table 7.13.[10]

Table 7.13   COMPARISON OF IMPACT MODIFIED PVC WITH UNMODIFIED PVC

Property	Unmodified	Modified (20%)
Hardness (Shore D)	78	80
Tensile strength (MN/m^2)	56.3	45.8
Elongation at break (%)	85	120
Tensile modulus (MN/m^2)	3080	1930
Flexural strength (MN/n^2)	96.8	63.5
Flexural modulus (MN/m^2)	3130	2050
Heat distortion temperature (18.2 kN/m^2 °C)	73	71
White light transmission (%)	78	72

7.5   PROCESSING AIDS

If PVC were chemically stable it might always be possible to raise the temperature of pro-cessing to a point where uniform flow was readily attained, although even this is doubtful. As it is, even with the best stabilising systems, it is sometimes difficult to get unplasticised compositions to flow smoothly through processing equipment and produce homogeneous products with good surface finish. This is particularly noticeable in the production of com-plicated sections, especially injection mouldings, but it is by no means uncommon in rela-tively straightforward extrusion. Processing aids are included in many unplasticised PVC compositions in order to reduce these problems, and thus generally to permit more tolerance in control of processing conditions. Specifically processing aids are claimed to reduce fusion time and melt viscosity, and improve melt strength, resistance to melt fracture and shear burning.[40]

The effects of processing aids alone and in combination with impact modifiers have been studied in torque rheometers and small-scale extrusion experiments,[37–40] and such experiments can be a useful guide to behaviour at full scale, though eventually it is only evaluation under production conditions that will ensure that a satisfactory formulation has been devised. The precise constitution of processing aids is rarely completely disclosed, but the best of them appear to be acrylic/methacrylic ester copolymers. Polymethyl methacrylate itself, as well as other acrylate polymers, of high molecular weight (i.e. higher than that of the PVC) are highly compatible with PVC and are effective as processing aids.[40] Concentrations employed are usually between 1 and 5 per cent. The mode of action is not completely understood, for while some appear to reduce melt viscosity, others increase it. Very often it is not possible to demonstrate the advantage of using processing aids in simple small-scale experiments, and their value may be revealed only after relatively lengthy production trials, when any advantage may appear in the form of a marked increase in the proportion of good quality product resulting from their inclusion.

7.6   ANTISTATIC AGENTS

Antistatic agents do not appear to be particularly effective in PVC, nor do they appear to be widely used. The vast majority of potential materials either lose their effectiveness or develop objectionable discloration during processing. The constitutions of those few materials that do work effectively are not disclosed. A few trade materials are undoubtedly effective in vinyl gramophone records, though the customer seems rarely to notice!

The problem of electrostatic charges does not appear to be so severe with PVC as it is with many other polymers.

7.7   COLORANTS

Much PVC, both plasticised and unplasticised, is processed and used in clear uncoloured form, but most of it is coloured in some way. It is not intended to discuss coloration in general here, since this is quite a big subject, and has been dealt with as adequately as possible elsewhere.[11–13] All that is attempted is to indicate the types of material and processes used to colour PVC, and in particular to point out some of the problems which are peculiar or are particularly liable to arise with PVC. In practice most suppliers have by now introduced adequate ranges of colouring materials for PVC, and practical coloration normally involves selecting from the established offers, and working to a few simple rules. The colouring of plastics in general requires attention to a number of factors, some of which are specific to plastics processing and some of which are shared generally with coloration of other products. The most important of these factors are:

(1) Dispersion of colorant throughout the plastic mass.
(2) Heat stability.
(3) Stability to the material environment.
(4) Migration.
(5) Light fastness.[11–17]

7.7.1 DISPERSION OF COLORANTS

The colorants used for PVC are normally insoluble powders, so that dispersion throughout the mass in solution form is not possible. While most pigments are fine particle powders, in practice the particles frequently tend to agglomerate more or less tightly together. Proper and efficient coloration of a plastics mass requires these agglomerates to be broken down, and for the primary particles of pigment to be distributed as uniformly as possible throughout the polymer matrix. Even where a mottled effect is required, as for example in flooring, agglomerates should be avoided and the non-uniform distribution should normally be on the gross scale of the pattern and not the micro or near-micro scale of the individual particles. Inadequate breaking down of agglomerates and dispersion of individual pigment particles results in gross cases in visible non-uniform colouring, but even when visible only under the microscope poor dispersion is likely to yield incomplete development of the full tinctorial power or intensity of the pigment, giving rise to dullness, haze, and lack of transparency. In gross cases agglomerates of pigment may be so large that they can readily be plucked from the surface of a finished product. The relative ease with which a pigment can be dispersed in a medium is known as its 'dispersibility', but it is difficult to quantify this property, which as might be expected shows a fair amount of variation. Some phthalocyanine, anthraquinone, perylene, and quinacridone pigments, and a few inorganic pigments such as titanium dioxide, tend to have hard particles which require particularly intensive mixing to achieve good dispersion.[14]

The dispersion of pigments in PVC may generally be regarded as neither easier nor more difficult than in other common plastics materials. Colorants are dispersed in PVC by most of the possible means, and as a general rule it will be found that if the processing homogenises the polymer or polymer/plasticiser mixture adequately, the colorant will also generally be adequately dispersed. There are, however, occasional occurrences of pigment separation either as 'plate-out' or in streaks in mouldings or extrudates.

Full compounding to produce self-coloured materials generally produces adequate colour dispersion without much trouble, and is commonly used for calendering, as well as extrusion and moulding.

For the latter two processes, concentrated masterbatches are frequently and effectively used. The masterbatches are generally fully compounded and granulated normal PVC compositions containing high concentrations, possibly up to as much as 50 per cent of pigment. They are produced by techniques similar to those used to produce normal self-coloured compounds (Chapter 10). In use, granules of natural uncoloured compound of the required composition are blended with the appropriate proportion of masterbatch granules, usually by simple tumbling, before extrusion or moulding, and it is relatively easy to obtain adequate dispersion in the finished product by this means. This procedure enables the processor to store a relatively small number of uncoloured compounds in fair quantity together with quite small quantities of a number of masterbatches, thus saving some of the storage space and capital investment involved in maintaining stocks of each compound self-coloured in each required shade.

Dry blending of pigment with granulated compound is not generally very satisfactory in dispersing pigment, but dry blending with polymer and additives, particularly in a 'high-speed' mixer developing a fair amount of frictional work, is very often practised satisfactorily.

As well as being available in powdered form and dispersed in compounded master-batches, many pigments are offered in dispersed form in a liquid medium, such as a plasticiser, or even a melt, such as a vinyl chloride/vinyl isobutyl ether copolymer.[14] Since the pigment has already been dispersed thoroughly in the medium, good dispersion throughout the finished product is usually readily achieved. Although these pigment dispersions may contain as little as 20 per cent of liquid medium, this may be sufficient to lower the softening temperature of unplasticised PVC, so they should only be selected for such compositions if evaluation has shown they are unobjectionable on this count.

7.7.2 STABILITY DURING PROCESSING

The majority of pigments are susceptible to chemical and colour changes on heating, and this excludes many because they fail to withstand the thermal treatment involved in processing. Specifically with PVC it is also necessary for a pigment to be unresponsive to other ingredients of the composition being processed and to hydrochloric acid which is bound to be produced during processing.

Several arbitrary tests are used, in which the pigments under examination are submitted to various heat treatment schedules; but these should not be regarded as more than preliminary screening tests. It is far more reliable to evaluate a prospective pigment by testing its behaviour in compositions in which it might be used and under conditions as near as possible to those to be experienced in production. In general it is found that inorganic pigments are better in yellow, red, and brown colours, while organic green and blue pigments are superior.[14]

Some pigments, notably some iron oxides, decrease the stability of the PVC itself, and special care must be exercised when contemplating the use of any pigment where there is a suspicion that this may occur.

7.7.3 MIGRATION PROBLEMS

PVC seems to be plagued rather more seriously than most other polymers by problems of pigment migration, but most of these can be avoided by proper attention to selection of nature and concentration to be employed. Migration can occur during processing and during later service life.

Separation of pigment from polymer during processing is normally associated with 'plate-out', usually resulting in deposition of pigment on the surfaces of the processing equipment. It is not clear whether or not pigments would separate by themselves. In practice the separation is usually associated with that of incompatible lubricants and stabilisers and appears to become more severe with increasing severity of shear conditions. It thus tends to become more of a problem as processing speeds (e.g. of calendering or extrusion) are increased. Sometimes partial separation of pigment and polymer becomes apparent in the form of streaks in continuously produced sections, such as calendered sheet or extrudates, which appear to be excessively heavily or lightly pigmented. It is not impossible that this, too, is an effect due to shear, arising for example during flow of melt around spider legs or through breaker plates. Usually the problem is only solved by trials under full processing conditions.

During service life migration of colorants can result in a number of apparently different but obviously related phenomena, known usually as 'blooming', 'bleeding', and 'chalking'. Blooming describes the migration of a material to the surface of a mass of polymer to form a deposit which is not only unsightly, but which is usually wiped off and picked up by objects rubbing against it. Chalking appears to be analogous to blooming but tends to be much more slow to develop. Whereas blooming very often becomes evident within a few days or weeks, chalking tends to appear more slowly, and may even take several years to show. Whether there are two phenomena or merely the same one taking place at different rates is not clear. It is suggested that high processing temperatures aggravate chalking.[15]

Bleeding involves migration of colorant from its intended site on to or into an adjacent material. In 'solvent bleed' the colorant passes into the adjacent material, which may be solid (for example, another plastic product, or liquid, such as an oil). In 'contact bleed' the migrating colorant stains the surface of an adjacent material.

In plastics materials generally, increase in concentration usually increases the tendency for blooming, and within limits bleeding, to occur, but with PVC the phenomena are in practice more likely to occur at excessively low concentrations. It is often found in plasticised PVC that bleeding or blooming are minimal at high concentrations, but that on reducing concentration a point is eventually reached below which both phenomena become excessive.[16, 18] At the same time there is often a dramatic change in hue. This behaviour is explained on the basis of partial solubility of a pigment in the PVC composition during processing.[16] When crystals of undissolved pigment are present in the plastic material after processing, any part of the pigment which dissolved crystallises out rapidly. At sufficiently low concentrations, however, the whole of the pigment might pass into solution if processing temperatures are sufficiently high, and on cooling a supersaturated solution of pigment remains. Crystallisation then occurs only at the surface of the PVC. Associated with the change of hue which coincides with this phenomenon, there is generally a reduced stability to light. This behaviour can obviously be obviated by increasing the pigment concentration or by using a less soluble pigment. This problem has particularly to be borne in mind when using masterbatches, lest using too low a concentration of masterbatch yields too low an overall concentration of pigment. Manufacturers' recommendations should always be followed for this reason if for no other. The severity of bleeding and blooming is also dependent on the size of the pigment molecules, and, as might be expected, the newer high molecular weight organic pigments are much less prone to these defects, partly owing to the lower solubility but possibly also to the greater resistance to diffusion which results from their relatively large molecular dimensions.

As well as the colorant itself, other factors such as the nature of the formulation as a whole, light, heat, and even humidity appear to affect the occurrence of blooming.[14] Increase in processing temperatures generally increases severity of blooming, but does not affect bleeding, whereas increase in service temperature promotes bleeding but not blooming.[11]

7.7.4 Colour stability

In most cases it is required that the original colour of a PVC product shall be maintained during its service life. Colour changes can, in fact, occur as a result of a number of different factors, mainly (1) instability of the pigments to light, (2) reaction of pigment with environmental chemicals, and (3) discoloration of the polymer due to degradation by radiation, generally only apparent in the most critical of pale pastel shades.

The ability of a colorant to maintain its colour under the influence of light is usually re-ferred to as its 'light fastness'. The light fastness of a colorant is difficult to predict, and, un-fortunately, accelerated tests using artificial radiation sources are rarely satisfactory guides to service behaviour. Consequently evaluation is best carried out by exposure of com-pounded samples in appropriate climatic conditions, and this is, of course, both expensive and time-consuming.[17] Light fastness does appear to be dependent on concentration,[11] though whether this is a true chemical effect or merely visual is not clear. Loss of colour appears more noticeably when the fading pigment is associated with a white pigment or stabiliser. In the absence of these, fading of the pigment reveals unfaded pigment deeper in the surface of the PVC matrix.

Reaction of a pigment with environmental chemicals is usually much more predictable, from a consideration of the basic chemistry of the pigment molecule. Thus lead compounds, such as chrome yellow or molybdate red, tend to darken in the presence of sulphur-containing substances, owing to the formation of lead sulphide, and ultramarine blues and violets fade rapidly in acidic environments.

Discoloration of the polymer itself is generally only noticed in very light shades of colour where the low pigmentation is insufficient to mask the slight changes which usually occur. Many pigments improve the light stability of PVC itself either by absorption of radiation or sometimes apparently by a genuine stabilisation.[14] This is particularly true of titanium dioxide.[42] On the other hand some pigments decrease the stability of PVC to light.

7.7.5   ELECTRICAL PROPERTIES

Particular care needs to be exercised in selecting pigments for PVC insulation applications. The major electrical properties which are affected by pigments are volume resistivity, per-mittivity, and breakdown.[16] The last-named is generally the result of inadequate dispersion, and its solution is therefore usually obvious. Electrical resistivity of PVC appears to be uni-versally reduced by the introduction of pigments, but the effect varies considerably from one to another. Thus, on selecting pigments with reasonable electrical behaviour and cost, black and white pigments are generally found to have the smallest effect on resistivity, fol-lowed by yellows, reds, browns, and some greens, while blues and violets are difficult to produce without appreciable loss. Effects, however, are very dependent on purity, and specially pure grades of appropriate pigments now permit the matching of most colours while retaining adequate resistivity.

Permittivity, or dielectric constant, is of particular importance in insulation in telecom-munication cables, such as telephone distribution and switchboard wires. Capacitance unbalance between adjacent wires tends to lead to cross-talk; a minimum difference in per-mittivity is thus required between the two insulants,[16] which are normally of different colours. A maximum permitted difference of 1 per cent between the permittivities of the colours with the highest and lowest values has been suggested.[20]

7.7.6   SELECTION OF COLORANTS[13, 14, 16] (TABLE 7.14)

Nowadays most suppliers offer a range of pigments which have been tested and found to be acceptable in PVC, and all that is necessary, except for the expert involved directly in the fundamentals of PVC coloration, is to be guided by the trade literature and service advice,

Table 7.14 PIGMENTS COMMONLY EMPLOYED IN PVC

Colour	Inorganic	Organic
Black	Carbon black; black iron oxide	Aniline black
White	Titanium dioxide (rutile)	
Red	Cadmium reds; lead molybdate and sulphochromate molybdate	Azo, perylene, quinacridine, pyrazolone, isoindolinone, and anthraquinome reds; diketopyrrolpyrrole
Red-brown	Indian red iron oxide	
Brown	Mn/Sb/Ti oxides	
Orange		Pyrazolone orange
Yellow	Chrome yellow (lead), and lead sulphochromates; cadmium yellows; nickel titanium yellows; bismuth vanadates; Cr/Sb/Ti and Cr/Sb/Ti/Zn oxides	Azo, anthraquinone, benzidine, isoindolinone, and flavanthrone yellows; azo calcium salts; diarylide-m-xylidide
Green	Chrome oxide; Co/Cr/Zn/Ti and Co/Ni/Ti/Zn oxides	Phthalocyanine greens; halogenated copper phthalocyanine
Blue	Cobalt, manganese and ultramarine blues; Al/Co oxide	Phthalocyanine, anthraquinone, and indanthrone blues
Violet		Dioxazine violet

with a minimum of screening tests. It is now possible to produce a very wide range of both transparent and opaque colours in PVC, as well as metallic and fluorescent effects. Optical whitening agents are also sometimes used to offset slight yellowing or to give brighter colours. These materials act by converting ultra-violet to visible blue or violet light, thus increasing reflectance, rather than reducing it as blueing agents do.

For reasons discussed in Section 7.7.3 it is clear that colorants for PVC should be pigments with as low a solubility in the medium as possible. Comprehensive recommendations have been made both in trade literature and in the journals.[11, 13–16, 19]

Inorganic pigments tend to be cheaper than organic, but since the tinctorial power of the latter is very often much the greater, much lower concentrations can be used for a given effect, and relative costs have to be evaluated for each individual case.

A number of pigments of intermediate shades are also commonly used. Apart from white and black, it is preferable to use a mixture of three or more pigments rather than a single pigment to meet a particular colour requirement. This permits minor adjustments of concentration to be made to balance relatively simply any variations in hue of the individual pigments. In addition to the coloured pigments, white or black or both may be included to control brightness.

Bearing in mind the requirements of Section 7.7.3 and the intensity of colour required, the concentrations of pigment required are likely to lie in the range 0.025–4 per cent, though considerably higher concentrations of titanium dioxide or carbon black are occasionally used for specific purposes related to properties other than colour.

Zinc oxide and lithopone should be avoided because more than minute proportions of zinc compounds catalyse thermal degradation of PVC.

As a consequence of increased awareness of toxicity problems pigments based on compounds of cadmium, their use in the European Union is now banned by an EU Council

Directive,[46, 47] and similar action has been taken or is contemplated in other parts of the world.

7.8   CHEMICAL BLOWING AGENTS

Cellular products can be produced from PVC compositions in the form of compounds for 'melt' processing or as pastes. While mechanical or gas injection methods can be used with the latter, chemical blowing agents are used in the former case, and are now more common with pastes as well. A chemical blowing agent will be required to decompose to release free gas at temperatures depending on the process being used, so several compounds of different behaviour are of interest in PVC.

Table 7.15   POSSIBLE CHEMICAL BLOWING AGENTS FOR CELLULAR PVC PRODUCTION

Agent	Decomposition temperature (°C)	Yield of gas (cm³/g)
1,1'-Azodicarbonamide (azobisformamide)	190–230	220
1,1'-Azohexahydrobenzonitrile	103–104	90
1,1'-Azobisisobutyronitrile	95–98	136
Benzene-1,3-disulphonhydrazide	146	170
Benzenesulphonhydrazide	90–100	130
Dimethyl 1,1'-azobisisobutyrate	95–200	97
Diphenyloxide-4,4'-disulphonhydrazide	175–180	120
Diphenylsulphon-3,3'-disulphonhydrazide	148	110
Dinitrosopentamethylenetetramine	160–200	240
N,N'-dinitroso-N,N'-dimethylterephthalamide	90–105	180
Substituted thiatriazole	112	130
Sodium borohydride (+ water)	Ambient	235

Contrasting with the somewhat exotic organic compounds listed in Table 7.15, the relatively simple sodium bicarbonate has been used quite widely, particularly in combination with citric acid, which lowers the decomposition temperature and increases the rate of gas evolution, thus permitting variation in melt characteristics, cell structure and density.[43] The dry blend technique is a useful one for preparing feed-stock for extrusion of cellular PVC, the blend being discharged from the mixer at 100–120°C to avoid premature decomposition of the blowing agent.[44]

Where the minimum density is required, so-called 'foam-kickers' can be useful. These act by accelerating the rate of decomposition of the blowing agent, and also reducing the decomposition temperature. They are mainly zinc oxide and lead, cadmium/zinc, barium/cadmium, or organotin compounds which are also stabilisers against degradation. A number of organic kickers is also available. To illustrate the effect of these compounds, the density obtained from a typical paste by chemical blowing was reduced from about 0.5 g/cm³ to 0.16–0.3 g/cm³ by additional inclusion of various combinations of commercial kickers. With properly selected materials and controlled process, densities down to 0.112 g/cm³ can be produced by chemical blowing.

Also used in formulations for the extrusion of cellular unplasticised PVC are 'foam stabilisers' which control the amount of expansion. They are thought to act by altering the rheological properties of the 'melt' so that the ultimate cell size is independent of the amount of gas released.[45]

Since the 1970s there has been a marked expansion in the manufacture of cellular PVC products, particularly in the construction and automotive industries.[44]

As a general rule open-cell cellular PVC is produced if decomposition of the blowing agent occurs before gelation of a paste, whereas closed-cell structure (or closed-cell with some open-cell structure) results if decomposition occurs after or at the same time as gelation.

7.9   BACTERIOSTATS AND FUNGICIDES

Where there is a possibility of bacterial or fungal growth it may be desirable to incorporate a bacteriostat or fungicide, but they have to be selected and used with care. This subject is discussed more fully in Chapter 16.

REFERENCES

1. U. JACOBSON, *Br. Plast.*, **34**, 328 (1961)
2. *Winnofil S; and Winnofil S in plasticised PVC*, I.C.I. Ltd., Mond Division
3. W. W. HARPUR, *Kunststoffe*, **56**, 10, 704 (1966).
4. P.I.A. MARTIN, *Br. Plast.*, **38**, 95 (1965)
5. W. SCHUBERT, *Plaste Kautsch*, **9**, 1, 28 (1962)
6. I. PHILLIPS and P. G. YOUDE, *Br. Plast.*, **30**, 3 (1957)
7. *'Geon' PVC resins*, British Geon Ltd. (1959)
8. *'Corvic' vinyl chloride polymers and copolymers*, I.C.I. Ltd. (1962)
9. J.H. GISOLF, *Plastica*, **15**, 10, 498 (1962)
10. *Resin Review*, **17**, 3, 18 (1967)
11. *The Colouring of Plastics*, I.C.I. Ltd. (1960)
12. J.M.J. ESTEVEZ, *J. Soc. Dyers Colour*, **77**, 300 (1961)
13. E. HERMANN, *Plastverarbeiter*, **10**, 586 (1963)
14. B. HIRSEKORN, *Gummi Asbest Kunststoffe*, **16**, 977 (1962)
15. C. MUSGRAVE, *Fibres Plast.*, **21**, 10, 291 (1962)
16. E.J.G. BALLEY, *Trans. J. Plast. Inst.*, **35**, 119, 707 (1967)
17. J.B. DECOSTE and R.H.HANSEN, *S.P.E. Jl*, **18**, 4, 341 (1962)
18. C.H. HALL, *Trans. J. Plast. Inst.*, **23**, 4 (1955)
19. C. MUSGRAVE, *J. Soc. Dyers Colour.*, **77**, 638 (1961)
20. *Bell Labs Rec.*, 291 (August 1959)
21. G.M. GALE, *Developments in PVC technology*, (J.H.L. Henson and A. Whelan, eds), Applied Science Publishers (1973) Chapter 3.
22. K.F. WORSCHECH, 'PVC Processing', *PRI Intl. Conf.*, Royal Holloway College, Egham Hill, Surrey, England (6–7 April 1978), Paper 3
23. S. KAUFMAN and M.M. YOCUM, *Plast. Rubb. Matls. Appns.*, **4**(4), 149 (1979)
24. M.M. O'HARA, W. WARD, D.P. KNECHTES and R.J. MEYER, *Flame Retardancy of Polymeric Materials*, Marcel Dekker, New York (1973) Chapter 3
25. J.W. LYONS, *The Chemistry and Uses of Fire Retardants*, Wiley, New York (1970)
26. M.M. O'MARA, *J. Polym. Sci.*, **9**, 1387 (1971)
27. G. MATTHEWS and G.S. PLEMPER, *Brit. Polym. J.*, **13**, 17 (1981)
28. Kabel-Und Metallwerke Gute A.G., Brit. Patent No. 1 260 533 (1969)
29. G. MATTHEWS and G.S. PLEMPER, *Brit. Polym. J.*, **15**, 95 (1983)
30. G. MATTHEWS and G.S. PLEMPER, *Brit. Polym. J.*, **16**, 34 (1984)

31. F.K. ANTIA, C.F. CULLIS and M M. HIRSCHLER, *Eur. Polym. J.*, **17**, 451 (1981)
32. I. SOBOLEV and E.A. WOYCHESHIN, *Fire Flamm./Fire Retard. Chem.*, **1**, 13 (1974)
33. I. SOBOLEV and E.A. WOYCHESHIN, *Coatings and Plastics Reprints*, **36**(2), 497 (1976)
34. W.V. TITOW, *Developments in PVC Production and Processing – 1*, (A. Whelan and J.L. Craft, eds) Applied Science Publishers (1977) Chapter 4
35. J. COWAN and R.T. MANLEY, *Brit. Polym. J.*, **8**(2), 44 (1976)
36. K.T. PAUL, *PVC Processing II*, PRI Intl. Conf., Brighton, England (26–28 April 1983) Paper 36
37. V. SAHAJPOL, *Developments in PVC Technology*, (J.H.L. Henson and A. Whelan, eds), Applied Science Publishers (1973) Chapter 4
38. A.P. WILSON and V.V. RAIMONDI, *PVC Processing*, PRI Intl. Conf., Royal Holloway College, Egham Hill, Surrey, England (6–7 April 1978) Paper 5
39. R.P. PETRICH, *ibid*, Paper 6
40. J. MOONEY, *PVC Processing II*, PRI Intl. Conf., Brighton, England, (26–28 April 1983) Paper 16
41. D.L. DUNKELBERGER, *Kunststoffe*, **80**(7), 816 (1990)
42. W. J. FERGUSON and B. E. HULME, *PVC Processing II*, PRI Intl. Conf., Brighton, England, (26–28 April 1983) Paper 17
43. R. BROWN, *ibid*, Paper 18
44. K.T. COLLINGTON, *Developments in PVC Production and Processing – 1*, (A. Whelan and J.L. Craft, eds), Applied Science Publishers (1977) Chapter 6
45. R.W. GOULD, *Developments in PVC Technology*, (J.H.L. Henson and A. Whelan, eds) Applied Science Publishers (1973) Chapter 9
46. European Communities Council Directive 91/338/EEC (18 June 1991): *Offl. J. Eur. Com.*, No. L186/59–63 12 July 1991)
47. U.K. Dept. of the Environment, Environmental Protection, (Controls on Injurious Substances) (No. 2), Regulations 1993; SI No. 1643 1993 (31 July 1993)
48. Martinswerk GmbH, *Flame Retardant Fillers in Plasticized PVC* (1993)

8

Principles of formulation

8.1 INTRODUCTION

Having discussed the nature of vinyl chloride polymers (Chapter 4) and the various additives which are commonly employed in PVC (Chapters 5, 6, and 7), we have now to consider the principles of formulation as a whole. As has been pointed out in previous chapters the selection of materials for a PVC formulation in industrial practice usually starts out from a position where a large amount of experience, both recorded and personal, is already available. Once the properties of a new compound are known, it is very often a relatively simple matter to formulate a composition on paper, having regard to this body of experience, and to find this composition confirmed as appropriate or very nearly so in the first practical evaluation. In these circumstances the principles of formulation are built into the prior experience and their application in practice is obscured. Nevertheless, these principles are being applied, even if subconsciously, and it is highly desirable for the novice to understand them. Some of the recommendations of Chapters 4–7 have therefore been drawn together and summarised below. Some typical formulations are set out in Section 8.9.

For the sake of logical presentation and clarity it is necessary to consider each component in turn, but it must be understood that each selected material may have some effect on the most appropriate selection for each other component. Some of the effects of formulation have been studied in fair detail.[1, 2]

8.2 PLASTICISATION

The first consideration to be faced is the rigidity or flexibility required in the final product. In the present state of advancement, if the product is required to be rigid, no plasticiser will be required. It should no longer be necessary to include plasticiser merely to lower the softening temperature or the melt viscosity of a PVC composition to render it processable. If the product is to be flexible, then the level of flexibility, taken in conjunction with the geometrical dimensions of the product will determine within fairly narrow limits the required proportion of any plasticiser or plasticiser combination. Unless there are some special requirements, in the current commercial situation the plasticiser will generally be DIOP or DOP, or a suitable mixture of one of these with an extender. If the application involves service at such an elevated temperature that volatility of octyl phthalates is a problem, a part or the whole may be replaced by higher phthalates such as DIDP or DTDP, a trimellitate, or by a polymeric plasticiser, the concentrations being adjusted according to the relative efficiencies. Similarly, if low temperatures are likely to lead to objectionable stiffening and brittleness, sufficient of an alkyl aliphatic dibasic ester, such as DIOZ, or a similar ester, should be included to give the required improvement in low temperature properties, again adjusting concentrations according to relative efficiencies. Where low extraction by oils, as for example in a package, is required the phthalates must be replaced by polyester plasticisers, or possibly, if the application is not too demanding, by epoxydised soya bean oil. For minimum inflammability TXP or TTP may be used in place of octyl phthalates. Most of the plasticisers not mentioned above will only be considered where special properties are required, or where the offered prices change relative to each other. Considerations such as the above lead to a specific plasticiser in nature and concentration. A decision to include a filler is the main thing likely to modify this choice, probably involving a slight increase in concentration to compensate for the hardening influence of the filler.

8.3   FILLERS

Apart from those cases where filler is included for specific technical effects, the decision whether or not to include filler will largely depend on whether the change in physical properties can be tolerated for the sake of reduced cost. If clarity is required, filler will be excluded, with the possible exception of certain aluminas which do not reduce clarity greatly at fairly low levels. Where fillers are required to reduce cost, natural calcium carbonates, china clay, or similar minerals are appropriate if mattness is required or acceptable. For high gloss, synthetic or finely ground natural calcium carbonates are necessary. For maximum electrical resistivity about 10 phr of a suitable calcined clay should be included, while for minimum flammability up to about 5 phr of alumina may be added. The concentrations of fillers employed depend on the purpose for which the composition is being designed, the maximum varying from about 60 phr in general purpose extrusion and moulding compounds up to as much as 500 phr in flooring.

8.4   STABILISER SYSTEMS

Any stabiliser system must be sufficiently effective to minimise decomposition during processing to an acceptable level. For many applications where clarity and non-toxicity are required, tribasic lead sulphate may be selected, though basic lead carbonate is still used occasionally where processing temperatures are not particularly high. Barium/cadmium/lead combinations are offered for unplasticised PVC profile and sheet extrusion.

At one time dibasic lead phosphite was recommended where maximum light stability was required, but it has now been replaced by mixed metal complexes.

Where maximum clarity is essential an organotin, or, where the odour is acceptable a thio-organotin, stabiliser will generally be required, although the best combinations of barium/cadmium/zinc stabilisers with epoxy compounds can be very nearly as good.

Selection of stabilisers for 'non-toxic' applications depends on the nature of the application and on local regulations. For water-pipes up to about 2 phr of tribasic lead sulphate can generally be used in unplasticised compositions. For aqueous liquids octyltin stabilisers may be acceptable, while for edible oils and fats choice was for a long time limited to combinations of barium, calcium, magnesium and zinc compounds, some epoxy compounds, and specific organic substances such as diphenylthiourea.

With the increased awareness of toxicity problems over the past few decades, efforts have been made to find effective replacements for lead, cadmium, and butyltin stabilisers. Helped by parallel improvements in machinery design and process control it is now possible to omit these in favour of more recently developed calcium, barium, magnesium, strontium, and zinc complexes.

8.5 LUBRICANTS

Since lubrication is so specific to composition, processing equipment, and processing conditions, little detailed guidance can be offered about selecting lubricants. For medium soft plasticised PVC around 1 phr of lead or calcium stearate will generally suffice, provided that high clarity is not required, in which latter case about 0.5 phr of stearic acid or a suitable wax are generally required. Sometimes higher concentrations of stearates or additions of other lubricants such as paraffin wax can be used to increase extrusion rates and possibly improve gloss, but care must be exercised to prevent excessive separation from the polymer phase.

Lubricant formulation for unplasticised PVC is now so specific that no attempt will be made to generalise. Starting from scratch it is usually necessary to begin with a simple system, such as a single metal stearate of wax, and work from there by trial and error. Extrusion rheometer experiments can sometimes be helpful. Specific formulations are often available from raw materials suppliers.

8.6 OTHER ADDITIVES

Little can be added to what has already been said about the inclusion of impact modifiers (Section 7.4) and processing aids (Section 7.5) in unplasticised PVC, or antistatic agents (Section 7.6) and colorants (7.7) in PVC compositions generally. A wide range of satisfactory colorants is now available commercially, and in practice selection is made from manufacturers' recommendations of those found satisfactory in PVC.

8.7 POLYMER

It might seem strange to leave selection of polymer until last of the components of a PVC composition, but in fact the characteristics of the most appropriate polymer are so dependent on the other components and on the processing that this is the most logical place to consider the subject. Some of the relationships between polymer properties and processing behaviour and product properties have been studied in detail.[1, 2]

Figure 8.1 gives a diagrammatic representation of the relationship between the various types of polymer and copolymer available. In the present commercial and technical situation generally a suspension homopolymer having porous particles (previously known as 'EP' polymers) will be selected for plasticised compositions, and one with particles of low or no porosity for unplasticised compositions, unless there is some particular technical reason to use alternative types of resin.

These polymers are available with a variety of different average molecular weights covering a wide range, represented by ISO Viscosity Numbers from about 50 to about 210, corresponding to weight-average molecular weights from about 40 000 to more than 500 000 and number-average molecular weights from about 20 000 to over 90 000. This is represented by the vertical line A–B in Figure 8.1. For the majority of applications average

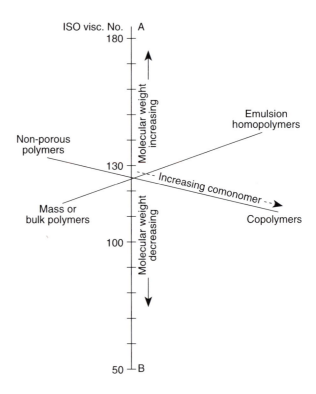

Figure 8.1   Relationship between different types of PVC resin

molecular weights corresponding to ISO Viscosity Numbers in the range 105–130 are usually high enough to attain adequate mechanical properties and low enough to ensure reasonable processing behaviour, at least in plasticised compositions. Since polymers of higher molecular weight are more expensive to produce, because of their relatively long polymerisation cycles, they are only used where the application is such that higher values of mechanical properties are required than are obtainable from polymers in the 105–130 ISO Viscosity Number range, for example in calendered sheet for magnetic tapes. These latter polymers, then, may be regarded as typical general purpose resins, used in most circumstances unless some property or processing requirement precludes them. They are essential for the production of plasticised dry-blends, but polymers having porous particles can be made by a two-stage mass or bulk polymerisation process.[3, 7] In the past mass or bulk polymers have competed successfully with suspension polymers, but the quantity produced has always been much smaller. Mass polymers were occasionally preferred where their high heat stability and low colour and haze were considered to be desirable, but the quality of suspension polymers nowadays is such that any advantage is very marginal. For this reason, and as a consequence of problems arising from the need to reduce residual monomer to a very low level, interest in mass polymerisation has decreased rapidly during the past couple of decades.

For the production of plastisols or pastes (Chapter 14) emulsion polymers, usually ground, or microsuspension polymers are essential.

Processing requirements which are difficult to meet with the so-called 'general purpose' polymers as described above are usually such that the temperatures required for adequate fluxing are so high that degradation reaches an unacceptable or undesirable level. This rarely if ever arises nowadays with unfilled plasticised compositions, but it is common with unplasticised or heavily filled compositions. A number of alternative solutions to the problem are available. These are to use instead of the general purpose polymer:

(1) Homopolymer of lower molecular weight.
(2) Suspension copolymer of similar molecular weight.
(3) Emulsion polymer of similar molecular weight.
(4) Suspension copolymer of lower molecular weight.
(5) Emulsion homopolymer of lower molecular weight.
(6) Emulsion copolymer of equal or lower molecular weight.

Wherever possible alternative (1) is preferred. This is because suspension homopolymers are cheaper than copolymers at present, and also cheaper than emulsion polymers; the heat-stability of suspension homopolymers is greater than that of the other alternatives; and the light-stability of homopolymers is greater than that of copolymers. The extent to which the molecular weight is lowered or to which the other alternatives are adopted depends on the requirements of the particular situation. Thus, for relatively straightforward extrusion of unplasticised PVC it is not necessary to move very far from the concept of the general purpose polymers, and homopolymers having ISO Viscosity Numbers as high as 105 are common, while even ISO Viscosity Numbers of around 125 have been used reasonably successfully. As the demands of the processing become greater, however, it is necessary to drop further in molecular weight. In injection moulding, where flow paths are generally longer, narrower, and more complex than in extrusion, ISO Viscosity Numbers of around 90 are more common. Similar values are common in bottle-blowing, though values as low as 70 have been used. Dropping much lower than this level usually brings unacceptable brittleness and it is then that copolymers become nearly essential. Thus vacuum-forming sheet, which requires a softening behaviour not offered by low molecular weight homopolymer, is usually based on copolymers of ISO Viscosity Numbers around 65–75 and vinyl acetate contents of 10–20 per cent. Probably the most difficult case is the moulding of gramophone records, where the processor is faced with the problem of accurately filling a very thin, very complex mould, and here it is common to use copolymers with ISO Viscosity Numbers as low as 55 with between 12 and 20 per cent of vinyl acetate.

The other important circumstance which reveals inadequate melt flow of general purpose homopolymers is where very high proportions of filler are included, as in many flooring compositions. Processing behaviour in the production of flooring depends not only on the composition, but also on the equipment and conditions employed, and it would be unrealistic to particularise too precisely, for apparently similar products can differ considerably in the characteristics of the polymers on which they are based merely because of different equipment and procedures used to manufacture them. In general however, it may be said that, as the filler content is increased, it becomes increasingly difficult for the polymer melt adequately to wet the filler particles, and a point is often reached where the particular polymer in use does not wet the filler sufficiently, and a dry crumbly 'melt' results, from which it is impossible to produce a smooth product. This necessitates action similar to that imposed by difficult processing of unplasticised compositions, namely changing to a homopolymer of lower molecular weight, or to a copolymer of equal or lower molecular weight. Thus, for flexible flooring, where the filler content may be said to be 'intermediate', homopolymers with ISO Viscosity Numbers between 70 and 105 are common, whereas very heavily filled vinyl tiles often use a copolymer of similar constitution to that used for gramophone records, i.e. ISO Viscosity Number around 60 and vinyl acetate content around 10–15 per cent.

A few other special circumstances require the use of alternatives to the general purpose type of polymer.

In general, homopolymers are not very soluble, and for solution applications, such as lacquers, low molecular weight copolymers are necessary. These usually need to be of very high purity, particularly with regard to contamination by homopolymer or other less soluble polymer. The molecular weights and comonomer contents are generally similar to those of the copolymers used for gramophone record moulding, i.e. ISO Viscosity Number around 60 and vinyl acetate content around 15 per cent.

In the extrusion of unplasticised dry blends the air in the space between and entrapped in the pores of the particles of porous suspension resins is liable to pass into the melt beyond the metering zone of the screw and thus to appear as porosity in the extrudate. The tendency for this to occur can be greatly reduced by using a suspension polymer with non-porous particles of high bulk and packing densities.

It is possible to increase the electrical resistivity of good quality insulation grades of PVC by special pre-treatment of the polymer, and a few manufactures offer special grades of this type, for use where the maximum possible resistivity is desired. In other respects these polymers are similar to normal suspension resins.

Suspension copolymers of low comonomer content (about 2 per cent) are sometimes employed where the processor requires a slightly easier fluxing, leading to marginally improved gloss and/or clarity over that obtainable with the corresponding homopolymer.

Occasionally it happens that the melt flow behaviour required is not met by a single available polymer but is intermediate between that of two different polymers. Provided that the two polymers are not too widely separated in softening behaviour it is usually possible to employ an appropriate mixture without undue difficulties. Indeed it is not unknown for a kind of eutectic to result where the flow behaviour of the mixture is better than that of either polymer singly. This can be the case for example with mixtures of suspension with emulsion homopolymers, a combination sometimes used in the past in unplasticised PVC extrusion.

8.8 PROPERTIES OF PVC COMPOSITIONS

8.8.1 GENERAL

A formulation is usually developed to meet a combination of requirements of end-product properties, processing behaviour, and cost. The first of these can usually be estimated fairly closely by consideration of the effects of all the possible components,[4] but final evaluation requires laboratory work. Processing behaviour, too, can be qualitatively estimated, but is more difficult to evaluate in the laboratory, often requiring full-scale plant trials. A processor may often use some form of extrusion rheometer in which he builds up an arbitrary body of experience related to production performance, but the fundamentals of processing behaviour are not yet completely understood. The overall cost depends on formulation, cost of the individual components, density of the composition, and conversion costs. The latter are important but difficult to estimate accurately for a new formulation without full plant-scale trials. Density is important because most end-product use is based on volume rather than weight, and so 'volume cost' is important.

It is not difficult to estimate the relative density of a composition with reasonable accuracy from its formulations, and thus, knowing the costs of the individual components, it is possible to estimate volume cost. Table 8.1 sets out relative densities for a number of typical components of a PVC composition.

Table 8.1 RELATIVE DENSITIES OF COMPONENTS OF PVC COMPOSITIONS

Material	Relative density
Vinyl chloride homopolymer	1.40
Vinyl chloride/vinyl acetate copolymer (84/16)	1.36
Di-n-heptyl phthalate	0.99
Di-'alphanyl' phthalate	1.00
Di-iso-octyl phthalate	0.98
Di-octyl phthalate	0.99
Di-isodecyl phthalate	0.97
Ditridecyl phthalate	0.95
Di-iso-octyl adipate	0.93
Di-iso-octyl azelate	0.92
Polypropylene sebacate	1.06
Tritolyl phosphate	1.17
Trixylyl phosphate	1.14
Epoxydised soya bean oil	0.99
Chlorinated paraffin	1.16–1.26
Basic lead carbonate	6.80
Tribasic lead sulphate	7.10
Dibutyltin dilaurate	1.04
Barium/cadmium complex	1.00–1.25
Calcium stearate	1.03
Lead stearate	1.10
Stearic acid	0.95
Mineral calcium carbonate	2.60
Precipitated calcium carbonate	2.50
Dolomite	2.80
Calcined clay	2.55
Asbestos	3.00
Acrylic impact modifier	1.11

The approximate relative density of a compounded PVC composition can be calculated from its formulation and information such as that typified in Table 8.1, by means of the following equation:

$$d_c = \frac{100 + P_1 + P_2 + P_3 + P_4 + \ldots}{71.4 + P_1/d_1 + P_2/d_2 + P_3/d_3 + P_4/d_4 + \ldots} \qquad (1)$$

where d_c is the relative density of the composition, P_1, P_2, P_3, P_4, etc., are the concentrations in phr of the additives to the polymer, and d_1, d_2, d_3, d_4, etc., are their relative densities. (Note: $71.4 = 100/1.4$.)

To illustrate the calculation consider a typical formulation for a cable insulation compound:

Suspension homopolymer	100 parts
DOP	36 parts
Chlorinated paraffin	18 parts
Tribasic lead sulphate	5 parts
Calcium stearate	1 part
Mineral calcium carbonate	20 parts
Calcined clay	10 parts

Taking the values of relative densities set out in Table 8.1, the following values for P/d are calculated:

DOP	26.36 (36/0.99)
Chlorinated paraffin	14.3 (18/1.26)
Tribasic lead sulphate	0.70 (5/7.10)
Calcium stearate	0.97 (1/1.03)
Mineral calcium carbonate	7.7 (20/2.60)
Calcined clay	3.92 (10/2.55)

Substituting these values in equation (1):

$$d_c = \frac{100 + 36 + 18 + 5 + 1 + 20 + 10}{71.4 + 36.36 + 14.3 + 0.70 + 0.97 + 7.7 + 3.92}$$

$$= 190/135.4$$

$$= \underline{1.40}$$

To calculate the materials cost and hence the volume cost of the compound, it is necessary to know the costs of the individual components. These fluctuate from place to place and from time to time, and the following values are merely taken to illustrate the calculation:

	£/tonne
Suspension polymer	400
DOP	520
Chlorinated paraffin	280
Tribasic lead sulphate	520
Calcium stearate	600
Mineral calcium carbonate	80
Calcined clay	120

The materials cost per unit of composition is given by the equation:

$$C_c = \frac{100\,C_0 + P_1 C_1 + P_2 C_2 + P_3 C_3 + P_4 C_4 + \ldots}{100 + P_1 + P_2 + P_3 + P_4 + \ldots} \tag{2}$$

where C_c is the materials cost per unit weight of compound, C_0 is the cost per unit weight of polymer, and C_1, C_2, C_3 etc., are the costs per unit weight of each of the additives. Substituting the values given above in this equation:

$$C_c = \frac{40\,000 + 18\,720 + 5040 + 2600 + 600 + 1600 + 1200}{190} \quad \text{£/tonne}$$

$$= 69\,760/190$$

$= £367.12$ per tonne or 36.7p/kg

Volume cost can now be calculated from the relative density and the cost per unit weight. For units of pence per cubic centimetre the materials volume cost is given by the expression:

$$C_v = C_c \times d_c/1000$$

$$= 36.7 \times 1.4/1000$$

$$= \textbf{0.051 p/cm}^3$$

for the example used above.

To obtain a true production cost for the end product, all other cost factors would have to be added. These would include conversion cost, dependent on production rate, capital investment in plant, labour, etc., packaging, overheads, and indeed any item contributing to the production of the end-product.

8.9   TYPICAL FORMULATIONS

The formulations which follow are intended merely for guidance as being typical of those that have been employed commercially, and it is certainly not claimed that any one is necessarily the best possible for its indicated purpose. In particular the simple stabiliser systems listed are likely in practice to be replaced by complex mixtures supplied by the manu-facturers.

8.9.1   COMPOUNDS FOR EXTRUSION

8.9.1.1 Rigid opaque sections, e.g. pipe

Suspension homopolymer of ISO Viscosity No. 85–105	100	parts
Tribasic lead sulphate	5–8	parts
Glyceryl monostearate	0–2	parts
Lead stearate	1	part
Precipitated calcium carbonate	0–10	parts

For extrusion direct from dry blend the polymer should be changed to one having high bulk and packing densities and non-porous particles. However, for good quality extrusion at reasonable rates the lubricant system would almost certainly have to be modified to match the machine in which it was being extruded, the main criterion being its effect on gelation behaviour rather than on melt flow behaviour as is the case in extrusion of fully com-pounded material.[5] A possible alternative lubricant system might be:

Lead stearate	0.65	phr
Dibasic lead stearate	0.1	phr
Montan ester wax	0.2–0.6	phr
Calcium stearate	0.25–1	phr

This may seem something of a hotch-potch concocted from the cookery book; but the function of each component has been qualitatively analysed.[5] In addition, processing aid, such as about 2 per cent of an appropriate acrylic copolymer, and/or an impact modifier, might be included.[8–11] There is a trend away from stabilisation based on TBLS alone, es-pecially for products for outdoor exposure applications, such as window frames, where combinations with barium/cadmium stabilisers at around 1.5 phr each are commonly used.

Pipe for water conveyance would require a stabilising system of acceptably low toxicity. In the UK and some other countries this can be achieved by reducing the concentration of tribasic lead sulphate to about 2 phr. The precise permissible maximum concentration would have to be determined in accordance with the requirements of the appropriate specifica-tions.[6] In some countries lead is not permitted, and recourse has to be made to fully non-toxic stabilisers.

Rigid foam extrusion requires slightly lower molecular weight, and, of course, blowing agent. Precise details of proprietry commercial formulations are not disclosed, but the fol-lowing is typical:[13]

Suspension polymer of ISO Viscosity No. 80	100 parts
Basic lead stabiliser	3–4
Lead co-stabiliser and lubricant	0.4–0.6
Lubricant	1–2
Processing aid	5–8
Sodium bicarbonate or azodicarbonamide	1.5–2.5

Titanium dioxide	3–4
Filler	2–4

8.9.1.2   Clear rigid sections, e.g. sheeting

Clarity demands the use of organotin or possibly efficient barium/cadmium/epoxy complex stabiliser systems. The problems associated with developing uniform flow across a wide die also demand better flow than is obtained with polymers of ISO Viscosity Number around 100, and it is generally necessary to use polymers of lower molecular weight.

Suspension homopolymer of ISO Viscosity No. 90	100	parts
Sulphur-containing organotin stabiliser	2–3	parts
Acrylic impact modifier	10	parts
Acrylic processing aid	2	parts
Montan ester wax	0.6	part
Cetyl stearyl alcohol	0.6	part
Mineral oil	0.1	part
Low molecular weight polyethylene	0.1	part

The sulphur-containing organotin stabiliser might well be replaced by a combination of a barium/cadmium/zinc complex (2 phr), an epoxy resin (2 phr), and an organic phosphite (0.5 phr), for which superior ageing behaviour has been claimed. For 'non-toxic' applications, such as food-packaging, the stabiliser system would have to be replaced by an octyltin compound or by a combination of calcium/zinc or calcium/magnesium/zinc compounds (2 phr) with epoxydised soya bean oil (1–2 phr), with the possible addition of 0.5 phr of α-phenylindole or diphenylthiourea.

For the production of sheet for subsequent vacuum-forming, the homopolymer would be replaced by a lower molecular weight copolymer, typically one of about 10–15 per cent vinyl acetate content and having an ISO Viscosity No. around 70.

8.9.1.3   Blown film for food packaging

Suspension copolymer of ISO Viscosity No. 90 and vinyl acetate content 6–10 per cent	100	parts
Epoxydised soya bean oil	3	parts
Ethyl palmitate	1	part
Calcium/magnesium/zinc complex	1	part
Diphenylthiourea	0.5	part
Stearic acid	0.5	part

8.9.1.4   Blown sheeting for sacks

Suspension homopolymer of ISO Viscosity No. 120–145	100	parts
A dioctyl phthalate	45–20	parts*
DIOZ	0–20	parts*
Mineral calcium carbonate	20–30	parts
Tribasic lead sulphate	4	parts
Lead stearate	1	part

* Total dioctyl phthalate plus DIOZ about 45 parts; ratio adjusted to required low temperature performance.

8.9.1.5   *Electric cable insulation*

Suspension homopolymer of ISO Viscosity No. 100–130	100	parts
A dioctyl phthalate	30–40	parts
Chlorinated paraffin	0–20	parts
Tribasic lead sulphate	5–8	parts
Lead stearate or calcium stearate	1	part
Mineral calcium carbonate	0–30	parts
Calcined clay	0–10	parts

8.9.1.6   *Sheathing for electric cables*

Suspension homopolymer of ISO Viscosity No. 100–130	100	parts
A dioctyl phthalate	40–60	parts
Chlorinated paraffin	20–30	parts
Tribasic lead sulphate	5–8	parts
Lead stearate or calcium stearate	1	part
Mineral or synthetic calcium carbonate	30–50	parts

8.9.1.7   *'High temperature' cable insulation*

Suspension homopolymer of ISO Viscosity No. 120–130	100	parts
DIDP, DTDP, or trioctyl trimellitate	50–70	parts
Tribasic lead sulphate or dibasic lead phthalate	5–8	parts
Antioxidant	0.5	part
Lead stearate	1	part
Mineral or synthetic calcium carbonate	0–30	parts
Antimony trioxide	5	parts

8.9.1.8   *'Low temperature' cable insulation*

Suspension homopolymer of ISO Viscosity No. 100–130	100	parts
A dioctyl phthalate (preferably linear)	25–50	parts*
DIOZ	40–50	parts
Basic lead carbonate	5–8	parts
Lead stearate or calcium stearate	1	part

* Total dioctyl phthalate plus DIOZ and ratio adjusted to required low temperature performance.

8.9.1.9   *'General Purpose' opaque flexible compound*

Suspension homopolymer of ISO Viscosity No. 100–130	100	parts
A dioctyl phthalate	40–80	parts
Tribasic lead sulphate	5–8	parts

Lead stearate or calcium stearate	1	part
Mineral or precipitated calcium carbonate	0–50	parts

8.9.1.10   *'General Purpose' clear flexible compound*

Suspension homopolymer of ISO Viscosity No. 100–130	100	parts
A dioctyl phthalate	40–80	parts
Epoxydised soya bean oil	3–5	parts
Barium/cadmium or barium/cadmium/zinc complex	2–3	parts
Organic phosphite	0–0.5	part
Stearic acid	0.5	part

8.9.1.11   *'Crystal' clear flexible compound*

Suspension homopolymer of ISO Viscosity No. 100–130	100	parts
A dioctyl phthalate	40–80	parts
Epoxydised soya bean oil	0–2	parts
Thio-organotin stabiliser	2	parts
Cadmium stearate	0.5	part

8.9.1.12   *'Low toxicity' flexible compound*

Suspension or emulsion homopolymer of ISO Viscosity No. 100–130	100	parts
A dioctyl phthalate	40–80	parts
Epoxydised soya bean oil	3–5	parts
Calcium/magnesium/zinc stabiliser	2	parts
Stearic acid	0.5	part

8.9.2.   COMPOUNDS FOR INJECTION MOULDING

8.9.2.1   *Rigid mouldings, e.g. pipe fittings*

Suspension homopolymer of ISO Viscosity No. 75–90	100	parts
Tribasic lead sulphate	2–8	parts
Glyceryl monostearate	1–2	parts
Calcium or lead stearate	0.5–1	part

8.9.2.2   *Flexible clear compound*

Suspension homopolymer of ISO Viscosity No. 85–125	100	parts
A dioctyl phthalate	40–95	parts
Epoxydised soya bean oil	3–5	parts
Barium/cadmium/zinc complex	2–3	parts
Organic phosphite	0.5	part
Stearic acid	0.25–0.7	part

8.9.2.3   Flexible opaque compound

Suspension homopolymer of ISO Viscosity No. 100–125	100	parts
A dioctyl phthalate	40–95	parts
Tribasic lead sulphate	5–8	parts
Lead stearate or calcium stearate	1	part

8.9.3.   COMPOUNDS FOR BLOW-MOULDING

Suspension homopolymer of ISO Viscosity No. 70–90	100	parts
Organotin or thio-organotin stabiliser	2	parts
Epoxy compound	1–2	parts
Glyceryl monostearate	1	part
Acrylic impact modifier	0–15	parts

For bottling aqueous beverages it would be necessary to avoid the toxicity of the organotin stabiliser by replacing it by an octyltin compound, or by a calcium/magnesium/zinc complex. For edible oils the latter alternative would generally be required. In either case the addition of 0.5 phr of diphenylthiourea or α-phenylindole would help by increasing stability. As Sahajpal has pointed out,[12] special care is required in selecting ingredients for beverage bottles. In particular high purity substances are necessary. The following two typical formulations were suggested as being suitable for bottles for mineral waters:

	Parts	Parts
Suspension homopolymer of ISO Viscosity No. 73	100	100
MBS impact modifier	10	10
Calcium stearate or isooctoate	0.05	0.3
Zinc octoate	0.15	0.12
Distearyl pentaerythritol diphosphite	0.4	—
α-phenyl indole	—	0.3
TNPP	—	1
Epoxydised soya bean oil	3	—
Acrylic processing aid	0–1	—
Calcium/zinc soap lubricant	—	—

8.9.4   COMPOSITIONS FOR CALENDERING

To a degree formulations for calendering follow those used for extrusion, the main difference being in the selection of lubricants, where considerable care is necessary to ensure that the optimum level of lubrication is achieved. Since this is usually dependent on process conditions as well as formulation little further guidance can be given.

For unplasticised sheeting it is generally necessary to use polymers of lower molecular weight than can be used for extrusion, i.e. in the ISO Viscosity Number range 85–90. The following formulations are of special interest in calendering.

8.9.4.1 *Flexible sheeting for oil packaging*

Suspension homopolymer of ISO Viscosity No. 100–400	100	parts
PPS	60–90	parts
Tribasic lead sulphate	5–8	parts
Lead stearate or calcium stearate	1	part

8.9.4.2 *Flexible flooring – continuous or tiles*

Suspension homopolymer of ISO Viscosity No. 55–105	100	parts
A dioctyl or butyl benzyl phthalate	35–50	parts
Epoxydised soya bean oil	3–5	parts
Barium/cadmium/zinc/ or barium/zinc stabiliser	2–3	parts
Calcium stearate	1	part
Mineral or precipitated calcium carbonate	100–300	parts

8.9.4.3 *Vinyl floor tiles*

Suspension copolymer of ISO Viscosity No. 55–70 and vinyl acetate content 10–15 per cent	100	parts
A dioctyl or butyl benzyl phthalate	35–50	parts
Epoxydised soya bean oil	3–5	parts
Barium/cadmium or barium/zinc stabiliser	2–3	parts
Mineral or precipitated calcium carbonate	100–250	parts
Alumina	50–250	parts

8.9.5 GRAMOPHONE RECORDS

As with other processing, particularly calendering, formulation of compositions for compression moulding of gramophone records (Section 13.2.2) requires considerable care in the selection of lubricants and the following can only be taken as a very rough guide.

8.9.5.1 *Using combined stabiliser/lubricant*

Suspension copolymer of ISO Viscosity No. 60 and vinyl acetate content 15–20 per cent	100	parts
Lead stearate or dibasic lead stearate	1–1.5	parts
Carbon black	0.5–1.5	parts

8.9.5.2 *Using lubricant plus separate 'low lubricating' efficient stabiliser*

(1) Suspension copolymer of ISO Viscosity No. 60 and vinyl acetate content 15–20 per cent	100	parts
Dibasic lead stearate	0.75–1	part
Tetrabasic lead fumarate or dibasic lead phthalate	0.75–1	part
Carbon black	0.5–1.5	part

(2) Suspension copolymer of ISO Viscosity No. 60
and vinyl acetate content 15–20 per cent 100   parts

Organotin stabiliser	0.5–0.75	part
Calcium stearate	0.5	part
Wax lubricant	0.25	part
Carbon black	0.5–1.5	parts

REFERENCES

1. R. Hammond, *Trans. J. Plast. Inst.,* **26**, 49 (1958)
2. B.S. Dyer, *Trans. J. Plast. Inst.,* **27**, 84 (1959)
3. J.C. Thomas, *S.P.E. Jl*, **23**, 10, 61 (1967)
4. W.J. Frissell, *Plast. Technol.*, **8**, 12, 32 (1962)
5. D.R. Jones and J.C. Hawkes, *Trans. J. Plast. Inst.*, **35**, 120, 773 (1967)
6. BS 3505
7. G.C. Marks, *Developments in PVC Technology*, (J.H.L. Henson and A. Whelan, eds), Applied Science Publishers, London (1973) Chapter 2
8. V. Sahajpal, *ibid*, Chapter 4
9. J.B. Press and D.A. Trebucq, *Developments in PVC Production and Processing – 1*, (A. Whelan and J.L. Craft, eds), Applied Science Publishers, London (1977) Chapter 9
10. R.P. Petrich, *PVC Processing*, PRI Intl. Conf., Royal Holloway College, Egham Hill, Surrey, England (6/7 April 1978) Paper 6
11. J. Mooney, *PVC Processing II*, PRI Intl. Conf., Brighton, England (26–28 April 1983) Paper 16
12. V. Sahajpal, *ibid,* Paper A.2
13. N.L. Thomas, R.P. Eastup and T. Roberts, *PVC '93 – The Future*, IOM Intl. Conf., Brighton, England (27–28 April 1993) Paper 25: *Plast. Rubb. Proc. Appn.* (in press)
14. E.J. Wickson, editor, *Handbook of PVC Formulating*, John Wiley & Sons, Inc., New York (1993)

9

General consideration of processing

9.1 INTRODUCTION

It is, of course, essential that the formulation used to produce a PVC product shall be such that the required properties are attained, which naturally has a prime influence on determining the formulation. It is also obviously essential that the formulation is such that the composition is amenable to the processing required to convert the components to the required product form, and this too, as was seen in the previous chapter, can have a profound influence on the nature of the individual components of a formulation.

The various possible ways of converting vinyl chloride polymers to useful end-products have been reviewed and described fairly fully as a whole[1-3, 39] and separately. Practically every kind of plastics processing can be applied to PVC in one form or another, and some PVC processing, e.g. paste processes, are not generally applicable to other polymers. The various different routes by means of which a vinyl chloride polymer or copolymer and additives can be converted to finished products are illustrated diagrammatically[1, 3, 39] in Figure 9.1. It will be understood from what has been stated in preceding chapters that not every combination of polymer and additives can be put through every possible sequence of processing treatments.

The various processes may be broadly classified into three classes, (1) mixing, (2) compounding, and (3) shaping; but it is not suggested that any one is mutually exclusive of the others, and indeed much of modern processing involves all three in one step. In the present context 'mixing' is seen as any process in which the various components of a composition are blended together mechanically, or physically, any softening or agglomeration of polymer occurring being insufficient to produce a continuum of gelled polymer. 'Compounding' is seen as apparently involving mixing on a molecular scale, usually with the formation of a continuous polymer matrix. 'Apparently' because it is now known that particle structures are commonly retained throughout processing, so that complete mixing on a molecular scale does not occur. 'Compounding' is thus seen as a particular form of mixing, but, although the differentiation between 'mixing' and 'compounding' may be fundamentally artificial, it is nevertheless a useful convention in practice. 'Shaping' comprises generally those processes which produce the final product. Some of these, e.g. extrusion and moulding, involve some amount of compounding, whereas others, e.g. thermoforming, do not include this at all. The object of processing is presumed to be the production of the required shape with the composition employed in as homogeneous a state as possible. Where more than one process is involved in the conversion each one may contribute towards the homogeneity. The greater the contribution from one process the smaller will be that required from the others. Thus

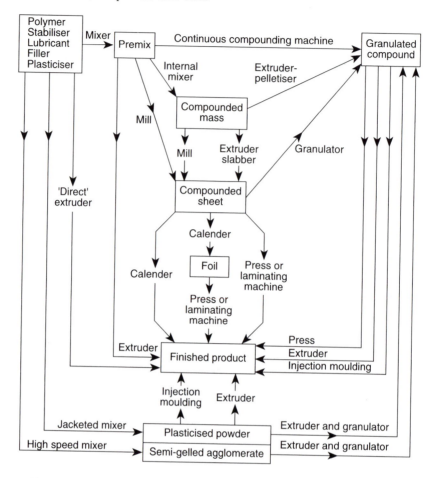

Figure 9.1.   Routes for conversion of PVC resins to finished products[1, 3, 39]

each process needs to be considered from the point of view of the effect it will have on the homogeneity of the final end-product.

9.2   MANUFACTURING PROCEDURES

For many years the most common procedures for extrusion and moulding involved the separate stages of (1) simple mixing of raw materials in some form of blending machine, (2) compounding, usually in an internal or intensive mixer, (3) sheeting on a two-roll mill, (4) granulating, and (5) extrusion or moulding. Sometimes all these operations are carried out by a single manufacturer, possibly all in one factory. More commonly, firms carrying out extrusion or moulding have purchased compounded granules from other manufacturers. These latter have mainly been the polymer producers, but there are also a few intermediate compounding concerns, purchasing polymers and other raw materials, and converting them to compounded granules for sale. As indicated in Chapter 7, Section 7.7.1, compound is produced either self-coloured ready for direct extrusion or moulding, or in natural unpigmented form which needs to be mechanically blended with an appropriate masterbatch before the final extrusion or moulding operation.

In calendering the granulating stage is almost invariably omitted, and for this reason, and because the scale of operations is usually sufficiently large to justify it, mixing and compounding is carried out in or adjacent to the calendering plant by the same company.

Over recent years the mixing and compounding processes have become more and more highly automated to reduce the originally very high labour content. This has been helped by the advent of so-called 'high-speed' mixers or 'fluid-mixers', or, to give them their full title, 'intensive non-fluxing mixers', and also by replacement of two-roll sheeting mills by 'extruder-slabbers'.

The use of continuous compounding machines, of which there are a number,[1, 39] has not developed as far as might have been expected, partly because of their relative inflexibility, even when the facility for changing mixing elements is available, but perhaps more because of the automation of batch compounding referred to already, and because of the growth in dry blend processing. This latter has become increasingly popular for extrusion and moulding of unplasticised PVC for a variety of large applications, such as pipe, pipe fittings, flat and corrugated sheet, window frame sections, and cladding. Dry blend techniques have also become more popular in the processing of plasticised compositions.

Of other processes, extrusion blow-moulding, particularly for the production of bottles for mineral waters and fruit squashes, developed to become a sizeable market during the 1970s and 1980s, but has received something of a set-back due to criticisms from environmental groups.

Powder coating by the fluidised bed technique[1] has usually required ground fully compounded material, but suitable dry blends can now be made.

Other processes involving homogenisation are those employing pastes, but the procedures used are relatively simple, and the degree of homogeneity achieved is dependent on the very small size of the primary particles of paste polymers. The processing of pastes is very specific to these materials and is discussed in detail separately (Chapter 14). Solution processes are in a somewhat similar category, in that the procedures involve only mild mixing conditions (Chapter 15).

Processes which do not in themselves involve homogenisation, at least not of chemical constitution, are thermoforming, including vacuum forming, and welding. PVC sheeting and other forms are fabricated by all these techniques (Chapter 15).

9.3   GENERAL MECHANISMS OF PROCESSING

9.3.1   INTRODUCTION

Some or all of the following separate processes are likely to be involved in the different processing procedures:

(1) Simple mechanical admixture of components of a formulation.
(2) Molecular admixture of ingredients.
(3) Thermal plasticisation of the polymer.
(4) Adhesion between plasticised polymer particles.

(5) Deformation of plasticised polymer particles.
(6) Degradation of the polymer and possibly other components under the action of heat and/or mechanical shear forces.
(7) Viscoelastic deformation and recovery, e.g. in passing through an extrusion die.
(8) Thermal expansion and contraction.

The theoretical bases of some of these have been considered earlier (Chapters 5, 6, and 7) and elsewhere,[39–43] while others are considered later (Chapters 10, 11, 12 and 13). A process which has not been clearly defined or understood is that of 'gelation' or 'fusion'. Indeed the two terms frequently appear to be used interchangeably. This is probably because the complex nature of the processes underlying the conversion of PVC powders to solid products has only become relatively clear during the past decade or so, and even now some aspects are not completely resolved.

9.3.2   GELATION

The word 'gelation' means different things in different contexts and to different people, but here it is envisaged as embracing the changes involved in the conversion of separate particles of polymer, or possibly compounded granules, to a more or less continuous poly-meric matrix. Clearly it must involve (3), (4), and (5) above (i.e. softening, deformation, and adhesion of the particles). For unplasticised PVC it would be imagined that this gelation is a relatively simple process. Its rate would clearly depend on the softening behaviour of the polymer (Chapter 4), on the rate of heating by externally applied means and by frictional work, and on any forces compacting the particles together. These forces might arise, for example, from the shearing action of the rotors of an internal mixer, or of an extruder screw, or from the pressure imposed by the ram of an internal mixer, or the die or nozzle of an extruder or injection moulding machine. It would be expected, too, that many additives, particularly lubricants, might profoundly affect gelation behaviour, and this is borne out in practice.[4, 5, 40–43] Processing of unplasticised PVC now appears to be dominated by powder feedstocks. While considerable investigation into gelation of unplasticised PVC has been carried out,[40–48, 67–75] it has been largely of an empirical nature, although the structural changes occurring are now fairly clear and generally agreed.

As discussed in Chapter 4, the morphology of PVC resin particles is complex, and it has long been known, not only that the morphological structure can have a major effect on processing behaviour, but also that structural features of the resin particles generally persist through to the finished product stage and have a profound effect on mechanical behaviour.[40–57] The pattern and degree of retention of particle structures depend on the intensity of processing, and vary according to processing equipment and procedures. Thus in dry blend extrusion under relatively mild conditions the gelation process involves compaction, densification, fusion, and elongation of 'grains' (Chapter 4), with little, if any, comminution of the grains. On the other hand under more intense shear conditions, as in an internal mixer or Brabender Plasticorder, considerable comminution of grains occurs. Two-roll mill compounding gives greater elongation of grains than extrusion, but no marked comminution.[40, 41] Processing aids accelerate fusion and aid break-down of particle structure.[43] Gotham and Hitch[61] demonstrated that poor removal of particle structure can lead to inferior mechanical properties; and in unplasticised PVC pipe extrusion Benjamin[62] found a 'gelation level' of 60 per cent gave optimum properties.

In torque rheometer experiments Gonze[70] and Faulkner[71] found that many PVC compositions showed three characteristic peaks in plots of torque vs temperature. Faulkner attributed these to the successive breakdown of 'Stage III' particles (100–150 μm diameter) or 'grains', 'Stage II' particles (Geil's 'agglomerates'?), and 'Stage I' particles, or 'microdomains'. The second peak is also associated with increase in particle density and decrease in porosity. At some temperatures beyond the third peak the composition exists as 'melt' with reducing torque due to falling melt viscosity as temperature rises.

Allsopp[40] reported that, unlike the progress of gelation in torque rheometers or internal mixers, extrusion of unplasticised PVC dry blends involves compaction, fusion, and elongation of grains with little comminution. The suggestion that there are two alternative routes for grain and primary particle fusion was supported by Gilbert and co-workers,[42, 67, 69] using differential interference contrast microscopy to examine changes in particle structure, differential scanning calorimetry to follow thermal changes, and hot-stage X-ray diffraction to follow crystallinity changes during processing. They found that most crystallinity disappears when PVC is heated to 180°C, but is partially recovered on cooling to 50°C, and extended Allsopp's proposal to include melting and recrystallisation leading to the formation of a network in the later stages. The secondary crystallinity formed on cooling is less well ordered than the initial crystallinity, leading to a lower overall level and lower density.[67]

All additives are likely to affect gelation and flow processes, whether they are employed for that purpose or not.

Lubricants and processing aids affect processing by modifying the gelation and forming processes. The behaviour of the former can be more complex than the simple division into 'external' and 'internal' lubricants implies (Section 7.2). Calcium stearate alone and in combination with a wax has been studied extensively.[68, 80–84] In general increase in concentration of lubricant was found to delay fusion, as would be expected, and the ultimate level of fusion can be restricted, but there were indications that calcium stearate can mildly promote fusion.[81]

In broad terms processing aids are said to reduce fusion time and melt viscosity, and increase melt strength and resistance to melt fracture[80] (Section 7.5). Acrylate processing aids are very compatible with PVC, and it has been suggested that they accelerate fusion by 'glueing' PVC particles together, thus increasing interparticle shear, and then mixing intimately with PVC molecules.[73]

With plasticised PVC there is a more complex situation arising from the specific interactions between polymer and plasticiser discussed in Chapter 6. Microscopy studies such as those referred to[6–10] in Chapter 6 show that swelling of individual particles by plasticiser occurs in the early stages. If the concentration of plasticiser is sufficiently high, e.g. if a few particles of polymer are 'floating' in a drop of plasticiser, swelling may reach a point where the external shape of the particle is lost and the polymer appears to go into solution (Figure 9.2). This behaviour is paralleled by determinations of critical solution temperatures by visual methods.[11] In dry-blending, plasticiser is absorbed rapidly into amorphous regions, and particle swelling by diffusion into the polymer network follows.[66] Swelling has also been demonstrated in actual compounding experiments.[12, 13] In systems of lower plasticiser concentration, as used in manufacturing compounding processes, it can be imagined that particles of polymer swollen by plasticiser impinge on each other and that their interfaces

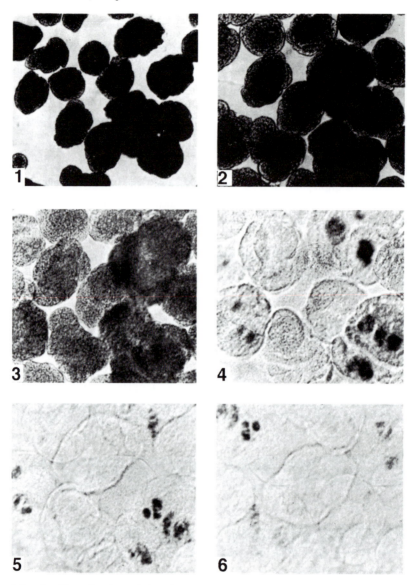

Figure 9.2.   Swelling of resin particles when heated with plasticiser (approx. × 100). (1) 70°C; (2)
90°C; (3) 105°C; (4) 110°C; (5) 112.5°C; (6) 116°C

merge. Diffusion aided by deformation arising from shearing or pressure during processing eventually results in a more or less homogeneous 'gelled' whole. These steps have been demonstrated in internal mixer compounding,[12, 13] but presumably are likely to occur in other processes as well.

The relative importance of the various factors on gelation behaviour has been analysed by using a Brabender Plastograph,[14] which showed that the nature and concentration of plasticiser had the greatest effect in determining the course of the gelation process. The type of polymer and the intensity of shearing were next in importance.

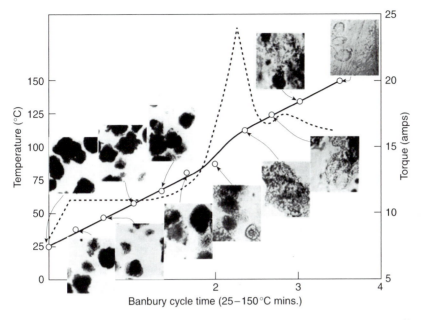

Figure 9.3. Behaviour of PVC plasticiser in an internal mixer. Solid curve refers to temperature and broken curve power consumption. Magnification approx. × 40

More recent studies using the 'Plasticorder' and capillary rheometry,[64, 65] suggested that fusion levels are controlled mainly by temperature, time and shear conditions being secondary. As might be expected, fusion was enhanced by increased plasticiser compatibility. The greater the compatibility the more rapid the fusion, the lower the peak temperature and the higher the torque at fusion peak.[79] Dynamic mechanical and tensile properties depend on the level of fusion, but once a high level of fusion is reached elastic modulus and viscosity depend on the viscosity of the plasticiser.[74] Under the relatively intense shear conditions of an internal mixer domains and primary particles are still observable after compounding to relatively low temperature (e.g. 124°C), although at a concentration of 80 phr of DIOP primary particle boundaries are beginning to blur. At higher 'drop' temperatures (e.g. 160°C) the primary particle structure is no longer detectable, but domains are still visible.[78] There is evidence that microdomain structures are retained in plasticised PVC compounded to temperatures up to 180°C.[81]

The amount of crystallinity present after processing may depend on the rate of cooling from the peak temperature. Horsley[76] showed that quenching a melt of PVC containing low concentrations of plasticiser (up to ca 15%) inhibited 'antiplasticisation', indicating prevention or reduction of recrystallisation. Annealing by heating to 65°C introduced the 'antiplasticisation', and presumably the crystallinity, characteristics of low plasticiser concentrations.

Investigations by Guerroro supported the view that 'antiplasticisation' is due to crystallinity,[87] but found that examination of the crystal structure by wide-angle X-ray diffraction yielded anomalous results.[88] Hobson and Windle[89] concluded that this supports Juijn's view[90, 91] that some isotactic units become included in the crystallites. Juijn and coworkers postulated that three different chain structures can occur, namely (1) syndiotactic zig-zag,

(2) isotactic helix, and (3) isotactic straight-chain; and that atactic chain segments occur which can fit into the syndiotactic orthorhombic lattice.[91] Using molecular modelling techniques Hobson and Windle calculated that a chain of isotactic PVC can emulate the shape and axial repeat of an equivalent syndiotactic molecule, and also that this conformation has its own energy minimum.[89]

To summarise, there appear to be two distinct possible sequences for the process of gelation as polymer is converted to end-product. Gilbert and co-workers[67] modified Allsopp's[40] suggested mechanism for twin-screw extrusion by postulating a gradual increase in melt homogeneity inside the extruder and the formation of a network structure on cooling. Which particular sequence of gelation is predominant in any particular case will depend mainly on the severity of the shear conditions through which the material has to pass, and presumably processes involving shear conditions intermediate between the two extremes are likely to exhibit features of both sequences. Since shear conditions depend on the geometry of the flow paths through the machinery, operating conditions, and formulation, all these will play some part in determining the progress of gelation, and each situation requires separate evaluation.

9.3.3   MELT FLOW BEHAVIOUR

It will be apparent from the previous section that the flow behaviour of a PVC composition through processing equipment is intimately bound up with gelation and particle structure, and it is scarcely possible to consider any one of these in isolation. While the extensive studies of melt flow behaviour of polyethylenes [15–17] may have some relevance to PVC, the 'peculiar' behaviour of PVC described in the previous section suggests that caution should be exercised in attempting to transfer the results to the latter material. The difficulty of defining accurately the physicochemical state of most PVC compositions, and the near impossibility of carrying out experiments at processing temperatures without some degradation of samples, make studies of PVC melt flow behaviour particularly difficult.

The Macklow–Smith Plastometer, introduced by Hayes in 1952,[20] has been widely used for routine evaluation and investigational wok with all kinds of PVC composition, but is also amenable to quite fundamental work. Most early work on this and other instruments was, however, of an arbitrary nature, although confirming expected trends of variations in flow behaviour with plasticiser content and with temperature. Thus at a fixed temperature, plots of pressures developed in extruding at constant rate against plasticiser contents yielded smooth curves of negative slopes. Other things being equal, the pressures required increased with molecular weight and decreased with comonomer content. Plots of pressures against temperatures also gave smooth curves having negative slopes.

Shear rate/shear stress relationships for PVC are markedly different to those for polyethylene, and a value of about 4 has been suggested for the slope of log/log plots vs shear rate against shear stress, compared to values of about 2 for polyethylene and 3 for polypropylene.[21] The velocity profile of a flowing stream of PVC 'melt' tends to be very steep near the walls of the confining channel, and almost zero near the centre of the flow path, i.e. PVC appears to exhibit 'plug' flow.[21]

The behaviour of plasticised PVC in a high shear rate rheometer has been related to processing conditions, the following shear rates being indicated:[22]

Compression moulding	$1-10s^{-1}$
Milling and calendering	$10-100s^{-1}$
Extrusion	$100-500s^{-1}$
Injection moulding	$1000-5000s^{-1}$

However these figures are generally much lower than those given by Riley, who has quoted values for shear rates in different extruder zones, at different speeds, and for calendering over a range of nip settings.[75]

As with polyethylenes, the tendency to produce extrudate distortions may be reduced by increasing melt temperature, but matters can be complicated by the onset of further distortions due to degradation,[21] or changes in flow regime.

Russian work,[23-25] using a combined rotary and torque rheometer, suggested that PVC compositions are processed under flow conditions of a 'breakdown-recombination' nature, which were visualised as involving actual rupture of chemical bonds during flow, but this suggestion is now discounted.

Other work also suggested that processing of PVC can involve rheological breakdown under shear,[26] and it is now generally recognised that particle structure and crystallinity, and changes therein during processing, play a major part in the flow behaviour of PVC. As discussed in the previous section the behaviour of PVC is complex, involving a number of regimes of structure, the retention or otherwise of which during melt flow depends very much on shear conditions and earlier treatment, so that apparent inconsistencies in rheological measurements are not surprising. An outcome of these observations is that methods of specimen preparation can have important effects on rheological behaviour. This not only points to a need for careful attention to this factor, but may also invalidate observation where it has been overlooked.

Even fairly detailed and comprehensive studies of melt rheology of PVC[27, 28, 40-57] can do little more than indicate general trends which need to be noted in practical processing, but which often cannot be applied quantitatively because of major effects arising from variations in additives and processing conditions.

Taking the Power Law[29, 30] in the form:

$$\eta = \eta_0 . \gamma^{n-1}/\gamma_0$$

where η = apparent viscosity at shear rate γ; η_0 = viscosity at an arbitrary standard shear rate γ_0 (usually one reciprocal second); n = the slope of the plot of log shear stress against log shear rate, then η_0 and n should describe the flow properties reasonably well. Tables 9.1 and 9.2 indicate the sort of results that have been obtained.

Table 9.1   MELT VISCOSITY RELATIONSHIPS FOR UNPLASTICISED PVC[28]

ISO Viscosity No. (Estimated)	188°C		199°C	
	n	$\eta_0 \times 10^5$ P	n	$\eta_0 \times 10^5$ P
57	0.533	0.849	0.656	0.22
79	0.413	2.06	0.477	0.981
94	0.378	3.74	0.380	2.7
107	0.362	5.49	0.356	4.17
116	0.358	7.94	0.323	5.83
125	0.361	13.2	0.307	6.96

Table 9.2   MELT VISCOSITY RELATIONSHIPS FOR PVC PLASTICISED WITH DOP[28]

DOP (phr)	173°C		193°C	
	n	$\eta_0 \times 10^5\ P$	n	$\eta_0 \times 10^5\ P$
0	–	–	0.418	5.88
8	0.301	13.2	0.425	3.58
20	0.313	8.33	0.465	1.88
30	0.348	4.75	0.532	1.23
50	0.426	1.175	0.600	0.29

As would be expected η_0 increases, while n decreases with molecular weight, values of the latter ranging from about 0.36 to 0.55 at 188°C and 0.45 to 0.8 at 215°C. Also, as would be expected, n increases with increase in plasticiser content. 'Energies of activation of viscous flow' at constant shear rate calculated from results like this indicate that a constant value of about 25 kcal (105 kJ) is attained once the molecular weight reaches a critical value (ISO Viscosity Number in the region of 100), and below this value the energy of activation increases with decreasing molecular weight. The idea that the critical value might be related to a critical value for chain entanglement was thought contrary to the smooth relationship between molecular weight and melt viscosity. The specimens for this work were prepared by two-roll milling with 2 phr of an organotin stabiliser, and it seems possible that these results are invalidated by lack of attention to the critical nature of the compounding conditions.

Unstable flow of forms similar to those that have been observed with other polymers[15–17, 31–33] has been noted in PVC.[15,28] Flow patterns of a plasticised PVC in a die with an entry angle of 180° were found to be markedly different from those of polyethylene, but distortion of turbulence still occurred at low extrusion temperatures, decreasing as the temperature rose.[15] At higher temperatures still, apparently similar distortion appeared, but this was attributed to degradation. More thorough study of PVC compositions revealed 'melt fracture' associated with critical shear stress and change in the slope of shear stress/shear rate curves, and a second form of unstable flow which occurred at much lower shear stress/shear rate values.[28] The latter was in the form of what was described as a 'herringbone' or 'fish-scale' pattern, and appeared to be similar to 'land fracture' observed in polyethylene,[31] but it is not always clear whether this is due to melt elasticity[31] or to a 'stick-slip' phenomenon at the die wall.[17]

Uneven flow can create severe problems in the processing of PVC, particularly in extrusion, blow moulding, and calendering. The most common forms encountered in industrial practice range from slight surface roughness to gross 'lumpiness', and include the phenomenon usually described as 'sharkskin'. 'Spiralling' and 'bambooing' are rarely if ever encountered in production, though they can be reproduced in the laboratory. The causes of these phenomena are often not obvious, though it is now generally agreed that most cases of melt instability are due to melt elasticity,[66, 75] and depend on the degree to which the particle structure is retained or destroyed, as discussed in the previous section.[40–57] Berens and Folt found that particle structure may not be completely destroyed up to temperatures of the order of 200°C.[44] They found that for emulsion polymers flow by particle slippage was favoured by relatively large particles, high molecular weight, and low processing temperatures, which minimised apparent melt viscosity, die swelling, and roughness of extrudate.[45] Clearly formulation must play a part,[4] and one of the functions of processing aids is to reduce the likelihood of these occurring.[43, 58] The heat/shear treatment of a PVC composition prior to its final shaping operation can markedly affect its rheological behaviour.[34, 35] In general the higher the temperature to which a composition is submitted the more likely it is for melt instability to arise in subsequent extrusion, and the higher the temperature is above that at which there is an onset of melt instability the more severe is the phenomenon likely to be.[34, 36, 66] Patel and Gilbert attributed this behaviour to network development. However the effect appears to reverse at high plasticiser contents.[66]

For the reasons just discussed it is usually desirable to compound to relatively low 'drop' temperatures. However if PVC is reprocessed at lower temperature under fairly intensive shear conditions, the melt flow behaviour characteristic of that lower temperature is recovered, attributable to the formation of a finer structure.[64] Diagnosis of uneven extrusion behaviour is occasionally complicated by the fact that superficially similar deformation can arise from quite opposite rheological behaviour, though closer examination of the extrudate and the melt behaviour within the head and die will usually reveal the true cause. Thus there have been cases where a die for a large section has been inadequate to provide the back-pressure necessary to compact PVC compounded diced feedstock sufficiently. Superficially the appearance of the extrudate may exhibit deformities associated with some forms of melt instability due to elasticity, but examination of a cross-section reveals that the dice have been only partially combined, leaving appreciable porosity within the mass. In such cases if it is impracticable to modify the die and/or head design, precompounding at relatively high temperature can sometimes provide a solution.

Like other thermoplastics PVC exhibits 'die swell', and some quantitative study of this has been made.[28, 59] Its behaviour appears to be contrary to theory in that the magnitude of the effect does not increase with molecular weight; indeed, is lower for high molecular weight polymers at low shear rates.[28] This is clearly associated with 'particle flow' behaviour, which is favoured by high molecular weight.[45, 48]

Because of the flow behaviour of PVC, if uniform flow from an extrusion die is to be achieved, special consideration has to be given to the design of heads and dies wherever there is a change in direction of flow or where the cross-section departs from the circular, coaxial with the extruder screw. Coupled with this the heat-sensitivity of the material demands that it shall flow continuously through the equipment.

For a comprehensive review and bibliography of the melt rheology of PVC see reference 63.

9.3.4   DEGRADATION

Enough has already been said (Chapter 5) about the degradation and stabilisation of PVC to indicate that they need consideration during design of processing equipment and operating conditions. Some degradation is almost bound to occur during processing, the amount depending on the inherent stability or instability of the composition being processed, the temperatures encountered during processing, and the period during which material is at these temperatures. For these reasons, if for no others, efficient temperature control is usually essential. This means that not only must heating and cooling arrangements be carefully selected, but that in any process where high shearing of 'melt' is likely to occur, care must be taken, through equipment design and operating conditions, to ensure that material temperatures do not rise too high. As indicated in the preceding section, the liability to degrade also requires that the PVC shall flow continuously through the equipment. Points of stagnation, such as corners, axial grooves between components, or even mere chips in surfaces, where melt is liable to reside indefinitely, lead eventually to decomposition which spreads to the flowing stream and appears in the product, to mar its appearance or even to ruin its mechanical properties. Clearly this problem becomes less intense the more stable the PVC composition, but it is advisable to design equipment so that all flow paths are stream-lined. Certainly for some of the very long running periods (several months) of many of today's operations, such a procedure is a *sine qua non*.

In addition to the design of equipment and careful control of processing conditions, it is essential that maintenance of equipment should be of a high standard. PVC compositions generally come away cleanly from metal equipment if stripped while still fairly hot, and acquisition of the slight skill required needs only a little practice – and a good pair of protective gloves! If tools are ever required to remove stubborn residues, they should if possible be of soft material such as brass. Any marring of the surface of metal contributing to flow channels is liable to result in trouble from degradation sooner or later.

One other aspect of processing in which degradation is an important factor is the subject of scrap material; but apart from stating that this is a matter requiring careful consideration little generalisation can be made. Obviously increased severity of processing conditions and decreased inherent stability will increase the danger of decomposition on reprocessing, and in extreme cases it may be desirable to reject scrap material. At the other extreme, however, with very well stabilised material and relatively mild conditions it may be possible to reprocess several times, even when unmixed with previously unprocessed material. It should be noted that degradation of a PVC composition may involve changes in or loss of other ingredients as well as the polymer, and it may be necessary to check that such changes will not adversely affect the material on reprocessing.

Though it does not strictly come within the scope of the word 'degradation', it is convenient to note here that, where PVC is reprocessed in a different composition, it is necessary to ensure that there will be no adverse reaction between any of the components of the two compositions involved, and that there will be no detriment to the properties of the final product arising from inclusion of some specific ingredient in the reprocessed composition. Examples of the former are the formation of lead sulphide and cadmium sulphide, respectively, when material containing lead or cadmium stabilisers is processed with material stabilised by sulphur-containing organotin compounds. An example of the latter would be involved in reprocessing a composition containing a dibutyltin stabiliser, in a composition destined for a 'non-toxic' application.

9.3.5 THERMOFORMING BEHAVIOUR

Much unplasticised PVC is thermoformed in some way, and clearly in this sort of process stress/strain behaviour over a range of temperatures below those at which flow occurs is of interest. Little if any fundamental information is available,[36, 37] though the requirements for different thermoforming operations and the behaviour of PVC sheet in particular have been briefly analysed.[38] Homopolymer compositions can be thermoformed reasonably well, and even vacuum-formed to fairly simple shapes, but for deep-drawn or complicated shapings low molecular weight copolymers are generally required.

REFERENCES

1. G.A.R. Matthews, 'Compounding Techniques', Chapter 5 in *Advances in PVC Compounding and Processing*, (M. Kaufman, ed.), Maclaren, London (1962)
2. W.S. Penn, *PVC Technology*, 2nd edn, Maclaren, London (1966)
3. J.A. Brydson, *Plastics Materials*, Chapter 9, Iliffe, London (1966)
4. D.R. Jones and J.C. Hawkes, *Trans. J. Plast. Inst.*, **35**, 120, 773 (1967)
5. G.M. Gale, *RAPRA Bull.*, **21**, 5, 78 (1967)
6. C.E. Anagnostopoulos, A.Y. Coran and H.R. Camrath, *J. appl. Polym. Sci.*, **4**, 11, 181 (1960)
7. F. Bargellini, *Materie Plast.*, **28**, 4, 372 (1962)
8. F. Bargellini, G. Fabris and F. Chiozzini, *Poliplasti*, **10**, 5 (1963)
9. F. Bargellini, *Materie Plast.*, **30**, 4, 378 (1964)
10. G. Matthews, D.H. Smith and J.A. Viney, 'Rate vs Temperature Relationships for Plasticiser Solvation of pvc Particles', *PVC Plasticiser Group Meeting*, Univ. of Technology, Loughborough (10 April 1981)
11. E.C.A. Horner, G.A.R. Matthews and W.W. Vincent, unpublished work (1956–1957)
12. R. Hammond, *Trans. J. Plast. Inst.*, **26**, 49 (1958)
13. B.S. Dyer, *Trans. J. Plast. Inst.*, **27**, 84 (1959)
14. T. Shiramatsu and N. Ueda, *J. Soc. Rubb. Ind. Japan*, **31**, 97 (1958)
15. P.L. Clegg, *Trans. J. Plast. Inst.*, **28**, 245 (1960)
16. E.R. Howells and J.J. Benbow, *Trans. J. Plast. Inst.*, **30**, 88, 240 (1962)
17. J.J. Benbow and P. Lamb, *S.P.E. Trans.*, **3**, 1, 7 (1963)
18. E B. Atkinson and H.A. Nancarrow, *Proc. Int. Rheol. Congr.*, Pt 2, 103 (1948)
19. B.A. Taylor and T.C. Tunbridge, unpublished work for award of B.Sc. at Borough Polytechnic (1964–65)
20. R. Hayes, *Chemy Ind.,* No. 44, 1069 (1952)
21. P. L. Clegg, *Br. Plast.*, **39**, 1 (1966)
22. E.H. Merz and R.E. Colwell, *A.S.T.M. Bull.*, No. 252 (TP 211) 63 (1958)
23. V.A. Khargin and T.I. Sogolova, *Zh. fiz. Khim.*, No. 6, 1328 (1957); *Dokl. Akad. Nauk S.S.S.R.*, **108**, 4, 662 (1956)
24. G.A. Slonimskii, *Khim. Nauka Prom.*, **4**, 1 (1959)
25. S.I. Klaz and E.C. Glukhov, *Soviet Plast.*, 26 (11 Nov. 1962)
26. S.K. Khanna and W.F.O. Pollett, *J. appl. Polym. Sci.*, **9**, 1767 (1965)
27. Y. Fukasawa, *J. chem. Soc. Japan*, **63**, 459 (1960)
28. C.L. Sieglaff, *S.P.E. Trans.*, **4**, 2, 129 (1964)
29. A.B. Metzner, 'Non-Newtonian Technology – Fluid Mechanics, Mixing & Heat Transfer', in *Advances in Chemical Engineering*, Vol. 1, (T.B. Drew and J.R. Harper, eds) Academic Press, New York (1956)
30. A.B. Metzner and R.E. Otto, *A.IO.Ch.E. Jl*, **3**, 13 (1957)
31. J.P. Tordella, *J. appl. Phys.*, **27**, 5, 454 (1956); *Trans. Soc. Rheol.*, **1**, 203 (1957); *J. appl. Polym. Sci.*, **7**, 215 (1963)
32. M.T. Dennison, *Trans. J. Plast. Inst.*, **35**, 120, 803 (1967)
33. F.N. Cogswell and P. Lamb, *Trans. J. Plast. Inst.*, **35**, 120, 809 (1967)
34. G.H. Burke and G.C. Portingell, *Br. Plast.*, **36**, 5, 254 (1963)
35. D. Dowrick, *Plastics*, **30**, 328, 63 (1965)
36. N. Platzner, *Mod. Plast.*, **31**, 3, 144 (1954)
37. N. Platzner, Chapter 8 in *Processing of Thermoplastic Materials* (E.C. Bernhardt, ed.), Reinhold, New York (1959)
38. J.M.J. Estevez and D.C. Powell, *Manipulation of Thermoplastic Sheet, Rod and Tube*, Iliffe, London (1960) P.I. Monograph
39. G. Matthews, *Polymer Mixing Technology*, Applied Science Publishers, London (1982)
40. M.W. Allsopp, 'Manufacture and Processing of PVC', (R.H. Burgess, ed.), Applied Science Publishers (1982) Chapter 8
41. M.W. Allsopp, 'PVC Processing II', *PRI Intl. Conf.*, Brighton, England (26–28 April 1983) Paper 4
42. M. Gilbert, D.A. Hemsley and A. Miadonye, *Plast. Rubb. Proc. Appn.* **3**, 343 (1983)
43. D.L. Dunkellberger, *Kunststoffe*, **80**(7), 816 (1990)
44. A.R. Berens and V.L. Folt, *Trans. Soc. Rheol.*, **11–1**, 95 (1967)

45. A.R. BERENS and V.L. FOLT, *Polym. Engg. Sci.*, **8**, 5 (1968); **9**, 27 (1969)
46. D. HANSON, *Polym. Engg. Sci.*, **9**(8), 405 (1969)
47. G.M. GALE, *Plastics & Polymers*, **38**, 183 (1970)
48. G. PEZZIN, *Pure and Applied Chemistry*, **26**(2), 241 (1971) also *Macromolecular Chemistry – 6*, Butterworths, London (1971)
49. T. HATTON, K. TANAKA and M. MATSUO, *Polym. Engg. Sci.*, **12**(3), 199 (1972)
50. G. MENGES and N. BERNDTSEN, *Kunststoffe*, **66**(11), 735 (1976)
51. G. MENGES and N. BERNDTSEN, *Pure & Appl. Chem.*, **49**, 597 (1977)
52. D.A. HEMSLEY, E. KATCHY, R.J. LINFORD and D.E. MARSHALL, 'PVC Processing', PRI Intl. Conf., Royal Holloway College, Egham Hill, Surrey, England (6–7 April 1978) Paper 9
53. J. GRAY, *ibid*, Paper 10
54. D.R. MOORE, *ibid.*, Paper 11
55. G. MENGES and N. BERNDTSEN, *ibid.*, Paper 12
56. D.E. MARSHALL, R.P. HIGGS and O.P. OBANDE, 'PVC Processing II', PRI Intl. Conf., Brighton, England (26–28 April 1983) Paper 13
57. G. MENGES, E. KRUGER, and J. PAREY, *ibid.*, Paper 11
58. R.P. PETRICH, 'PVC Processing', PRI Intl. Conf., Royal Holloway College, Egham Hill, Surrey, England (6/7 April 1978) Paper 6
59. V. SAHAJPAL, *Developments in PVC Technology*, (J.H.L. Henson and A. Whelan, eds), Applied Science Publishers, London (1973) Chapter 4
60. D.A. TESTER, *Manufacture and Processing of PVC*, (R.H. Burgess, ed.), Applied Science Publishers, London, (1982) Chapter 9
61. K.V. GOTHAM and M.J. HITCH, *Brit. Polym. J.*, **10**, 47 (1978)
62. P. BENJAMIN, 'PVC Processing', PRI Intl. Conf., Royal Holloway College, Egham Hill, Surrey, England (6–7 April, 1978) Paper B5
63. D.W. RILEY, *Encyclopedia of PVC*, vol. 3, (L.L. Nass, ed.), Marcel Dekker, New York (1992) Chapter 9
64. S.V. PATEL and M. GILBERT, *Plast. Rubb. Proc. Appn.*, **5**, 85 (1985)
65. S.V. PATEL and M. GILBERT, *Plast. Rubb. Proc. Appn.*, **6**, 321 (1985)
66. S.V. PATEL and M. GILBERT, *Plast. Rubb. Proc. Appn.*, **8**, 215 (1987)
67. J.A. COVAS, M. GILBERT and D.E. MARSHALL, *Plast. Rubb. Proc. Appn.*, **9**, 107 (1988)
68. O.P. OBANDE and M. GILBERT, *Plast. Rubb. Proc. Appn.*, **10**, 231 (1988)
69. O.P. OBANDE and M. GILBERT, *J. Appl. Polym. Sci.*, **37**, 1713 (1989)
70. A. GONZE, *Plastics*, **24**, 49 (1971)
71. P.G. FAULKNER, *J. Macromol. Sci.-Phys.*, **B11**, 251 (1975)
72. G.C. PORTINGELL, *Particulate Nature of PVC*, (G. Butters, ed.), Applied Science Publishers Ltd, London, (1982) Chapter 4
73. D.L. DUNKELBERGER, K.P. ROZKUSZKA and V.K. SAHAJPAL, *Kunststoffe*, **80**, 816 (1990): *Kunststoffe German Plastics*, **80**, 33 (1990)
74. L. RAMOS-DE VALLE and M. GILBERT, *Plast. Rubb. Comp. Proc. Appn.*, **15**, 207 (1991)
75. D.W. RILEY, *Encyclopedia of PVC*, (I. Nass, ed.), Marcel Dekker (1992) Chapter 9
76. R.A. HORSLEY, *Plastics Progress 1957*, (P. Morgan, ed.), Iliffe & Sons, London (1957) pp. 77–88
77. J.K. SEARS and J.R. DARBY, *The Technology of Plasticizers*, John Wiley & Sons, Inc., New York, (1982)
78. M. BOTTRILL and R.C. STEPHENSON, 'PVC Processing II', PRI Intl. Conf., Brighton, England (26–28 April 1983) Paper 6
79. D.L. RAMOS-DE VALLE and M. GILBERT, *Plast. Rubb. Proc. Appn.*, **13**, 157 (1990)
80. J.D. BOWER and M.H. HEFFNER, *J. Vinyl Tech*, **5**, 116 (1983)
81. E.B. RABINOVITCH, E. LACATUS and J.W. SUMMERS, *J. Vinyl Tech.*, **6**, 98 (1984)
82. A. YU, P. BOULIER and A. SANDHU, *J. Vinyl Tech.*, **6**, 110 (1984)
83. R. KRZEWKI and E.A. COLLINS, *J. Macromol. Sci. Phys.*, **B20**, 465 (1981)
84. J. MOONEY, 'PVC Processing II', PRI Intl. Conf., Brighton, England (26–28 April 1983) Paper 16
85. E. FOLDES, T. PAZONI and P. HEDVIG, *J. Macromol. Sci. Phys.*, **B15**, 527 (1978)
86. P.L. SONI, P.H. GEIL and E.A. COLLINS, *J. Macromol. Sci. Phys.*, **B20**, 479 (1981)
87. S.J. GUERRERO, *Macromolecules*, **22**, 3480 (1989)
88. S.J. GUERRERO, H. VECOSO and E. RANDON, *Polymer*, **31**, 1615 (1990)
89. R. HOBSON and A.H. WINDLE, Polymer, **34**, 3582 (1993)
90. J.A. JUIJN, J.H. GISOLF and W.A. DE JONG, Kolloid-Z.u.Z., *Polym.*, **235**, 1157 (1969)
91. J.A. JUIJN, J.H. GISOLF and W.A. DE JONG, Kolloid-Z.u.Z., *Polym.*, **251**, 456 (1973)

10
Mixing and compounding

10.1 INTRODUCTION

A wide variety of different machines is used for the mixing and compounding of PVC compositions, and it seems almost as if every possible design has at some time or another been pressed into service for the purpose. The simple mixing machines used widely throughout the industry were adopted directly from those used in other industries, particularly the food and confectionery and the rubber industries. Compounding machines were adopted directly from those used for rubber, with relatively small changes in design. During the 1950s and 1960s several machines were introduced, designed with the processing of plastics, especially PVC, in mind. Most if not all of the mixing and compounding machines have been designed largely on empirical bases, mainly because the processes are still at best only imperfectly understood.

10.2 THEORETICAL CONSIDERATIONS

Simple mechanical admixture of components has received a fair amount of theoretical treatment, backed up by limited experimentation.[1-10, 44] For present purposes it is very doubtful if it is worth taking much account of the mathematical equations that have been derived and postulated, since they do not take into account several highly significant features inherent in mixing of most PVC compositions. At most they can be applied directly to such simple instances as the tumble-blending of masterbatches with similar granules of natural compound, or the blending of several batches of nominally identical granules or powder for the purpose of unification of any minor departures of compositions or properties from the mean. Nevertheless the qualitative concepts arising from the theoretical considerations are important in any form of mixing process.

10.2.1 SCALE OF SCRUTINY

In considering the visual appearance of a pigmented surface it will be realised that if the particles of pigment and any unpigmented areas between them are too small for separation by the eye, the distribution of pigment amongst the polymer will appear to be uniform, assuming, of course, that there are no gross fluctuations in concentration across the surface being observed. If the particles of pigment are just too small to be picked out by the naked eye they will be observed as separate particles under the microscope. If, now, one imagines the particles becoming smaller and being dispersed more uniformly until even the microscope cannot reveal them, the non-uniformity will still exist; even in the ultimate it will always be there on the molecular scale since one molecule cannot occupy the space occupied by another! These observations illustrate the concept of the 'scale of scrutiny', namely, the idea that the more closely a mixture is scrutinised the more likely is it to appear non-uniform. Conversely one can conceive that for any mixture there will be a scale of scrutiny which is just insufficiently fine to reveal lack of uniformity. The scale of scrutiny which it is necessary to meet will depend on the components being examined. Thus sufficient uniformity would exist for a pigment if the visual appearance was uniform (forgetting for the moment that this state of affairs might involve the use of unnecessarily high concentrations of pigment). For stabiliser, however, it may be assumed that mixing on a molecular scale would be desirable, even if not attainable. Evidence has, in fact, been obtained to suggest that overall stability of a given composition is increased the smaller the scale of scrutiny that can be applied to it before non-uniformity is detected.[11]

In its original concept[2, 10] 'scale of scrutiny' was seen as being the minimum size of the regions of segregation which would cause the mixture to be imperfect for the purpose intended, and could be expressed as a length, an area, or a volume.

10.2.2   DEGREE OR GOODNESS OF MIXING

It is possible to visualise a mixture of a number of components as being completely uniformly distributed, i.e. in a repeating constant three-dimensional pattern, as might be achieved for example with a mixture of equal numbers of black and white building cubes by stacking them manually, placing black and white cubes alternately next to each other. At first sight this would appear to correspond to the ideal of a 'perfect' mixture. In practice, however, a mixture of this nature can generally not be obtained. The best that is theoretically possible in a normal process is a statistically random mixture, and conventionally such a mixture is defined as 'a perfect mixture'. This concept can be confusing in that it is possible to visualise a more uniform, albeit an unattainable, state, but it is mathematically convenient as seen below. Indeed, mixing of a completely uniform composition might be expected to produce 'unmixing' to approach the random from the opposite direction to that of normal mixing. For the simple case of mixing of particles of two different materials, the state of admixture at any particular time can be determined by extracting a sufficiently large number of samples and determining their deviations from the mean or overall composition. The deviations can be expressed in terms of the differences between the concentrations of one component and its mean concentration. From the values of these deviations the standard deviation s and variance s^2 can be calculated by normal statistical means. If P is the mean or overall proportion of one of the constituents and n is the number of particles in each sample, the variance s_r^2 of a completely random mixture is given[3] by:

$$s_r^2 = P(1 - P)/n \tag{1}$$

For the completely unmixed state, the variance s_o^2 is given[1] by:

$$s_o^2 = P(1 - P) \tag{2}$$

For incomplete mixing the variance s^2 will lie between the values of s_o^2 and s_r^2, and its value gives a measure of the degree of goodness of mixing.[10] Various expressions have been proposed to define M, the degree of mixing, of which the following[3] is perhaps the most useful:

$$M = (s_o^2 - s^2)/(s_o^2 - s_r^2) \qquad (3)$$

In this expression it will be noted that at completely random mixing: $s^2 = s_o^2$ and M reduces to unity.

At complete unmixing: $s^2 = s_o^2$ and M reduces to zero

Intermediate degrees of mixing are characterised by values between zero and unity. A number of other indices and ways of describing 'goodness' of mixing have been proposed.[47]

10.2.3   RATE OF MIXING

Clearly a mixing process occupies a finite period of time and involves rate, i.e. it is a 'rate process'. It would be expected that the rate of approach to the random state would decrease as the gap narrowed, and that the approach would be asymptotic. It has been suggested[3] that the change in degree of mixing M with time t is given by the expression:

$$M = 1 - e^{-kt}$$

where k is a rate constant for the mixing process.

This expression yields a linear plot of $\log_e (1 - M)$ against t, the slope being k, the rate constant, and thus a measure of the speed of the mixing process.

10.2.4   PRACTICAL COMPLICATIONS

In some mixing processes some unmixing may occur, and equation (4) should be rewritten:[10]

$$M = 1 - e^{-kt} - e^{k't}$$

where k' is a rate constant for the umixing process.

Apart from the blending of batches of like material or blending natural compounds with masterbatches, the particles of the components being mixed are usually of different sizes, and one component is very often made up of particles having a wide size distribution. Such systems have been treated theoretically,[12] but practical application is limited, partly because the liability of small particles to pack into the spaces between large particles reduces the trend to approach a random distribution.

Other complications which can arise have not been analysed. They complicate not only mathematical analysis of mixing processes but also the actual processes themselves, and include the following:

(1) There are frequently as many as 6, 7, or 8 components to be mixed.
(2) There are usually big differences between the concentrations of the various components.
(3) There are frequently big differences between the relative densities of the various components, and this can lead to a tendency for the heavier components to concentrate to-

wards the bottom of the composition. For this reason it is desirable to pay some attention to the order in which the ingredients of a composition are charged to mixing equipment.

(4) Many of the additives to PVC are of a waxy or sticky nature, and thus difficult to disperse.

(5) Liquid additives such as plasticisers usually have rather high viscosities.

(6) Liquid additives can often be absorbed by or have specific interactions with polymer and filler particles, as a result of which initial uneven distribution of the liquids during addition can be impossible or difficult to disperse. For this reason it seems a sound policy to introduce liquid additives by fine spraying on to the solid components while the latter are being thoroughly agitated.

(7) In mixing processes involving high shear, and particularly in compounding, particles of solid may be disintegrated, and if the temperature becomes sufficiently high particles of polymer may agglomerate and eventually flux to a more or less continuous melt.

10.3   MIXING MACHINES

Almost every conceivable mixing machine seems to have been used for mixing of PVC compositions at some time or another. The various types are considered below.

10.3.1   TUMBLE-MIXERS

Tumble-blending varies from the simplest form, in which the components to be mixed are merely placed in a drum which is then turned end over end by some simple mechanical means, to quite large sophisticated mixers with chambers of complicated shape (e.g. double conical) with internal baffles, and often eccentrically rotated.[9, 13, 47] The Patterson-Kelley 'Zig-Zag' continuous mixer is essentially a continuous tumble blender. It comprises a feed chute leading to an eccentric drum which controls residence time in the six zig-zag section that follows.[48] The main use for tumble blenders with PVC is for blending of different batches of granulated compound, or of 'natural' compound with masterbatch.

10.3.2   PADDLE MIXERS

This title is intended to embrace those mixing machines in which the components being mixed are agitated in a vessel by means of blades, rotors, or the like, rotating usually on vertical shafts. In some the axis of the shaft follows a circular path concentric with the walls of the mixing chamber at the same time as the stirrer rotates. Planetary mixers are of this type. In some cases the mixing chamber itself also rotates. A wide variety of mixers of this type exists with all kinds and shapes of stirrers, including the familiar anchor-type, flat blades, propeller-like blades, impellers, etc.[13, 47] A number of models have two sets of shaft and stirrer blades very often rotating in a 'figure-of-eight' chamber so that the central part of the chamber is swept alternately by the blade of each stirrer.[9, 13, 47]

These types of mixer are sometimes used for large-scale mixing of ingredients before compounding, but they are not really satisfactory for this purpose, and their main use in in the production of pastes.

10.3.3 RIBBON BLENDERS

For many years this type of mixer was the most commonly employed machine for the production of blends or 'premixes' of PVC compositions. They appear to be a development from the Archimedean conveying screw, and indeed many of them have mixing blades resembling the appearance and at least in part having the effect of an Archimedean screw. The mixing blades or agitators are usually in the form of strip metal mounted more or less in spiral fashion around a horizontal shaft rotating in a trough-shaped mixing chamber. Many different designs of mixing blades exist, some machines having more than one set attached to the same shaft, the main function being to lift the materials and drop them under gravity in such a way as to produce a complex flow path to ensure good mixing. Discharge is effected by tilting the whole mixing chamber or through a port in the base of the chamber. Continuous blenders are available, in which components of the mixture are metered in at one end, mixed and conveyed along the mixing trough, and discharged at the other end.

Very little frictional heat is developed in ribbon blenders, but the mixing chamber is often jacketed for heating and cooling, so that they may be used to produce plasticised dry blends. Heat exchange is very inefficient, however, and dry blending cycles can be very long, as much as several hours for larger machines.

Another type of mixer which might be classified as a ribbon blender has an Archimedean type of screw rotating in such a way that it tends to convey materials up the inclined wall of the mixing chamber, which has the form of an inverted cone. The conveyor is mounted on an inclined shaft parallel to the walls of the mixing chamber, and its upper bearing is carried on the outer end of a radial arm rotating about the centre of the base of the cone (the top of the mixing chamber). Thus the conveying screw moves around the whole of the inclined surface of the mixing chamber at the same time as it is conveying material upwards. This action appears to give very efficient mixing, and is suitable for premixing plasticised PVC compositions for subsequent compounding. A subsequent development of this design of mixer involves two inverted conical chambers, joined and partially overlapping towards the top like Siamese twins.

Other variations include machines with plough- or spade-like members mounted on a central rotating shaft and rotating fairly near to the cylindrical wall of the mixing chamber, thus introducing a smearing action. In one variety of this type of machine the cylinder is perforated so that material is squeezed out through the walls of the cylinder into an outer mixing chamber. This variety of blender is also available in a continuous processing form.

10.3.4 Z-BLADE OR SIGMA-BLADE MIXERS OR KNEADERS

Like ribbon blenders this type of mixer is widely used in the plastics industry, not only to produce relatively dry blends, but also for pastes and other viscous compositions. They comprise a double U-shaped mixing chamber with two rotors or kneaders, usually resembling the letter Z, hence the name. The rotors usually rotate at different speeds thus yielding a complex flow pattern, and in addition to the normal lifting function of the rotor blades there is compression where the two blades interact near the central ridge of the mixing chamber. The action of these machines has been described as 'folding and compressing the mix much as a baker might do in kneading dough'.[9] Like ribbon blenders, this type of mixer may be jacketed for heating and cooling, and can thus be used to produce plasticised powders or dry blends, but here, too, poor heat transfer leads to rather long cycle times, which increase with size, and may be as long as several hours.

10.3.5   'AIR' MIXERS

This type of mixer employs air to 'fluidise' the components to be mixed, thus offering very efficient dispersion and rapid mixing. The fluidisation is usually effected in a vertical cylindrical chamber, and can be achieved by the conventional porous plate principle of the fluidised bed process, or by appropriate air entry channels and baffles to give a cyclone effect upwards near the walls of the mixing chamber with a return path downwards through the centre. In both types fluidisation is usually alternated with settling by applying the air in controlled intermittent blasts.

10.3.6   INTENSIVE NON-FLUXING MIXERS ('HIGH-SPEED MIXERS)

It is unfortunate that the term 'high-speed' is commonly used for these mixers because it can be misleading. While it is generally true that they have much higher rates of mixing than most other types of mixer, the term is also often applied to more conventional types which are claimed to mix more rapidly than their competitors. For this reason, and because the motion even of solids in in this type of mixer may be likened to that of fluids, the term 'fluid mixer' is possibly preferable. These machines were introduced in the mid-1950s,[14, 15] and have now effectively ousted other types of mixer for all kinds of PVC mixing, especially of plasticised and unplasticised dry blends, because of their efficient and rapid mixing, and because they can be operated to produce powders or agglomerates in ideal form for subsequent extrusion or moulding.

The essential feature of an 'intensive non-fluxing' or 'fluid' mixer is that movement of the components to be mixed is achieved by a rotor at the base of the mixing chamber, rotating on a vertical axis at speeds ranging from a few hundred to a few thousand rev/min. Under these conditions the components are thrown out by the centrifugal action of the rotor and up the walls of the mixing chamber, to return down the central zone, thus exhibiting a fluid-like movement with a vortex. Depending on the precise design of the rotor and its speed, more or less intensive frictional heat is developed and the temperature rises continuously within the mixing mass. Indeed the difficulty or impossibility of removing the heat completely is one of the few drawbacks of this type of mixer, necessitating careful control and possibly imposing an upper limit to the practicable size. The frictional heat developed is sufficient to induce rapid plasticiser absorption by porous particle resins, and so plasticised powders can be produced without the application of external heat, although heating jackets are often used to minimise mixing cycles. If mixing is allowed to continue, the temperature of the mixed material continues to rise and agglomeration will ensue, but by careful operation the agglomerates can be produced in controlled particle form and compounded state eminently suitable for subsequent processing. Commercial fluid mixers differ in the shapes of the mixing chambers and the rotors. Often a selection of rotor shapes is available to suit different mixing requirements.[47]

Owing to the large amounts of heat generated by the mixing rotors of fluid mixers, cooling is an inefficient or near impossible process, so it is customary on any scale of production operation to discharge the mixed material, as soon as it has reached the required state, into an associated low-speed cooling mixer, preferably with a large surface/volume ratio, where the mixture is cooled to a temperature at which it will not agglomerate further.

Most fluid mixers are fitted with lids carrying feed ports, thermometer probe, sight glass, and vacuum-line for removal of volatiles. Complete units comprising interconnected fluidising and cooling mixers with a high degree of automation are now commonplace. The complete enclosure is a big advantage in processing unplasticised PVC dry blends stabilised with solid tribasic lead sulphate, since it eases the problem of avoiding release of the toxic powder into the surrounding atmosphere.

Capacities of fluid mixers range from laboratory sizes taking something like 2.2–3.2 kg of material to machines of 2500 l capacity with batch sizes of the order of 500 kg. Cycle times vary from as little as 90 s up to about 20 min, depending on the mixer and the nature of the composition.

10.3.7   PRODUCTION OF DRY BLENDS

10.3.7.1   Dry blend formulation

To produce a true plasticised powder containing appreciable proportions of plasticiser, a porous particle resin is essential, though plasticised dry blends of sorts can be produced from other types of resin if partial gelation and agglomeration is effected in a fluid mixer.

For unplasticised dry blends polymers with non-porous particles are preferred because they permit faster extrusion than porous particles before air entrapment in the melt occurs.

Other requirements of the polymer are that it should have relatively high packing density, to decrease the air space between the particles, that it should flow freely when in the dry blended state, and that it should gel readily. The first of these is generally opposed to the other two, and a compromise is usually necessary.

For unplasticised dry blends irregular shaped particles generally gel more readily than regular particles and therefore process better in extruders having relatively low compression. Moreover, polymers with regular shaped particles usually have higher packing densities and their use in extruders of low compression can lead to over-feeding of the screw and disturb the pressure conditions in the melt zone.[63] On the other hand, in extruders of relatively high compression, the denser regular particle polymers are advantageous, because their shape leads to relatively little friction between the particles.[62] Generally the larger the machine and the higher the output required, the more the irregular type of polymer is preferred.[64]

Possibly even more important than the choice of polymer is that of lubricant, since this seems to have a greatly disproportionate effect on gelation behaviour, as well as the usual effects on flow behaviour. In general, increase in lubricant concentration decreases the ease of gelation, and variation in lubricant type and concentration can be used to control the progress of gelation along an extruder screw. The effect is most marked with incompatible materials of low melting point, for example stearic acid, which can be in excess at a concentration as low as 0.33 phr.[64] A combination of lubricants can often offer better control of the gelation process than a single lubricant alone, and for this reason lubricant systems for unplasticised PVC dry blends can be quite involved. A lubricant formulation might contain typically a mixture of normal and dibasic lead stearate, a montan wax derivative, and calcium stearate, the main function of the latter being as external lubricant in the head and die.[64]

Processing aids help in unplasticised dry blend extrusion, apparently by increasing melt-viscosity and so increasing mixing of the material leaving the screws.[64]

Fine particle fillers, used to improve impact strength, appear to increase melt viscosity since they result in increased back pressure and reduced outputs, apparently by increasing the distribution of gelled material along the screw.[64]

10.3.7.2   Dry blending techniques

Simple mixtures of the ingredients of an unplasticised PVC composition can sometimes be directly extruded satisfactorily if the lubricant system is adjusted to allow adequate homogenisation and gelation; but control of such extrusion is delicate. It has indeed been claimed that there are distinct advantages in 'cold blending', including increased extrusion outputs, provided that the extruder design and processing conditions are appropriately selected.[65]

Plasticised dry blends can be made reasonably well in simple heated mixers, such as ribbon blenders and Z-blade mixers, but 'fluid' or 'high-speed' mixers are preferable, as they are also with unplasticised dry blends. Dry blending in fluid mixers can produce some absorption of additives into the particles of the polymer, some attrition, and improved flow, and more prolonged heating and shearing can lead to increased bulk density and ultimately to 'semi-gelled' agglomerates.[64] Absorption of additives, such as lubricant, can presumably only occur if the temperature reaches a point sufficiently high to melt them, and the temperature to which a blend is allowed to rise appears to be critical in this respect. Optimum maximum temperatures for unplasticised dry blends are typically in the region of 120–140°C,[64] but even higher temperatures have been proposed.[65] Failure to allow the blend to reach its optimum temperature is likely to result in porosity and extrudates of inferior strength.

10.3.8   BALL MILLS

These machines are rarely if ever used to mix PVC compositions, their main use being to grind pigment agglomerates.

10.3.9   COLLOID, DISC, AND PIN MILLS

These mills impose intense frictional work on a relatively very small bulk of material at a time. Hence they achieve good dispersion throughout limited volumes, but do not counteract any variations in composition of material feed. Their main use is for dispersing pigments, stabilisers, and blowing agents which need very fine dispersion, though some have been claimed to be suitable for the production of plasticised PVC dry blends. This seems hardly practicable. Mechanical foam mixers (Section 14.4.8.2) are essentially of this type.

10.3.10   MULLERS

Mullers usually consist of a cylindrical working chamber, with a circular base, in which one or more usually two solid cylinders are rotated either about horizontal co-axes (edge-runner mill) or about parallel vertical axes (side-runner mill). Frictional work is imposed on the mixing materials between the faces of the rotating cylinders and the walls and base of the chamber. In some models the materials can be heated by circulating air, and it has been claimed that such machines are suitable for the production of plasticised dry blends.[16] As with other machines where an intense grinding action is developed, the main use for mullers appears to be in dispersion of pigments and stabilisers in media such as plasticisers. Good dispersion is usually obtained rapidly, in as short a time as a few minutes, but compositions with coarse hard particles (e.g. polymer) do not mix well in these machines.

10.3.11   ROLL MILLS

'Roll' or 'roller' mills are of two types, with rather different function, the most obvious difference being that one type is heated and the other is not.

The non-heated roll machines are mainly employed for dispersing solids such as pigments, stabilisers, blowing-agents, and paste resins in plasticisers, which is achieved by passing the mixture several times through the nips of the machine, which are adjustable and are set at clearances of only a few thousandths of an inch, the precise gap depending on the materials being dispersed. The most common form is the 'triple-roll' mill, but mills with as many as five rolls are also available. Heated roll mills are more conveniently considered as compounding machines.

10.4   COMPOUNDING MACHINES

10.4.1   ROLL MILLS

10.4.1.1   Constructional features

Heated roll mills are usually two-roll mills, with one roll in fixed bearings and the other in bearings which can be moved towards and away from the fixed ones. Usually the movement of the moveable bearings towards the fixed ones is effected by two independent adjusting screws operated by 'lug' bars, while movement outwards depends on pressure of material being processed within the nip.

Commonly the rolls are of chilled cast iron, but alloy steel and chrome-plated models are obtainable. Heating and cooling is usually effected by passing steam, hot water, cold water, or heat-exchange fluid through the rolls, which are drilled for the purpose. Where steam is used temperatures are controlled by steam pressure valves with associated gauges, preferably independent sets for each roll. With heat-exchange fluid temperature is controlled by external heat-exchange units. Some machines have rolls heated by electrical resistance heaters.

The size of the rolls varies with requirements, the smallest generally being about 150 mm long and 50–75 mm in diameter, obviously for laboratory purposes only, for which, however, the most common size is about 300 mm by 150 mm. Also in common use in laboratories are 450 by 230 mm and 600 by 300 mm sizes. Sizes extend up to 2 m by 0.66 m for development and production.

The rolls are contra-rotated downwards into the nip, either at the same speed or at different speeds, the ratio of which is known as the 'friction ratio'. This usually has a value between 1 and 1.4. The actual speeds of the rolls vary with the size and are usually fixed, but variable speed roll machines are available. Higher speeds give more rapid mixing than low, but make handling of the compounded material more difficult. For a small laboratory mill, with rolls 300 mm long and 150 mm in diameter, speeds are usually in the region of 30 rev/min, corresponding to surface speeds of the order of 13.5–15 m/min. Rotational speed decreases with increase in roll dimensions. A 1.5 by 0.5 m mill will generally have speeds of about 20 rev/min, corresponding to surface speeds of about 30.5 m/min.

Batch size can be adjusted within limits by adjustable V-shaped chucks which limit the spread of material along the length of the rolls, but the maximum charge is clearly approximately proportional to the length and diameter of the rolls and to the nip setting. Table 10.1 gives typical batch capacities for a number of standard commercial machines.

Table 10.1

Size (m)	Batch capacity (kg)
0.3 × 0.15	0.9–1.8
0.46 × 0.23	2.2–4
0.6 × 0.3	5.5–8
0.9 × 0.4	14–18
1.05 × 0.4	16–23
1.2 × 0.4	18–27
1.5 × 0.55	40–70
1.8 × 0.6	70–110
2.1 × 0.65	135–180

An important feature of all roll mills is the hazard presented by the nips, into which the rotation of the rolls naturally tends to take not only material being processed, but any adventitious objects such as spatulas, spoons, and human fingers! Every machine must have reliable quick-acting stop mechanisms operable from any working position, but not requiring deliberate hand operation. Guards must be fitted around the upper parts of the rolls making it impossible for an operator to put his hands in a dangerous position, but means have to be provided for the introduction of materials to the nip, so it is impracticable to bar the possibility of inanimate objects being dropped into the nip. For this reason only plastic or wooden tools should be used for materials handling to the mills, as these will not damage the rolls as metal tools would.

10.4.1.2   Operation

In operation the rolls are first heated to the required temperatures with the nip controls slack to allow for expansion. The temperatures will depend on the formulation of the composition to be processed, particularly on the nature and molecular weight of the polymer, and the nature and concentration of the plasticiser, but will usually be in the region of 130–180°C for compounding purposes. The front, working roll is usually heated to a temperature about 3–5°C higher than that of the back roll. Temperatures can be checked by means of bow thermocouple pyrometers. On reaching the required temperatures the rolls are set in motion and the nip is reduced to a small clearance. Mixed material to be compounded is charged to the nip in small quantities at a time, allowing each quantity to gel to a continuous crêpe before adding the next. The ease with which the composition will gel is also dependent on the formulation, especially on the nature of the polymer. Thus, other things being equal, emulsion polymers will gel very rapidly, much more so than suspension polymers, and small particle suspension polymers will gel more rapidly than large particle suspension polymers. As more and more material is added the nip controls may be slackened off cautiously to permit the material to open it and increase the quantity of material which can be carried. Any material which falls through the nip can be returned, provided that care is taken to avoid contamination. Eventually the whole charge will have been fed to the mill, forming a continuous crêpe or blanket of gelled material around the working roll, with a more or less cylindrical 'rolling bank' of material in the entry region to the nip. Mixing across the face of the rolls is limited, so mixing is usually aided by cutting the crêpe periodically and allowing it to fold across the face of the working roll. This can be done by hand, using a mill knife or scraper, or by means of doctor blades fitted to the machine. Cross-mixing can be achieved mechanically by locating plough-like separator guides adjacent the roll carrying the compounding material,[47, 49] or by means of a 'stock-blending' unit.[47] After adequate compounding, which will usually correspond to a period of about 10–15 minutes from the time the whole batch has gelled, the nip is adjusted to give the required thickness of sheet, and the crêpe is removed. On small mills this is usually done by means of the working roll doctor blade, the crêpe being taken off manually as a sheet. On larger machines the same procedure may be used, but often small strip doctor blades are fitted by means of which the crêpe can be taken off as a strip, fed through a cooling water bath, dried, and granulated.

For compounding purposes continuous operation, in which mixed material is fed to the nip at one end of the mill and strip is taken off at the other, is not generally practicable on a single mill, because the dwell times are too short to achieve adequate homogenisation.

Owing to the rather low output rates and high labour usage of compounding on two-roll mills, they are now rarely employed for this purpose in production, but they are still commonly employed for 'sheeting' compounded material from internal mixers ready for granulation. For this purpose continuous operation is feasible, and once set up requires little labour. They are also extensively used in laboratories to examine the compounding behaviour of different components of PVC formulations, and for the preparation of specimens.

The effectiveness of a two-roll mill is clearly largely dependent on the interactions between the surfaces of the two rolls through the material in the nip, which will be dependent on the gap-setting, the surface speeds, and the viscoelastic behaviour of the material. The loads on the bearings, and hence the forces tending to separate the rolls, can be measured by replacing the standard adjusting screws by a modified form fitted with strain gauges,[17] and an apparent viscosity of processed material can be calculated from the bearing loads.[18] The process of compounding plasticised PVC on a two-roll mill has been studied,[19] and is often used to assess the behaviour of different polymers and other ingredients of PVC compositions in the laboratory. Some of the reported conclusions are surprising. For example, it has been stated that the higher the molecular weight of the polymer the more rapid is the disappearance of the particles and the homogenisation of the composition.[19] This proposition, however, was based on observations of milling experiments at a constant temperature of 150°C, since above this temperature the rate of particle disappearance was too high to be measured accurately, and below it the higher molecular weight polymers did not gel properly. It should be noted that comparisons carried out at the same temperatures need very careful interpretation. On the other hand results obtained with variation of temperatures to match the polymers may also require careful thought if any theoretical conclusions are to be drawn. Penetration of polymer articles by plasticiser is more rapid during milling than in static tests, which has been attributed to crushing of the particles by the rolls.[19]

10.4.2 INTERNAL MIXERS

10.4.2.1 Introduction

The internal mixer was originally designed by Banbury[20] for compounding rubber stocks, and machines of this type are frequently referred to as 'Banbury mixers' or 'Banburys', though this is not strictly correct as many other makes and designs are now available, and the name should be applied to the machines made by the limited number of manufacturers whose designs have descended directly from Banbury's original.

A high proportion of PVC is still compounded in internal mixers, but the use of continuous compounding machines and especially dry blend techniques has expanded considerably during the past twenty years. At the same time the operation of internal mixers and compounding plants as a whole has become more sophisticated with automatic and computer control.

10.4.2.2 Constructional features

In general arrangement[15] internal mixers resemble Z-blade or 'Sigma' mixers, but they are much more intensive in their action and of necessity much more robust in construction. The mixing chamber is a hollow regular cylinder with a cross-section shaped rather like a prone figure of eight. The upper V-shaped section is closed by a ram which can be raised and lowered in a hollow vertical shaft, usually by hydraulic means, thus affording a charging entry for the mixing chamber and a means of compressing the material being processed by the application of a load to the ram. The discharge exit can be in the form of a hinged door forming part of one of the sides, but is nowadays more often constituted by a hinged or sliding unit forming the inverted-V section at the bottom of the mixing chamber.

Within each circular section of the mixing chamber is a sturdy rotor or kneader, driven by a powerful motor and reduction gear. The rotors usually contra-rotate in similar fashion to the rolls of a two-roll mill, downwards into the centre region between the rotors. Different types of internal mixer differ mainly in the shapes of the rotors. They are usually of what may be called roughly convoluted shape, but sometimes they are more like smooth circular cylinders carrying lugs. Speeds of rotation vary from about 30–150 rev/min for small laboratory models to about 15–60 rev/min for large production machines.

The walls of the mixing chamber and the rotors are cored for circulation of heating and cooling media such as steam, water, or oil, though, as in fluid mixers, the amount of frictional heat developed during compounding is so great that additional heat is not essential, and a circulating heat-exchange medium is mainly used to increase rates of heating or to avoid over-heating where prolonged compounding cycles are necessary.

Usually the rotors are mounted in bearings at each end, but one type of machine uses the so-called 'overhung' arrangement in which only one bearing per rotor is employed, the two bearings being side by side at one end of the mixing chamber, leaving the other end free to form a discharge door hinged at the top.[15, 21]

Temperature is usually recorded by means of a thermocouple probe mounted in the inverted-V section at the bottom of the mixing chamber. Some idea of the power consumption can be obtained by observation of an ammeter in the motor circuit. For more precise investigation a dynamo-meter has to be incorporated.

Apart from general improvements in construction, notably in respect of bearings, and increases in rotor speeds, the main advance in internal mixers over the past few decades has been the provision of automatic control facilities, which permit not only a reduction in labour, but also a more uniform repetition from batch to batch.[22]

10.4.2.3 *Operation of internal mixers*

In production it must be very rare indeed that a single batch of a particular composition is compounded, and the aim is always to organise production schedules so that runs of consecutive batches of the same composition (including colour) are as long as possible. This is to maintain 'rhythm' which aids smooth running and maintains output at as high a level as possible, and to keep to a minimum the number of stoppages for changing and cleaning. Once a batch or two have been processed the internal mixer will usually have reached an equilibrium temperature, and externally applied heat is often then not necessary, but it is desirable at least to heat the mixer to around the equilibrium temperature before starting, in order to reduce the risk of variations in treatment between the first few batches and the majority that follow.

With the discharge door closed, the ram raised, and the rotors in motion a charge of previously mixed material ('premix') is fed to the mixing chamber through the vertical chute – the shaft in which the ram operates.

The size of the charge depends on the bulk and packing density of the premix, but should generally be large enough to maintain a positive resistance to the downward movement of the ram throughout the compounding cycle.

After charging, the ram is lowered to compact the material into the mixing chamber, and possibly raised and lowered a few times to ensure that all the premix enters the chamber.

One of the major problems with internal mixer compounding arises from the tendency of premix to stick to the sides of the chute and settle on top of the ram. On discharge such uncompounded material may fall into the compounded mass, and almost certainly can contribute to the formation of 'fish-eyes', 'nibs', etc., in processed material.

Once the ram is finally down the temperature of the batch will be seen to climb steadily. At around gelation point a more or less steep peak in power consumption may be seen to occur and the temperature will continue to rise. When the temperature reaches a point predetermined to yield the required degree of compounding, the ram is raised, the door opened, and the compounded mass discharged in a state varying from a quite soft continuous shapeless mass to an agglomerated crumb-like material, depending on the composition and the compounding cycle.

Where the reactivity of the plasticiser is relatively low, as for example with polyesters like PPA and PPS, it is often necessary to prolong the compounding cycle in order to achieve adequate homogenisation. In such cases external cooling will generally be required to prevent the temperature from rising so high that excessive decomposition occurs or the mass becomes too fluid and sticky.

The mass dropped from the internal mixer is usually fed to a two-roll mill where it is sheeted. Where the compound is to be calendered, small quantities are removed from the mill to feed to the calender as required. In the production of compound for extrusion and moulding, a strip is removed continuously from the mill, cooled by passing through a water-bath, dried, for example by 'air-knives', and granulated. A metal detector is usually interposed somewhere in the line to reduce the risk of contamination. From the mill onwards the process is virtually continuous. The equivalent to several batches of granulated compound may be thoroughly homogenised by tumble blending, though with the precision of control systems nowadays this should not normally be necessary.

The two-roll mill may be replaced by an 'extruder-slabber', which is essentially a large diameter (0.2–0.3 m) short length extruder, with a very large throated hopper which accepts a whole charge from an internal mixer and extrudes it in the normal way. Usually the slabber is fitted with a die producing a slit tube which is opened out between rollers to form a sheet as it is extruded, the sheet being cooled and granulated in the same manner as strip from a mill. Although this type of system seems attractive superficially, in practice control of gelation levels can be very difficult.

10.4.2.4   Behaviour of PVC in internal mixers

The course of events occurring in compounding PVC compositions in internal mixers has been studied in some detail. Much work has been done using the Brabender Plastograph or 'Plasti-Corder', and some has been carried out in laboratory internal mixers, which can be correlated reasonably well with production machines.[23–29, 44, 46, 51–55] The shear conditions arising during compounding in an internal mixer have been calculated.[29]

Compounding of plasticised compositions appears to follow a course in which a sequence of three fairly distinct phases can be distinguished,[29] as illustrated in Plate 9.2 (Section 9.3.2). During the first phase the temperature rises steadily owing to frictional work and heat conducted from the machine if it was previously hot. Plasticiser penetrates the polymer particles and swells them at an increasing rate as the temperature rises. Meanwhile the power consumption by the machine remains constant. The onset of the second phase arises from agglomeration due to the swollen polymer particles sticking together and is characterised by an increase in the rate of temperature rise. The latter is due to the fact that the agglomerating mass offers much more resistance to rotation of the kneaders; hence, there is a rapid rise in power consumption and frictional work up to a point where the agglomerates fuse rapidly to form a substantially homogeneous gel. This is followed by a return of the rate of temperature rise to a similar order to that of the first phase, and a steady decrease in power consumption as the temperature rises, corresponding, to all intents and purposes, to a normal viscosity/temperature relationship for the 'melt'. Within the melt there will generally remain the vestigial traces of a small proportion of incompletely plasticised polymer particles, and these disappear more or less rapidly as the compounding process is continued. Clearly the fewer undispersed polymer particles remaining after the power peak as passed the better, but in many applications using opaque compositions a few incompletely dispersed particles are of no importance, and then the important feature of the compounding cycle is the time taken to reach the peak, or time to gelation.

In many other applications, such as clear sheeting, however, incompletely dispersed particles cannot be tolerated, and the time taken to achieve complete dispersion becomes more important. In comparing a number of different polymers, ranking according to gelation time frequently does not correlate with a ranking according to time to achieve complete dispersion. Even if all particles of a sample to be compounded are identical, statistical considerations indicate that the particles will soften, agglomerate, and gel in a random fashion, so that by the time the bulk of the polymer has passed into a gelled, viscous fluid state, there will still be a proportion of particles in various stages of incomplete dispersion. It seems likely that the change in state from agglomerated mass to viscous fluid occurring around the power peak will result in slower dispersion of those particles that remain. How much worse must the situation be if the particles are not identical. Precise identity of size and shape even in a single batch of polymer is unlikely to be achieved, but it is doubtful if minor variations would be important if no other variations were present. Adventitious particles of higher molecular weight, different comonomer content, or completely different type are obviously likely to give trouble.

Amongst suspension polymers of constant molecular weight there appears to be a definite correlation between compounding cycle time and the packing densities and surface areas of the polymers,[29] a linear relationship between the former and compounding time having been demonstrated, i.e. increase in packing density resulted in a related increase in cycle time, the main effect being on the times to reach the power peak.[29] This is, perhaps, a little surprising. It certainly is important to establish the truth or otherwise of the relationship if a change from one polymer to another of higher packing density is being considered as a means of increasing batch size and hence production rate. The relationship between cycle time and surface area was non-linear, but reduction in surface area led to increased cycle times.[29] It should be pointed out that these observations were based on a particular composition (100 parts polymer: 33 parts DAP), and it seems highly likely that a critical factor is the wetness or stickiness of the mixture, and the type of polymer giving optimum performance may well vary with the nature and concentration of the additives.

As an instance of this it was sometimes found that in compounding vinyl/asbestos floor tile compositions, a non-porous particle polymer would lead to faster gelation than a porous particle suspension resin, possibly because its use leaves more plasticiser available to wet the filler and make it sticky. In comparing different types of polymer it will generally be found that porous particle suspension polymers yield shorter times to power peak than suspension polymers having more regular and smaller particle size, but there are considerable variations with different compositions.

For unplasticised compositions it might be imagined that the compounding process would be relatively simple in that no plasticiser is involved. As discussed in the previous chapter, the intensive shear conditions in an internal mixer tend to produce more comminution of polymer particles than occurs in extruder compounding. In the early stages of a compounding cycle additives are located on the surfaces of the 'grains', which latter remain discrete, become comminuted, and form loosely bound small agglomerates. The additives become more uniformly distributed, but the 'primary particle' structure tends to remain until the mass densifies and becomes more like a true 'melt'.[53, 54]

In addition to mixing and dispersion of ingredients the nature of the compounding cycle can have a profound effect on the rheological properties.[31–33, 55–57] In general there is a tendency for the apparent melt viscosity of any given composition to increase on compounding at relatively high temperatures. The effects that this may have in subsequent processing such as extrusion cannot be generalised, however, and all that can be said is that the effects on rheological behaviour of PVC compositions and the relationships of these to subsequent processing need careful attention when determining what compounding cycle should be used.

10.4.3 OTHER BATCH COMPOUNDING MACHINES

A number of compounding machines of rather different design to internal mixers have been introduced from time to time, but few find general acceptance for PVC, usually because the intensive shearing is too fierce for the thermal sensitivity of the polymer.

One possible exception[34] which appears to have received fairly wide acceptance in Germany comprises essentially a cylindrical mixing drum, in which a shaft carrying mixing arms rotates at a peripheral speed of 25–40 m/s, thus creating conditions of high shear between the arms and the walls of the mixing drum. Mixing cycles with this machine are short, and can be as low as 25 s, giving outputs of over 2040 kg/h.

10.4.4   CONTINUOUS COMPOUNDING

10.4.4.1   Introduction

Any process involved in the conversion of raw materials to finished product may contribute to the overall homogenisation of the composition. Thus an ordinary conventional extruder may add to the compounding previously received by granules fed to it, and ideally the separate compounding process used to produce the granules should stop at the point where the extruder is able comfortably to complete the required homogenisation. Ordinary conventional extruders, however, are not very effiecint compounding machines and will only cope with relatively simple homogenisation requirements, being generally inadequate to cope alone with the homogenisation of a PVC composition. Nevertheless, they have the advantage over internal mixers of operating in continuous fashion. Ordinary extruders have occasionally been used to compound PVC by passing the material through the machine several times, but this is clearly an inefficient procedure. Several continuous compounding machines have been developed, and most of these are essentially extruders with screws designed to ensure that adequate homogenisation will be achieved.

Continuous compounding of thermoplastics has been reviewed,[35, 36, 47] but special considerations arise in the compounding of PVC compositions.[15] The tendency of even the most stable PVC compositions to decompose on heating is sufficiently great to require that compounding temperatures should not be too high for too long a time, so that increasing frictional work to increase homogenisation has to be done with care, and stagnation of material for any length of time must be avoided if extended compounding runs are expected. For these reasons compounding machines which introduce very intensive shearing or which have complicated working parts, such as planetary gears,[35] are generally advisable for PVC.

In addition to the problem of decomposition it is sometimes found that conditions in a continuous compounding machine, designed to increase homogenisation, are such as to change the rheological properties of a PVC composition and thus its behaviour on subsequent processing, particularly extrusion. It is difficult to be specific on this point, because the requirements vary considerably from one situation to another, and it may be necessary to establish by trial whether or not a particular machine is appropriate to produce compound for a particular extrusion procedure. In many cases the most favourable rheological state of the compound is obtained by a combination of high shear and low temperature, which are seen to be to some extent mutually incompatible.

Feeding simple premixes direct to screw machines can be difficult, particularly with plasticised compositions, but even unplasticised simple blends can feed haphazardly, and it is often preferable to prepare true dry blends rather than attempt to process simple premixes direct. Some machines are fitted with 'force-feed' rams or screws to reduce the problem.

Automatic dosing and mixing units are available for attachment directly to the feed port of an extruder, so that the whole operation of mixing and compounding, and indeed also of extrusion to finished product, can be carried out continuously.[15, 58, 66] This possibility does of course require that machine design and processing conditions are adequate and matched to the particular composition to be processed and to the finished product.

For a fuller description and discussion of continuous compounding machines the reader is referred to the author's *Polymer Mixing Technology*.[47]

10.4.4.2 *Single-screw compounding machines*

The normal single-screw extruder is generally inadequate to homogenise any PVC dry blend in a single pass.

One of the best known continuous compounding machines (the 'Ko-Kneader' of Buss A.G.),[15, 37, 38] achieves the required additional homogenisation by slotting the thread of the screw (about three slots per revolution), fitting the barrel wall with corresponding 'kneading teeth' or lugs, and giving the screw a reciprocating motion during which the slots pass over the lugs, in addition to the normal rotary movement. This introduces more intensive shear than is obtained with a normal screw all the way down the barrel, so that the required additional homogenisation is obtained relatively gradually during the whole of the passage of material. Where the compounded material is to be used for subsequent extrusion or moulding the machine is usually fitted with a multi-rod die, and the extrudate is cut into granules of appropriate length by means of a die-face cutter. For calendering, a strip die can be used feeding direct to the calender, or over an intermediate conveyor. This type of compounding machine is available in a range of sizes from a small laboratory model to a large machine capable of producing as much as about 2200 kg/h of unplasticised PVC for extrusion, and 4000 kg/h of a typical plasticised PVC extrusion grade.

A compounding machine (the 'Plastificator' of Werner & Pfleiderer A.G.),[15, 39] which is particularly adapted to the production of compounded granules from plasticised PVC compositions, especially cable insulation and sheathing materials, achieves the required homogenisation by introducing a conical compounding zone in which intensive shearing is effected between the internal wall and fillets on the conical section of the screw. The clearance between the fillets and the wall can be adjusted to balance the shear conditions to the requirements of particular compositions. Once conditions have been established no external heat is applied to the compounding zone, the required temperature being maintained by the frictional heat developed. Originally it was thought that the high friction present in the compounding zone limited the size of machine possible to one having an output of around 180 kg/h, but developments in design have made it possible to introduce a model which is capable of producing up to between 900 and 1350 kg/h. With these machines the compound is produced as die-face cut granules, but this does not appear to be inherent in the working principle.

Another single-screw machine used for compounding PVC or extruding it directly from dry blend, particularly for cable covering, is essentially similar to a vented barrel extruder. Some way down the screw the flights are interrupted for several turns, and a breaker plate is fitted around the smooth section, offering resistance to forward flow of material, hence increasing shear mixing, and producing a decompression zone on the die side. The second section of the screw picks up the melt, mixes it further and conveys it to the die for direct extrusion or granulation. Machines of this type with outputs up to about 270 kg/h are available, although there seems to be no theoretical limit to the size to which such machines could be made.

Other single screw compounding extruders are available, but their main interest is for the direct extrusion of unplasticised PVC dry blends, and a discussion of such machines is deferred until consideration of this latter process.

10.4.4.3   Twin-screw compounding machines

In an extruder with two or more screws there exists the possibility of increasing homogenisation of material by interaction between the screws,[15, 47] and a number of twin-screw extruders has been used as compounding machines, particularly for unplasticised PVC. Indeed there has been a considerable development in twin-screw machines for the processing of unplasticised PVC, but the main interest has been in the direct extrusion of dry blend to finished product, which is considered in more detail in Chapter 11.

In twin-screw compounding machines, zones resembling two-roll mills or internal mixer kneaders are sometimes introduced to achieve the required extra mixing.[15, 47] In another rather versatile design, screw and kneader sections are mounted on parallel driven shafts and can be interchanged to suit the requirements of different materials.[15, 47] One of the problems with this machine, and with many corresponding machines, is that although they are very often capable of compounding unplasticised PVC, speeds have to be kept low because of the high pressures and temperatures which are otherwise likely to develop. Attempts to operate too fast can result in deterioration of the PVC and mechanical failure of the machine. This problem appears to have been overcome in some installations by limiting the restriction to flow at the head of the twin-screw compounding section, and transferring the compounded material to a more or less conventional single-screw extruder section which conveys the material to the granulating head.[42, 47, 59–61] Special designs of this type of machine have been developed for the production of compounded material for gramophone record pressing.

Mention should also be made of an approach 'from the opposite direction', adapting an internal mixer for continuous operation by extending the rotors to include a twin-screw extruder section.[43, 47]

A useful feature of some continuous compounding machines is that incorporation of some ingredients can be deferred and venting of volatiles can be effected by locating feed hoppers at appropriate points along the flow paths.

REFERENCES

 1. W.D. MOHR, Chapter 3 in *Processing of Thermoplastic Materials* (E.C. Bernhardt, ed.), Reinhold, New York (1959)
 2. P.V. DANCKWERTS, *Chem. Engng Sci.*, **2**, 1 (1953); *Research, Lond.*, **6**, 355 (1953)
 3. P.M.C. LACEY, *Trans. Instn chem. Engrs*, **21**, 53 (1943); *J. app. Chem., Lond.*, **4**, 257 (1954)
 4. W.D. MOHR, R.L. SAXTON and C.H. JEPSON, *Ind. Engng Chem. ind. Edn*, **49**, 1855 (1957)
 5. R.S. SPENCER and R.M. WILEY, *J. Colloid Sci.*, **6**, 133 (1951)
 6. K. STANGE, *Chemie-Ingr-Tech.*, **26**, 3, 150 (1954); **26**, 6, 331 (1954)
 7. J.F.E. ADAMS and A.G. BAKER, *Trans. Instn chem. Engrs*, **34**, 1, 91 (1956)
 8. R. SHINNAR and P. NAOR, *Chem. Engng Sci.*, **15**, 220 (1961)
 9. E.G. FISHER and E.D. CHARD, *Int. Plast. Engng.* **2**, 2, 54 (1962); **2**, 3, 113 (1962)
10. W.R. MOORE, *Trans. J. Plast. Inst.*, **32**, 247 (1964)
11. G.B. GROOM, unpublished work (1960)
12. D. BUSLIK, *Bull. Am. Soc. Test. Mater.* 66 (1950) 92T
13. A. SCHNEIDER, *Kunststoff-Rdsch.*, **7**, 10, 333 (1963)
14. P. PILZ, *Kunststoffe*, **47**, 64 (1957)

15. G.A.R. Matthews, 'Compound Techniques', Chapter 5 in *Advances in PVC Compounding and Processing*, (M. Kaufman, ed.) Maclaren, London (1962)
16. J.L. Foster, *Powder Mixing Techniques for Polyvinyl Chloride Resins*, B.F. Goodrich Chem. Coy (1953)
17. R.H. Carey, *S.P.E. Trans.*, **2**, 265 (1962)
18. F.D. Dexter and D.I. Marshall, *S.P.E. Jl*, **12**, 17 (1956)
19. B. Uriu and H. Yamamoto, *J. chem. Soc. Japan*, **63**, 3, 435 (1960)
20. D.H. Killefer, *F.H. Banbury the Master Mixer*, Palmerton, New York (1962)
21. Anon., *Plastics*, **26**, 282, 79 (1961)
22. W.H. Barclay, *Plast. Wld*, 24 (1962)
23. T. Shiramatsu and N. Ueda, *J. Soc. Rubb. Ind. Japan*, **31**, 97 (1958)
24. F. Koji and H. Daiman, *J. Soc. chem. Ind. Japan, Ind. Chem Sec.*, **64**, 11, 2053 (1961)
25. Omiccioli-Zenella, *Materie Plast..*, **29**, 6, 409 (1963)
26. N.W. Touchette. H.J. Seppala and J.R. Darby, *Plast. Technol.*, **10**, 7, 33 (1964)
27. H.S. Bergen and J.R. Darby, *Ind. Engng Chem. ind. Edn*, **43**, 10, 2404 (1951)
28. R. Hammond, *Trans. J. Plast. Inst.*, **26**, 49 (1958)
29. B.S. Dyer, *Trans. J. Plast. Inst.*, **27**, 84 (1959)
30. W.R. Bolen and R.E. Colwell, *S.P.E. Jl*, **14**, 24 (1958)
31. S.K. Khanna and W.F.O. Pollitt, *J. appl. Polym. Sci.*, **9**, 1767 (1965)
32. G.H. Burke and G.C. Portingell, *Br. Plast.*, **36**, 5, 254 (1963)
33. D. Dowrick, *Plastics*, **30**, 328, 63 (1965)
34. *Brit. Pat.* 954 366
35. D. Grant, *Rubb. Plast. Age*, **39**, 6, 481 (1958)
36. C. Prat, *Revue gén. Caoutch*, **39**, 10, 1561 (1962); **40**, 5, 737 (1963)
37. J. Aeschbach, *Kunststoffe*, **45**, 10, 456 (1955); *Poliplasti*, **7**. 33, 77 (1959)
38. H. List, *The Ko-Kneader in the Plastics Industry*, Buss, A. G.
39. A. Vogt, *Kunststoffe*, **44**, 151 (1954); *Brit. Pat.* 863 521
40. Anon. *Rubb. Plast. Age*, **37**, 11, 789 (1956)
41. Anon., *Int. Plast. Engng*, **2**, 5, 218 (1962)
42. K. Becker, *Kunststoffe*, **54**, 533 (1964)
43. *P.T. News*, 13 (May 1963)
44. J.M. Funt, 'Mixing of Rubbers', *RAPRA* (1977)
45. S. Middleman, *Fundamentals of Polymer Processing*, McGraw-Hill, New York (1977)
46. Z. Tadmor and C.G. Gogos, *Principles of Polymer Processing*, Wiley, New York (1979)
47. G. Matthews, *Polymer Mixing Technology*, Applied Science Publishers, London (1982)
48. F. Roesler and A.C. Shah, *Encyclopedia of PVC* (L.I. Nass, ed.), Marcel Dekker, New York, Volume 3 (1992) Chapter 1
49. United States Rubber Company, Brit. Patent No. 847 588; USP No. 2 976 565 (1960)
50. J. Brown, *Rubb. Plast. Age*, **37**(6), 400 (1956)
51. J.T. Bergen, *Processing of Thermoplastic Materials*, (E.C. Bernhardt, ed.), Van Nostrand Reinhold, New York (1959) Chapter 7
52. H. Ellwood, *Eur. Rubb. J.*, **159**(1/2), 17 (1977)
53. M.W. Allsopp, *Manufacture and Processing of PVC* (R.H. Burgess, ed.), Applied Science Publishers, London (1982) Chapter 8
54. M.W. Allsopp, 'PVC Processing II', *PRI Intl. Conf.*, Brighton, England (26–28 April 1983), Paper 4
55. R.C. Stephenson and M. Bottrill, *ibid.*, Paper 6
56. K.V. Gotham and M.J. Hitch, *Brit. Polym. J.*, **10**, 47 (1978)
57. P. Benjamin, 'PVC Processing', *PRI Intl. Conf.*, Royal Holloway College, Egham Hill, Surrey, England, (6–7 April 1978) Paper B5
58. M. Weber, 'PVC Processing II', *PRI Intl. Conf.*, Brighton, England (26–28 April 1983) Paper 33
59. F. Hensen and E. Gothmann, *Kunststoffe*, **64**, 343 (1974)
60. P. Rice and H. Adam, *Developments in PVC Production and Processing* (A. Whelan and J.L. Craft, eds), Applied Science Publishers, London (1977), Chapter 5
61. F. Hensen, 'PVC Processing', *PRI Intl. Conf.*, Royal Holloway College, Egham Hill, Surrey, England (6–7 April 1978) Paper B1
62. N.T. Flathers, R.E. Johnson, V.R. Pallas and W.M. Smith, *Mod. Plast.*, **38**, 210 (1961)
63. H. Marhenkel, *Plastics*, **30**, 334, 57 (1965)

64. D.R. JONES and J.C. HAWKES, *Trans. J. Plast. Inst.*, **35**, 120, 773 (1967)
65. A.J. BOULTON, *Developments in PVC Technology*, (J.H.L. Henson and A. Whelan, eds), Applied Science Publishers, London (1973) Chapter 6
66. REIFENHAUSER GmbH, 'Direct Extrusion with the Co-Rotating Twin Screw Extruder – Reitruder.RZE85 (1993)

11

Extrusion of PVC

11.1 INTRODUCTION

Plasticised and unplasticised forms of PVC are extruded in a great variety of different sections. Wire covering and unplasticised PVC extrusions such as window profiles, rainwater goods, and rigid pipe for conveyance of water, effluent, etc., account for around half of total PVC consumption.

The general principles and practice of extrusion are dealt with very adequately elsewhere,[1, 2, 34, 35, 61] and there is no point in covering the same ground here.

11.2 GENERAL CONSIDERATIONS

11.2.1 FEEDSTOCK

PVC feedstock for extrusion may be in the form of fully compounded granules, a blend of uncoloured ('natural') granules with highly pigmented granules (masterbatch), or a dry blend. It should be clear from the preceding chapters that the requirements in extruder design vary with the forms of feedstock as well as with formulations.

PVC compositions, particularly in compounded granule form, do not absorb sufficient water from the atmosphere to necessitate drying before extrusion, provided that reasonable care is taken. Water may well be condensed from the atmosphere of the extrusion shop if bags or bulk containers from a cold store are opened in the shop before the contents have been allowed to warm up to the ambient temperature. For this reason materials to be extruded should be brought into the extrusion shop sufficiently in advance of the time they are needed (something like 2 h). The higher surface area of dry blends makes moisture more of a problem with them and special care is necessary to keep the dry. Vacuum and heated hoppers should help, but do not yet seem to have attained widespread acceptance in the UK. A 'vented' extruder can be advantageous in removing small amounts of water, but does not remedy any deterioration in powder flow arising from dampness.[3] Small traces of moisture usually lead to porosity in the extrudate, but sometimes also produce severe surface roughness.

11.2.1.1 Rework
It is naturally desirable and often essential to be able to regranulate scrap material and re-extrudate it. Three main problems affect the possibility of this being done successfully,

191

namely (1) degradation of the polymer and possibly some of the additives during initial compounding and first pass extrusion, (2) effects of processing on the rheological behaviour of the composition,[3, 4] and (3) different particle shape, size, and size distribution of regranulated compared with original material. From the point of view of degradation alone, the possibility of re-using scrap depends on the stability of the composition and the severity of the extrusion conditions. A really well stabilised PVC composition, containing say 8 phr of tribasic lead sulphate, might be reprocessable alone (i.e. at 100 per cent), especially if the colour is a fairly dark or deep one which masks any slight yellowing. At the other extreme it might be inadvisable to reprocess at all an unplasticised composition stabilised by an inefficient non-toxic stabilising system. The effects of the extrusion on rheological behaviour and therefore on performance during re-extrusion are difficult to predict, and all that can be said is that it must be ensured that any regranulated material does have appropriate extrusion behaviour. Provided that stability is adequate there is no reason why scrap should not be recompounded alone or with additional raw materials, even into a different formulation, provided also that the total overall composition is correct. An advantage of recompounding rather than direct granulating of scrap is that the granule size and shape can then be made identical to those of the original compound.

11.2.2   MACHINE DESIGN

11.2.2.1   General

PVC can be extruded successfully using single- or multi-screw extruders and all kinds of sections can be obtained, though it has been claimed that twin-screw extruders have almost entirely superseded single-screw extruders for processing unplasticised PVC.[36] Whatever machine is selected particular attention must be paid to the design of the extruder itself, to its attachments (e.g. head and die), and to the way in which these components fit together, to ensure that there will be continuous flow of material at all points of the flow path, with none where stagnation of material can occur. These latter include sharp corners, inaccurately abutting surfaces of separate components, and chips, scratches, and abrasions in the surfaces of the metal constituting the boundaries of the flow path. All surfaces should be maintained in as smooth a state as possible.

Nitrided steel or high-boron cast iron are preferred as materials for barrels, screws, etc. Adhesion between PVC and the surfaces of the extrusion equipment is important,[5] but little work appears to have been done to measure adhesion and friction, or to apply any measurements to practical machine construction.

11.2.2.2   Screws

Screw design should ideally be matched to the requirements of the material and to the die, but even if the information is available to determine the optimum design in each case, it is clearly uneconomic to use the large number of screws required. It is rarely worth using a twin-screw machine for plasticised PVC, a single-screw machine almost always being adequate, whereas there are practical advantages in twin-screw machines generally for the extrusion of unplasticised PVC, particularly for dry blend extrusion (see Section 11.2.4.2). For single-screw machines, single-start screws are preferable, with constant pitch and compression achieved by increasing root diameter (i.e. decreasing depth of flight). Screws with decreasing flight depth over the whole length are often inadequate, but it is preferable to have a metering zone of constant depth over some three or four flights. The ideal compression ratio depends on the material to be processed, and the design of the die, and is dependent on the sometimes conflicting requirements of the amount of shearing a material requires for homogenisation and the amount it can withstand without developing excessive frictional work and heat, and hence degradation. Because of their lower melt viscosities under processing conditions, plasticised PVC compositions can be processed on screws having somewhat higher compression ratios than can be used for unplasticised PVC. For compounded granules of plasticised PVC compression ratios in the region of 2–2.5 are generally satisfactory, but higher compression ratios can often be used, provided that care is taken to avoid overheating, *L/D* ratios as low as 12:1 or 13:1 are usually adequate, though longer screws (e.g. 15:1 or 20:1) are generally preferred. For plasticised dry blends, longer screws with higher compression ratios are essential if comparable outputs and quality are to be achieved, *L/D* ratios should be in the region of 18:1 to 20:1 with compression ratios between 3:1 and 4:1. The ideal compression ratio falls as the plasticiser content is reduced, but this is of little practical significance until the change is made to unplasticised PVC, when lower compression ratios are generally necessary to avoid excessive frictional heating, and screws with non-compression (i.e. constant depth) can often be used successfully. To compensate for the lack of shearing long screws are preferable (e.g. 22:1). As with plasticised material, however, un-plasticised dry blends require higher compression ratios and longer screws than compounds in order to achieve adequate homogenisation (e.g. 2:1 to 3:1, and 25:1 respectively). In Europe the tendency has been to use twin-screw extruders in order to achieve the required homogenisation and also to ensure a constant positive forward flow of material. A twin-screw extruder is in effect a laterally extended gear-wheel pump and has therefore a positive take-up of the feedstock. Single-screw extruders are not as good as twin-screw at the feed inlet, but are superior at the die end because it is easier to design them and their associated head attachments free of dead spots where stagnation can occur. It is difficult to avoid these in the head of a twin-screw extruder. Twin-screw extruders with conical screws are now common.

Flight depth should not be too great. A couple of decades ago a depth of about 6 per cent of the screw diameter at the metering zone was recommended for unplasticised PVC, but as a result of subsequent developments in material and machine design much deeper screw channels are now common. Screw/barrel clearance should be in the region of 75 to 125 µm.

The tip of the screw should not be square, as this design leads to local stagnation; a hemispherical tip, or a conical tip with a curved end, should be used.

Removal of the screw from its drive mechanism should be possible without recourse to pulling at the delivery end, as any hole or other arrangement intended to facilitate pulling is likely to provide a region in which stagnation and degradation can occur. It is preferable for the machine to be designed so that access to the screw is possible at the rear, through the drive mechanism, and it should be possible to force the screw out by gentle tapping with a brass rod.

It is useful for the screw to be bored for internal cooling to increase shearing of the melt, but this should not be necessary if the machine and material have been properly matched. In any case, screw cooling should be very carefully controlled, as excess is likely to cause stagnation of material in the flights, leading to loss of output and degradation. Modern extruders are usually provided with screw-cooling facilities incorporating accurate control systems.

11.2.2.3   Sizes of machines

The main dimension used to define the size of an extruder is the screw or barrel diameter, though clearly the length is also important. Potential output obviously increases with size of machine, but the effects of variations in materials and dies are so great that generalisations about output are not much help, and can be positively misleading. Some idea of the variation in published figures can be gained from the list in Table 11.1 of minimum and maximum outputs taken from manufacturers' literature. These figures illustrate the difficulty of estimating the minimum size of machine necessary for a required output of any particular extruded product, and caution is obviously necessary when selecting a machine.

Rather surprisingly the power requirements for PVC extruders do not appear to be greatly different from those for polyethylene, as the figures in Table 11.2 indicate.[6]

Table 11.1

Size of extruder (Screw diameter in mm)	Output (kg/h)
25	2.3–25
38	6.8–68
64	25–136
89	63–181
114	86–238
152	227–544

Table 11.2

	g/W/h	W/kg/h
'Rigid' PVC	4.2–6.0	165.8–232.1
Plasticised PVC	6.0–7.8	132.6–165.8
Low density polyethylene	4.2–6.0	165.8–232.1
High density polyethylene	2.4–4.8	215.5–414.5
Polypropylene	3.0–6.0	165.8–331.6

11.2.2.4   Head and die units

Particular care is needed in the attachment of head and die units to PVC extruders to ensure that abutting surfaces mate well and that the surfaces of the flow channels are not interrupted by shoulders, gaps, etc. With unplasticised PVC in particular there is need for special attention to the strength of the attaching devices (e.g. bolts or clamps), to ensure that the high melt-pressures encountered do not distort the alignment of the various parts or even fracture the attachment devices.

Like all other flow channels the holes of the breaker plate should be streamlined by tapering at both ends. The cross-sectional area of the holes should total about a quarter to a third of the total cross-sectional area of the plate. It is common practice to support a single stainless steel gauze on the screw side of the breaker plate, a 40 BSS mesh (c. 400 µm aperture) being appropriate, the main functions of which are to add to the effect of the breaker plate in smoothing out flow of melt from the screw, and to act as a filter to prevent particulate contamination from passing through the die. If increased back pressure is desired, to increase the homogenising action of the screw, one or two finer gauzes (say 100–180 BSS mesh, i.e. 140–85 µm) may be supported on the coarser gauze. On no account should the finer gauzes be used alone as they will be liable to failure. Sandpacks are occasionally used in PVC extrusion but are not generally recommended. What has been said about the streamlining of breaker plate holes also applies to any other items interposed in the flow channels. Thus, if a mandrel or torpedo is mounted separately from the breaker plate, for example on 'spider legs', the surface of the mandrel an its supports must all be stream-lined.

After the final point at which the stream of melt is split into a number of streams (e.g. through a breaker plate or around spider legs), it is necessary to compact it in order to ensure homogenisation across the interfaces between the separate streams. If this is not done adequately flow lines will appear in the extrudate and these may be associated with weakness. It is generally desirable to achieve this compaction by progressive reduction in cross-sectional area up to the die entry,[5] something in the region of 5/1 to 7/1 being appropriate, though the value can be reduced as the die parallel land length is increased. The optimum die parallel land length varies with the material and section being extruded, but should generally be about 15 times the thickness.

Design of heads and dies to achieve the required flow of extrudate from the die lips, though still difficult, is now possible on rational and fundamental bases,[7–10] the main problem lying in adequate characterisation of the flow behaviour of the material to be extruded.

Prime requirements of die design and construction are good control of temperature and uniformity of heating of the melt. Control of heating has received a good deal of attention,[11–15] but barrel cooling facilities are also often useful.[16] Uniformity of heating is not so easy to achieve, but it can be aided by keeping the streams of melt as thin as possible and by fitting heaters in mandrels and any similar items of equipment. Examination of extrudates with complex sections, such as window profiles, demonstrates the remarkable precision now obtainable with good die design and controls.

11.2.3   OPERATION

11.2.3.1   General
The main general points to be observed in PVC extrusion are concerned with thermal stability and high melt-viscosity of unplasticised grades. Generally no attempt should be made to start extrusion until the machine has reached the required temperature. When ready the screw should be started at low speed and the material fed to the screw cautiously if the material is unplasticised. Some machines have controlled feed rate arrangements and these should be started at a low rate. Once melt is issuing from the die in a reasonable state the speeds of rotation and material feed can be increased steadily to the required levels.

It is sometimes desirable to stop extrusion while material is still in the machine. There is no particular problem in doing this as long as the formulation being processed is one of reasonable stability. If the machine is to be stopped for any length of time, the heaters should be switched off. On restarting, the material in the machine should be allowed to reach its processing temperature before starting the screw at a low speed, and increasing the speed to the required value as before. Where this procedure is employed it may be desirable to use so-called 'freezing compounds', which are highly stabilised and lubricated.

Temperature should be controlled accurately and attention should occasionally be given to the temperature indicators or recorders and to the state of the extrudate, to ensure that overheating is not occurring. If degradation is encountered it is best to cut off the feed, switch off or reduce the heaters, and run the machine as free of material as possible.

Temperatures for extrusion depend on the composition and form of the feedstock, the design of the screw, head, and die, the rate of extrusion and, perhaps not so obviously, the temperature recording and controlling equipment, in particular the location of the thermocouples in relation to the melt and the heaters. The actual temperature of the material at any point will not, in general, be the same as any temperature indicated by a nearby thermocouple, partly because the latter is in the metal of the barrel, not in the PVC itself, and partly because of the frictional heat developed within the PVC which will tend to raise its temperature. The data in Table 11.3 are therefore to be taken only as a very rough guide.

Table 11.3

		Feed	Metering zone	Head zone	Die
Unplasticised PVC		130°C	170°C	165–170°C	175°C
Plasticised PVC of BSS	10	140	170	175–180	185
	20	135	165	175	180
	30	130	160	170	175
	40	130	150	160	165
	50	125	145	155	160
	60	120	140	150	160
Filled PVC of BSS	40	115	150	155	175

Stripping of the machine for cleaning or die-changing is best carried out while the material is still fairly hot. With a little practice it is usually possible to pull PVC cleanly from the metal surfaces while it is in the partially softened state. Even the holes in breaker plates or spider supports can be cleared in this way, by pulling the PVC away from one side first, and then removing the plugs with the material on the other side. Little if any subsequent cleaning should be necessary. Indeed, a well stabilised filled grade of plasticised PVC can be a very good cleaning compound for the removal of other more sticky thermoplastics from the equipment.

11.2.3.2 Surface gloss

Inherent gloss or mattness of a composition is mainly determined by the presence or absence, the concentration, and the nature of filler, but even an inherently glossy compound can yield a dull or rough surface if extruded incorrectly. Obtaining maximum gloss can be difficult because in general it requires the surface to be heated to relatively high temperatures. This can sometimes be achieved by setting a high die temperature, but this may cause sticking at the die lips, and may also lead to excessive heating of the adjacent die mounting. Where the desired gloss cannot be achieved by die heating the extrudate can sometimes be heated as it leaves the die, preferably by means of appropriately arranged radiant heaters.

11.2.3.3 Transparent compositions

One of the problems with transparent materials is that any defects in the surfaces or the body of an extrudate show up very readily, and more care is usually necessary than with opaque compositions. This, of course, applies to some of the possible material defects as well as to defects in machine and procedure. Maximum clarity of an intrinsically clear composition will generally require thorough homogenisation and relatively high die temperatures. Tube extrusion often results in dull slightly haze products, because the mandrel or torpedo is at a lower temperature than the die surface forming the external wall of the extrudate. A heater fitted in the mandrel can often solve this problem, but great care needs to be exercised in heating it, for it is difficult to include means for cooling should overheating occur.

11.2.4   DRY BLEND EXTRUSION

11.2.4.1   Introduction
The literature of PVC dry blend extrusion has been reviewed,[17] and a fair amount of other useful information is available.[18–22, 36–39] Special attention has to be paid to formulation, dry blending conditions, extruder design, and extrusion conditions, all of which play important parts in determining whether or not satisfactory extrudate can be produced at acceptable rates. These four factors are to a large extent interdependent, and changes in one often require changes in the others for continuance of satisfactory performance.

11.2.4.2   Equipment and procedure for dry blend extrusion
The special requirements of dry blends as compared with compounded granules are for greater mixing to compensate for the lack of homogenisation on a molecular scale of the dry blending process, and for greater compaction to compensate for the lower bulk densities. Plasticised dry blends can usually be processed satisfactorily in suitable single-screw extruders, but twin-screw machines are generally preferred for unplasticised dry blends.[37, 38] It has however been claimed that a cascade arrangement of two single-screw machines has distinct advantages over twin-screw machines.[40] Vented extruders can be advantageous in reducing air pressure and in removing small amounts of moisture.

For plasticised dry blends a single-screw extruder with a fairly long screw (i.e. *L/D* ratio 18/1 and above) with gradual but high compression (3/1 to 4/1) is preferred. Relatively shallow flights give a greater chance of achieving homogenisation but give low outputs. Inadequate homogenisation is manifest by undispersed particles of polymer in the extrudate, appearing as bumps in the surface, by porosity, and by failure to realise the full potential tensile properties of the composition being extruded. Clearly, the earlier the dry blend fluxes and gels, the greater the chance of achieving adequate homogenisation, and a relatively steep temperature gradient from the feed zone is therefore indicated; but care must be taken to avoid gelation too near the hopper zone, interfering with feeding. There is also a risk that unstable early gelation will entrap air in the melt, and once this occurs the air is almost certain to be carried forward into the extrudate, since it is not in communication with the outside air. A number of things can be done to increase homogenisation. These include:

(1) Slight and very carefully controlled screw cooling with circulating water.
(2) Increasing the screen packs carried on the breaker plate, e.g. use of two 140 BSS mesh screens supported on a 40 BSS mesh.
(3) Use of a die with increased length of die land parallel.
(4) Increase of the back pressure by adjustment of the restrictor, if a 'valved' extruder is used.

It is also possible to obtain an improvement by turning a small amount, i.e. 25 to 100 μm, off the screw lands, though this is a rather drastic measure to take.

Extruders for unplasticised dry blends have been critically reviewed, and design features and process control have been described in some detail.[36, 38, 40]

For single-screw extruders the required extra homogenisation can be aided by having a relatively large clearance between screw and barrel in the melt section so as to give relatively high leakage flow in that region. Relatively high back-pressure from the die also helps, but caution is necessary in designing dies with this in mind as there is a danger of inducing excessive shearing with consequent overheating and degradation. Melting should take place in the compression zone which should be gradual and lead to a metering zone equal to about one-fifth of the length of the screw. Two-stage screws with a decompression zone appear to be advantageous, and are widely used in the US; but their performance is said to be very susceptible to changes in density and formulation of the dry blends, and feed screws and valved extruders are used to cope with this.[22] What amounts to a variation on the two-stage screw arrangement is a machine with a vertical extruder which delivers melt to a second extruder in the conventional horizontal position. The cascade extruder arrangement referred to previously[40] is based on the same principle. Such an arrangement permits separate but coordinated control of processing in two stages, so that delivery of 'melt' from the first stage can be more accurately matched to the optimum requirements of the second. In addition, degassing is relatively simple between the two stages. The principle is not unlike that of the Farrel 'MVX' integrated mixing, venting and extrusion machine, introduced in 1977, in which raw materials are metered into a continuous internal mixer which discharges into the feed zone of a single-screw extruder.

The preference for twin-screw machines is based on a number of considerations, namely that the high shear rates in the single-screw extruders are liable to break down the polymer structurally or chemically, and that interaction between the screws of a twin-screw machine offers the chance of more efficient mixing. Even conventional twin-screw extruders of the earlier types are often not capable of homogenising unplasticised PVC dry blends adequately at reasonable output rates, and machines have been specially developed for this purpose in which the required extra mixing is achieved by modifying the basic design in some way. Thus, leakage flow is often increased by providing some of the screw lands with flat sections, or by having exceptionally large clearance (e.g. 2.54 mm) between the screws in the metering zone. Other designs employ kneading zones, rather on the same principle as the compounding machines referred to in Section 10.4.4.3. Another arrangement is to use conical screws, or screws with conical sections at the feed end. The original and very important purpose of this device was to provide more room for bearings, which could thus be made more robust and permit transmission of more power to the material being processed. As with single-screw machines, units are also available in which a twin-screw extruder with a gelling function feeds melt to a second unit whose main function is mixing at high output.[22] The main disadvantage of such an arrangement is relatively high cost.

11.2.4.3   Conclusion

There can be little doubt that use of dry blend techniques will continue to expand, and the tendency may well be for obsolescent compounding equipment not to be replaced. Systems in which individual ingredients can be metered in at one point and final extruded product delivered at another[24] are usually essentially combinations of matched units performing in sequence the separate operations of metering, blending, compounding and extrusion. Reifenhauser have introduced a twin-screw extruder which is designed to perform all these operations, but it is not known if it would be satisfactory for all PVC compositions.

11.2.5   EXTRUSION FAULTS

11.2.5.1   Introduction
PVC is no more or less prone than other materials to extrusion faults that can arise from inherent poor quality material, incorrect machine design, incorrect procedure, or failure to balance the material and its processing treatment against the machine design. Failure to achieve the optimum level of gelation or fusion can produce a variety of faults as well as failure to obtain maximum mechanical properties. Factors affecting gelation and fusion are considered in Section 9.3.2. The more common faults encountered are discussed below.

11.2.5.2   'Nibbiness' or 'bittiness'
This is a marring of the extrudate surface by 'nibs', or small specks of material which often appear to be small lumps of polymer which have not been fully plasticised in the thermal or chemical sense. The fault can arise in normal material of good quality owing to inadequate homogenisation of the composition. It can also arise from contamination from harder materials or from over-heating of part of the composition. In the latter case the nibs are usually irregular in shape, whereas undispersed polymer particles generally have a circular or elliptical shape.

11.2.5.3   Porosity
Fine porosity in the body or on the surface of an extrudate can arise from decomposition of additives (Section 5.4.2), the presence of small amounts of water in the composition, or from entrapment of air in the melt beyond the tip of the screw. The solution of the first two problems is obvious. Porosity due to air entrapment is usually dependent on extrusion rate. Once satisfactory conditions for extrusion at low speeds have been established, it is usually desirable to run the machine as fast as possible. As the speed is increased a point is often reached where porosity appears in the extrudate, and this may be the only factor limiting output to this particular level. The onset of porosity is here associated with movement of the gelation region along the screw until a point is reached where the material leaving the tip of the screw is no longer in the fully gelled melt state. This type of fault is much more likely to occur with dry blends than with granulated compound, owing to their lower bulk density. For this reason dry blends are best based on polymers of relatively high bulk density, and unplasticised dry blends on polymers with non-porous particles, because air entrapped within the pores of porous particles can aggravate the problem. Some alleviation can sometimes be obtained by increasing back pressure on the material in the screw. Ideally this should be done by appropriate die design; but this is often uneconomic since it demands long die lands and consequently large dies.

11.2.5.4   Transverse weakness
This fault is rarely encountered if machine and conditions are anywhere nearly matched to the material. It arises from inadequate gelation and is mainly met when attempting to extrude dry blends in inadequate machinery, when it is invariably associated with porosity.

11.2.5.5   Longitudinal markings and weakness
Lines in the direction of flow of the extrudate are usually associated with incomplete homogenisation of the separate streams of flowing melt after passing through a breaker plate

or around spider legs, etc. This can be due to inadequate back pressure on the material imposed by restrictions to flow of the channels in the head and die, but can also arise from decomposition or plate-out of lubricant at the points of division of the flow path, resulting in an intermediate layer of poorly miscible material between the various streams. In gross cases actual weakness results along the lines.

Reduction in head and die temperatures, to increase melt viscosity and hence back pressure, can sometimes alleviate this problem, provided that this can be done without disturbing other features such as gloss.

Longitudinal lines can also result from defects in the die lips or from marring due to hard particles deposited on to die lips.

11.2.5.6   Plucking
This fault may appear at intervals along the line of extrusion, at spacings usually related to the rotation of the screw. It bears a superficial resemblance to stitching by a needle. The cause is usually overheating of material in the region of the screw tip arising from overheating of the latter, and is consequently more likely with unplasticised than with plasticised PVC. Ideally the cure lies in redesign of the screw tip, but internal cooling can sometimes help.

11.2.5.7   Lumpiness
This fault takes the form of lumps in the surfaces, randomly distributed in mild cases, but almost indistinguishable from ripple in severe cases. It is usually due to inadequate mixing in the screw, and in gross cases can even be associated with incompletely thermally plasticised individual granules. It is frequently accompanied by high porosity. Though the material properties are sometimes partially responsible for the onset of this fault, its main occurrence is due to insufficient back-pressure from the head and die, which is liable to arise with dies for the extrusion of thick sections. Increase in melt viscosity can help, and this may be achieved by changing the formulation or by decreasing the head and die temperatures.

11.2.5.8   Ripple
Ripple is a form of unstable flow encountered in extrusion of other thermoplastics, and referred to previously (Section 9.3.3). Its form is of a regular surface unevenness or undulation which if severe is almost indistinguishable from 'lumpiness'. Generally it can be attributed to overheating of the material at some stage, leading to changes in rheological behaviour. The remedy lies in extruding at relatively slow rates or modifying the nature of the feedstock.

11.2.5.9   Sharkskin
This, too, is a form of unstable flow encountered with other thermoplastics, and is generally attributed to a slipping-sticking alternation in the die, usually arising from too low die temperatures or inadequate lubrication of the composition.

11.2.5.10   Orange peel
The term 'orange peel' is applied to a surface dimpling which can result from breakdown

of the film of lubricant in the die due to too low a die temperature or incorrect formulation. It is much more likely to occur with unplasticised than with plasticised PVC.

11.2.5.11   Lack of gloss

This has already been dealt with (Section 11.2.3.2) and no further comment is called for here.

11.2.5.12   Plate-out

'Plate-out' has had considerable publicity as a troublesome phenomenon in calendering, but it is by no means unknown in extrusion. It appears to be due to separation of lubricant under conditions of shear, the lubricant carrying with it pigment, filler, stabiliser, etc., and depositing on adjacent surfaces, such as breaker plate, spider legs, screw tip, and die entry. As the deposit builds up it begins to be carried forward by the flowing melt, to form a defect in the extrudate or to deposit latter on the die lips, where it is again liable to mar the surface of the extrudate, either by interfering with the flow, or by being plucked away in the surface. To avoid this behaviour it is necessary to reduce both temperature and shear, which are rather difficult to accomplish concurrently without changing the design of the machine components or altering the formulation.

11.2.5.13   Ringing

This term applies to the fault seen as rings of thicker or thinner wall thickness in extruded pipe, but a similar fault can occur with other extruded sections. It is attributable to slipping in sizing and haul-off equipment, unsteady haul-off, or to screw pulsation. The latter may be due to mechanical defect, to uneven feeding of material to the screw, or to overheating.

11.3   SPECIFIC EXTRUSIONS

11.3.1   ROD AND SOLID PROFILES

A fair quantity of PVC, both unplasticised and plasticised, is extruded into solid sections. For unplasticised extrusion such as window frame profiles, contra-rotating twin-screw machines are preferred. The main problems arise in designing dies which will produce the required profiles, and in handling the extrudates so as to avoid deformation. The melt flow of PVC is such that flow through relatively restricted channels is disproportionately reduced, and thin parts of a section of unequal thickness require proportionately broader channels than thick sections. However, most of these problems are shared with other polymers, and are dealt with adequately elsewhere.[1, 2]

11.3.2   SHEET

Plasticised PVC sheeting and fairly thick film can be produced relatively easily by extrusion through a flat slit die, but commercial use of the process for this purpose is limited by wide establishment of calendering. Extrusion of unplasticised sheet, for use as such or for corrugation or subsequent vacuum-forming, is more difficult, but has nevertheless developed quite rapidly. The difficulties of obtaining uniform flow from a wide slit die without degradation are being overcome by careful design of die and formulation,[25, 26] although extrusion of a circular tube which is slit at the die offers advantages in design and running.[5]

After leaving the die the sheet is usually passed through highly polished steel cooling rolls, usually in a bank of three, and the sheet may be corrugated transversely or longitudinally by passing through appropriate corrugating rollers.

Really thin unplasticised film is extremely difficult to produce by flat film extrusion, and the thickness range rarely goes below about 0.5 mm, rising to about 6 mm at the thicker end.

There is a trend to extrude flat unplasticised sheet from dry blends, and this requires some attention to selection of formulation and extruder, as discussed in more detail in Section 11.2.4.

11.3.3   PIPE AND TUBING

Plasticised PVC tubing is extruded in a wide variety of sizes, from a wide range of different compositions, for a wide range of applications, but there are no major problems specifically associated with these operations. Support and transport of the extruded tubing during cooling can sometimes require attention, but it is usually sufficient to transport the extrudate on a fairly simple conveyor belt, and to maintain a very slight pressure of air inside the tubing by sealing the leading end and using a bored mandrel or torpedo connected to an accurately controlled source of low pressure air. Very often even this is not necessary, provided that the leading end of the extruded tubing is kept open so that there is constant access between outside atmosphere and the inside of the extrudate leaving the die.

Extrusion of unplasticised PVC pipe in many sizes for various different applications has expanded rapidly to become a major use of PVC, particularly in the larger sizes for conveyance of water and waste. This rapid expansion has coincided with, and indeed has stimulated, a parallel growth in extrusion of dry blends (Section 11.2.4), and much unplasticised PVC is now extruded or moulded from such starting material. Many of the applications for unplasticised PVC pipe are rather demanding in respect of dimensional tolerances, and special provision has to be made to meet these requirements.

Machinery, plant lay-out, and processing conditions for UPVC pipe extrusion have been described in some detail.[36-39, 47]

Figure 11.1 illustrates a typical arrangement for handling extruded unplasticised PVC pipe. The essential features are (1) a sizing die mounted close to the extruder die, (2) a cooling water bath, (3) a steady and firm gripping haul-off, and (4) a travelling saw. The sizing die consists of a smooth hollow cylinder whose internal dimensions define the external dimensions of the pipe, these being slightly larger than the die orifice, thus allowing a little

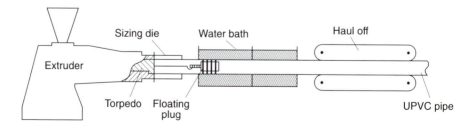

Figure 11.1. Arrangement for extrusion of unplasticised PVC pipe

die swell. The cylinder is surrounded by a circulating cooling water jacket. Close location of the external surface of the extruded pipe with the internal surface of the sizing die may be maintained by vacuum applied through holes in the inner surface of the die or by air pressure within the pipe. The former type is more expensive and vacuum may not be enough to maintain adequate contact between pipe and die in larger diameter sizes. It is, however, valuable for a small pipe, such as electrical conduit, or for complicated hollow profiles where the second type of sizing die can scarcely be operated. When using air pressure, to avoid sealing the end of the pipe and disturbing the pressure every time the pipe is cut, it is convenient to use a 'floating plug'. This is essentially a means of effectively sealing the air within the pipe until the latter has cooled sufficiently to retain its shape without support. A convenient form comprises a number of circular flexible rubber washers mounted coaxially on a rod which can be held to the mandrel by means of a 'hook-on' chain device. With such an arrangement it is relatively easy to maintain a controlled constant pressure within the pipe as it is being 'sized'. The pipe passes into and out of the cooling bath through flexible rubber seals so that water can be maintained around the pipe, and the bath may be divided into two or more zones by partitions fitted with similar seals, so that a controlled cooling sequence can be applied. The cooling water should be circulated through the bath, both to ensure uniformity and to prevent a steady rise in temperature as heat is absorbed from the extrudate. The haul-off should be one which exerts a positive grip on the pipe and should be free from backlash. Any slight departure from a steady pull will be transmitted back to the die and produce variations in dimensions. Caterpillar-type haul-off machines are preferable, two, three, or four belts being desirable, depending on the size of the pipe, but bevelled roller-type haul-offs can give quite satisfactory performance. Since unplasticised PVC pipe cannot be reeled it has to be cut into appropriate lengths, and this is best achieved without interrupting the steady movement of the pipe by using a circular saw mounted on a table which moves forward at the same rate as the extrudate, while a cut is made.

Production rates and productivity in the manufacture of unplasticised PVC pipe have been reviewed.[20]

11.3.4   TUBULAR EXTRUDED BLOWN FILM

The potential properties of tubular extrusion blown PVC film are of considerable interest, particularly for packaging applications, but in spite of the fact that the feasibility of its production was demonstrated well over thirty years ago, commercial production of blown PVC film has not developed to the same extent as other products such as 'rigid' pipe. It must be admitted that this process is one of the most if not the most difficult operation of the plastics industry, but there is no doubt that it can be carried out successfully if reasonable care is applied to selection and design of equipment and formulation, and to procedure. As usual, the main problems arise from the melt flow behaviour and the thermal instability of the polymer, and these necessitate particular attention to head and die design.

The film can be produced by horizontal extrusion with a horizontal bubble,[26] by vertical extrusion using a cross-head die,[27, 28] or by vertical extrusion using a vertical extruder. There can be little doubt, surely, that the last-named process should be best for PVC if not for other materials, having the advantages and not the particular disadvantages of the other two processes; but it has not been developed or used widely, the main difficulty apparently being in

achieving satisfactory feeding of material to the screw. It seems unbelievable that this problem cannot be solved by a little judicious design.

Horizontal extrusion and blowing (Figure 11.2) has the advantage that a straight-through
tube-type die can be used, thus avoiding the complication involved in conducting the melt
around a right-angle.[26] With this arrangement, however, it is quite difficult to produce film
of uniform thickness and free from wrinkles. This is due to the tendency of the bubble to
sag under gravity and for its upper surface to become hotter than the lower owing to con-
vection of hot air within the bubble.

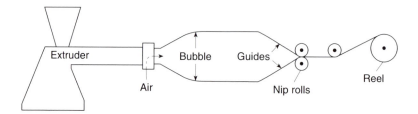

Figure 11.2. Arrangement for horizontal blown film extrusion

Vertical extrusion, usually upwards, using a cross-head die, involves the problem of converting the horizontal flow from the extruder to a vertical tubular flow. Many conventional
dies for ethylene are quite unsuitable for PVC, either because they provide regions for
stagnation or because the mushroom-type mandrels are not sufficiently rigid and strong to
withstand the stresses developed, particularly with unplasticised PVC.[27] It is not particularly
difficult, however, to design a cross-head die in which although the mandrel is solidly supported in the head structure the melt is nevertheless led smoothly through its right angle
bend and delivered uniformly to the die lips[27, 28] (Figure 11.3).

A wide range of PVC compositions can be blow-extruded for a wide range of applications, from clear unplasticised film as thin as 0.125 mm to translucent filled plasticised tubular sheeting from 0.15 to 0.31 mm thick for the manufacture of sacks. Highly plasticised
PVC compositions are difficult to blow-extrude because of the increased tendency for
blocking to occur between the two opposite layers as the lay-flat film passes through the nip
rollers.

11.3.5   WIRE COVERING AND RELATED EXTRUSION OPERATIONS

Electric cable insulation and sheathing accounts for a goodly proportion of PVC consumption. In 1968 the UK usage amounted to some 41 660 t, about 14.5 per cent of the total PVC
usage.[29] By 1975 the usage had risen to 46 000 t, 16 per cent of the total PVC usage.[48] More
recent figures are difficult to obtain, partly because of the extensive changes in company
structures and other commercial developments that have taken place in recent years. Quite
large quantities are also used for similar products such as covered wire for chain-link and
similar fencing, covered cords for stays and washing lines, and PVC-covered metal pipes.
The basic technique for producing all these items is similar, though obviously there are considerable differences in detailed design of dies and procedure. Design and manufacture of
PVC power cables has been described in some detail.[30, 49, 50]

Head and die design is in principle superficially like that for blown film extrusion, the

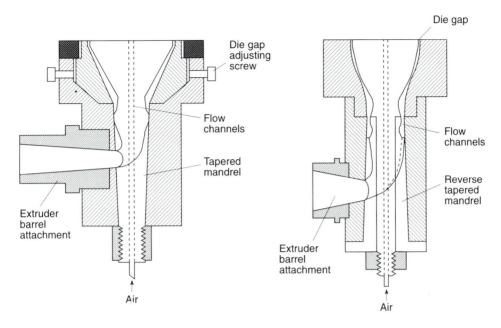

Figure 11.3 Cross-head blown film dies for PVC

article to be covered taking the place of the inflating air. Extrusion is usually horizontal with a crosshead at 90° to the screw axis, though where the item being covered permits the angle may be reduced to 45° or even 39°, making the achievement of uniform flow of melt around the mandrel or torpedo rather easier.

A typical installation for the application of PVC to a wire conductor (Figure 11.4) consists of a conventional single-screw extruder with cross-head and die, wire 'pay-off' and tensioning units feeding the wire to the cross-head, cooling-bath, capstan, tension-control unit, and reeling gear. Two types of die may be used, one in which wire issues from the mandrel within the melt (Figure 11.5(a)), thus ensuring close contact between the wire and its PVC coating, and the other in which the mandrel exit is adjacent to the die lips (Figure 11.5(b)) so that an extruded tube of PVC is drawn down on to or 'tubed on' to the wire. The latter technique is particularly useful where it is required that the PVC should be easily stripped from the wire, as when making connections.

It is often advantageous to interpose a preheater for the wire between the tensioning unit and the die, as this reduces the tendency for strains to be introduced into the PVC leading to deterioration in physical properties, particularly at low temperatures, a problem es-

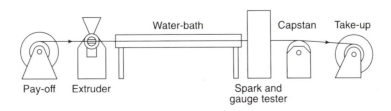

Figure 11.4. Typical installation for wire covering with PVC[1, 30]

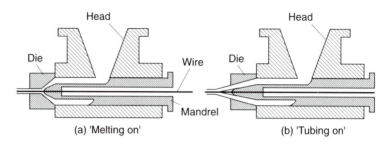

Figure 11.5. Wire covering dies

pecially likely to arise in the production of thin coverings, and when extruding 'high temperature' or non-migratory compositions (Sections 6.3.4.5 and 6.3.4.6). For best results the wire should be heated to around 150°C. Higher temperatures may result in excessive adhesion between the PVC and the metal, and may also cause drawing of the wire to occur. Wire pre-heating not only helps to avoid the introduction of strains where this is proving troublesome, but may also permit the use of higher speeds which would otherwise lead to the trouble.

With the ever-increasing speeds being used for wire-covering, particularly for thin-wall insulation, there is an increasing likelihood that friction developed in the metering zone and in the die will lead to overheating and consequent porosity, degradation, and change in rheological behaviour. It has therefore become necessary to pay particular attention to the formulation[30] (Section 5.4.2) and processing (Section 9.3.3) of compositions, and the design of equipment (Section 11.2.2).

Sheathing of single- or multi-core cables is carried out by what is essentially the same technique, though obviously head and die design differs in detail. The same may be said about the covering of cord with a PVC sheath or even for the sheathing of metal pipes. The latter are almost invariably too large and rigid for reeling and must therefore be fed to the cross-head one after the other to be sheathed continuously and separated later, after the sheathing has cooled. Where strong adhesion is required, the pipe may be primed with heat-activable adhesive just before it enters the cross-head.

11.3.6   RIGID FOAM EXTRUSION

A sizeable market now exists for expanded UPVC profiles with smooth skins, particularly in the construction industry. Three different processes are employed,[51-54, 62] all of which generally use chemical blowing agents in dry blend or compounded granules.

In the 'free foam' process extrusion is usually from a single-screw machine, and expansion occurs after leaving the die and before entering a vacuum shaping and sizing unit. In this process sodium bicarbonate, in combination with citric acid as an activator for the decomposition, is commonly used, because it tends to form dense surface layers on rapid cooling.

In the 'controlled foam' process expanding extrudate flows directly into a vacuum shaping unit, with no gap between this unit and the die. Twin-screw machines are normal for this process, but the choice of type of extruder seems to depend more on the size of the section to be extruded, the free foam process being used for production of relatively small section profiles, and the controlled expansion process for larger sections.

The third method for manufacturing foamed PVC profiles is by 'coextrusion' of two or more compositions from different extruders. This system can be used to produce multi-layered sheets including a central foamed layer, and profiles with foamed cores, with and without metal reinforcement.[54, 63]

The properties of extruded foam depend on the characteristics of the extruders used,[62] the formulation,[64, 65] and processing conditions.[63] Foam density tends to decrease with increasing temperature up to a 'critical' temperature, above which density increases with temperature.[66]

A typical co-extrusion arrangement includes a twin-screw extruder and a single-screw extruder, the blowing agent being a combination of azodicarbonamide and sodium bicarbonate.[63]

11.3.7   EXTRUSION BLOW MOULDING OF PVC

The consumption of PVC for blow-mouldings in the UK for 1968 was a mere 2800 tons.[29] By 1975 the consumption of PVC for blow-moulding bottles had risen to 12 000 t,[55] representing some 4.2 per cent of the total UK PVC consumption. Later figures are difficult to obtain but the proportion of PVC consumption going into bottles appears to have risen to about 8 per cent by the 1990s. The future is somewhat uncertain due to attacks on environmental and health grounds made against PVC in general and PVC bottles in particular.

The intrinsic attractions of unplasticised PVC for bottles compared with other polymers of similar cost are listed below:

(1) Potential high clarity if properly formulated and processed.
(2) High Young's modulus, permitting thin walls and light weight.
(3) Resistance to alcohols, oils, and greases.
(4) Low permeability to oxygen and other gases.

Apart from the production of some parts for automobiles there is not a great deal of interest in blow-moulding of plasticised PVC at the present time.

There are four different techniques for extrusion blowing of unplasticised PVC, namely:

(1) Conventional cross-head extrusion of a vertical parison.
(2) Horizontal extrusion and blowing in a continuous sequence.
(3) Vertical parison production from a vertical extruder.
(4) Conventional horizontal extrusion of parisons, which are cooled and blown in the vertical position in a separate machine.[32]

The main problem with the first (conventional) procedure is the same as that in any cross-head extrusion of unplasticised PVC (e.g. blown tubular film extrusion), namely, that of converting the horizontal plug flow into a uniform tubular flow in a direction at right angles to the original, without degradation. One way around this problem is to conduct the melt through a curved uninterrupted channel from the extruder to the rear end of the parison-forming mandrel.

The other three methods are all attempts to avoid this problem. The third method, using a vertical extruder, shares with the first the problem associated with all vertical parison extrusion-blowing, that the parison tends to thin at its upper end owing to the weight of the lower end. This problem can be overcome or reduced by using dies with movable mandrels, but care is necessary to ensure that the mechanisms employed do not offer points for stagnation and subsequent degradation of polymer.

The second method listed above appears to have much to commend it for PVC, though there appear to be some limitations on the minimum size of bottle that can be produced economically with machines designed on this principle.

The fourth method has the obvious advantage over the first and third listed, that it enables parisons with very uniform thickness and length to be made. Another possible advantage in some situations is that parisons can be produced at very high rates at one central location and then transported to blowing units adjacent to several different packaging stations. This would be cheaper than transporting already blown empty bottles. A disadvantage is that the material is processed through two separate heating and cooling cycles, an may therefore require more efficient stabilisation than the other procedures. After considerable development and some commercial success, this process seems to have been abandoned, largely on economic grounds, once problems with conventional blow-moulding systems had been overcome.[56]

Injection blow-moulding of PVC is also used to a limited extent.[26, 56, 57]

During the mid 1970s special techniques for bi-axial orientation of PVC blow-mouldings during production were introduced,[56-58] some of which are based on extrusion and others on injection systems. The orientation increases stiffness, tensile strength, impact strength, burst strength, gas and vapour resistance, and transparency, also enabling bottle weight to be reduced. This makes PVC bottles suitable for packaging lightly carbonated beverages and the like that normally require PET, heavy glass bottles or metal cans.

Good control of temperature of blowing moulds is particularly necessary with unplasticised PVC, and materials of relatively low thermal conductivity, such as aluminium alloys, are therefore inadvisable. Tool steel, highly polished and chromium plated if high gloss surface finish is required, is adequate. There should be no corrosion problems in the mould, so that expensive stainless steels are not necessary. A mould inner surface temperature of about 30°C will usually be found to give best results.

Formulation of PVC compositions for extrusion blow-moulding follow similar principles to those discussed for PVC generally (Chapters 5 and 7), though the effects of formulation variations in PVC for bottle blowing specifically have been examined.[34] The main problems in formulation are concerned with attaining adequate melt flow together with toughness, while retaining high clarity where required. For adequate flow it is usually necessary to employ polymers of relatively low molecular weight (e.g. ISO Viscosity Nos. 70–90), and thus impact modifiers are needed to boost toughness. Many of the latter tend to impart translucence, but acrylic compounds are now available which permit bottles of very high clarity to be produced.

Re-use of scrap follows a similar pattern to other PVC processing, but the high quality demands of many bottle applications often limit the proportion that can safely be worked away in virgin material, and no more than 20 per cent should be used in any clear composition.

Machinery, formulation, and procedures for blow-moulding PVC have been described in some detail.[56-59]

REFERENCES

1. E.G. FISHER, *Extrusion of Plastics*, Iliffe, London (1964) P.I. Monograph.
2. J.B. PATON, P.H. SQUIRES, W.H. DARNELL, F.M. CASH and J.F. CARLEY, Chapter 4 in *Processing of Thermoplastic Materials* (E.C. Bernhardt, ed.) Reinhold, New York (1959)
3. G.H. BURKE and G.C. PORTINGELL, *Br. Plast.,* **36**, 5, 254 (1963)
4. D. DOWRICK, *Plastics*, **30**, 328, 63 (1965)
5. A. KENNAWAY, 'Trends in PVC Extrusion', Chapter 6 in *Advances in PVC Compounding and Processing,* (M. Kaufman, ed.) Maclaren, London (1962)
6. H.R. SIMONDS (ed.), *The Encyclopedia of Plastics Equipment*, Reinhold, New York (1964)
7. D.J. WEEKS, *Br. Plast.*, **31**, 4, 156 (1958)
8. P.L. CLEGG, *Trans. J. Plast. Inst.*, **28**, 78 (1960)
9. J.R.A. PEARSON, *Trans. J. Plast. Inst.*, **30**, 230 (1962); **31**, 125 (1963); **32**, 239 (1964)
10. J.F. CARLEY, *S.P.E. Jl* **19**, 977 (1963)
11. R.F. TAYLOR and J.L. HOPLEY, *Br. Plast.*, **39**, 599 (1966)
12. A. ROMANOWSKI and G. VON BENNIGSEN, *Gummi Asbest Kunststoffe*, **18**, 5, 607 (1965)
13. R. BERGER, *Kunststoffe*, **55**, 6, 516 (1965)
14. D. GRANT and P. WILKINSON, *Plast. Polymers*, **36**, 124, 333 (1968)
15. D.E.A. BROOKS and P.R. LEVER, *Plast. Polymers*, **36**, 124, 343 (1968)
16. N.C. WHEELER, *Barrel Cooling Effectiveness in Decreasing Melt Temperature,* Davis-Standard
17. G.M. GALE, *RAPRA tech. Rev.*, **28** (1966)
18. N.T. FLATHERS, R.E. JOHNSON, V.R. PALLAS and W.M. SMITH, *Mod. Plast.*, **38**, 210 (1961)
19. G.M. GALE, *RAPRA Bull.*, **5**, 78, Sept./Oct. (1967)
20. T.J. DANIEL and D. MATTHEWS, *RAPRA Res. Rep.*, 163 (1967)
21. D.R. JONES and J.C. HAWKES, *Trans. J. Plast. Inst.*, **35**, 120, 773 (1967)
22. D.R. JONES, Lecture to Plastics Institute, Midland Section, 24 Jan. (1968)
23. H. MARHENKEL, *Plastics,* **30**, 334, 57 (August, 1965)
24. G.A.R. MATTHEWS, 'Compounding Techniques', Chapter 5 in *Advances in PVC Compounding and Processing*, (M. Kaufman, ed.) Maclaren, London (1962)
25. D.J. KIEL, *Int. Plast. Engng*, **1**, 8, 382 and **1**, 9, 438 (1961)
26. H. DOMINGHAUS, *Plastverarbeiter*, **14**, 12, 775 (1963)
27. Anon., *Plastics*, **26**, 287, 119 (1961)
28. A.E. PARKER, *Br. Plast.*, **32**, 456 (1959)
29. Anon., *Br. Plast,* **42**, 1, 58 (1969)
30. D.H. BOOTH, P.M. HOLLINGWORTH and W.H. LYTHGOE, *I.E.E. Conf. Rep.* Series No. 6, 1 (1963)
31. C.M. THOMAS, *Trans. J. Plast. Inst.*, **35**, 120, 793 (1967)
32. Anon., *Plastics*, **29**, 316, 81 (1964); **29**, 319, 97 (1964)
33. W.B. SISSON, *Plast. Polymers*, **36**, 125, 453 (1968)
34. Society of the Plastics Industry, *Plastics Engineering Handbook*, (J. Frados, ed.), Van Nostrand Reinhold (1976)
35. Z. TADMOR and C.G. GOGOS, *Principles of Polymer Processing*, Wiley, New York (1979)
36. U. SCHEIBLBRANDNER, 'PVC Processing II', *PRI Intl. Conf.*, Brighton, England (26–28 April 1983) Paper 8
37. J.W. HUMPHREYS, *Developments in PVC Technology*, (J.H.L. Henson and A. Whelan, eds), Applied Science Publishers, London (1973) Chapter 7
38. J.B. PRESS and D.A. TREBUCQ, *Developments in PVC Production and Processing*, (A. Whelan and J.L. Craft, eds), Applied Science Publishers, London (1977) Chapter 9
39. J.B. PRESS, 'PVC Processing', *PRI Intl. Conf.*, Royal Holloway College, Egham Hill, Surrey, England (6–7 April 1978) Paper B3
40. F. HENSEN, *ibid*, Paper B1
41. FARREL BRIDGE, *Production Development News* (1977)
42. Anon, Plast. Tech., **23**(11), 143 (1977)
43. D.M. SALDEN, 'Melt Compounding and Compounding Machinery', *1st Major Conf. on Thermoplastic Compounding*, PRI, London (1978)
44. T.A. MURRAY, *Plast. Tech.*, **24**(11), 83; **24**(12), 65 (1978)
45. R. WOOD, *Plast. Rubb. Intl.*, **4**(5), 207 (1979); **5**(1), 25 (1980)
46. G. MATTHEWS, 'Polymer Mixing Technology', Applied Science Publishers, London, (1982). 180

47. P. BENJAMIN, 'PVC Processing', *PRI Intl. Conf.*, Royal Holloway College, Egham Hill, Surrey, England (6–7 April 1978) Paper B5
48. D.A. TESTER, *Developments in PVC Production and Processing*, (A. Whelan and J.L. Craft, eds), Applied Science Publishers, London (1977) Chapter 1
49. G.P. BARNETT, *ibid*, Chapter 8
50. V.A.C. BURTON and T.J. CLARKE, 'PVC Processing', *PRI Intl. Conf.*, Royal Holloway College, Egham Hill, Surrey, England (6–7 April 1978) Paper B2
51. Produits Chimiques, Ugine Kuhlmann, *Brit. Pat.* Nos. 1 184 688 and 1 233 968
52. R.W. GOULD, *Developments in PVC Technology*, (J.H.L. Henson and A. Whelan, eds), Applied Science Publishers, London (1973) Chapter 9
53. K.T. COLLINGTON, *Developments in PVC Production and Processing*, (A. Whelan and J.L. Craft, eds), Applied Science Publishers, London (1977) Chapter 6
54. R. BROWN, 'PVC Processing II', *PRI Intl. Conf.*, Brighton, England (26–28 April 1983) Paper 18
55. D.A. TESTER, *Developments in PVC Production and Processing*, (A. Whelan and J.L. Craft, eds), Applied Science Publishers, London (1977) page 4
56. J. PICKERING, *ibid*, Chapter 7
57. J. PICKERING, 'PVC Processing', *PRI Intl. Conf.*, Royal Holloway College, Egham Hill, Surrey, England (6–7 April 1978) Paper A4
58. M. WEISS and R. WHITEHEAD 'PVC Processing II', *PRI Intl. Conf.*, Brighton, England (26–28 April 1983) Paper 25
59. J. BURRIDGE, *Developments in PVC Technology*, (J.H.L. Henson and A. Whelan, eds), Applied Science Publishers, London (1973) Chapter 8
60. S.V. PATEL and M. GILBERT, *Plast. Rubb. Proc. Appn.*, **8**, 215 (1987)
61. F. HENSEN, ed., *Plastics Extrusion Technology*, Hanser Publishers, Munich (1988)
62. P. KLENK and H.P. SCHNEIDER, *ibid*, pp. 471 to 488
63. N.L. THOMAS, R.P. EASTUP and T. ROBERTS, 'The Effect of Processing Conditions on the Density and Morphology of Rigid PVC Foam', *PVC '93 – The Future*, IOM Intl. Conf., Brighton, England, England (27–29 April 1993), 176–196
64. G. SZAMBORSKI and J.L. PFENNIG, *J. Vinyl Tech.*, **14**, 105 (1992)
65. K.U. KIM, T.S. PARK and B.C. KIM, *J. Polym. Eng.*, **7**, 1 (1986)
66. B.C. KIM, K.U. KIM and S.I HONG, *Polymer (Korea)*, **10**, 215 (1986)

12
Calendering PVC

12.1 INTRODUCTION

Calendering is the major process for the production of sheet and film from plasticised PVC, and the same is probably true of unplasticised PVC, in spite of the growth in the sheet extrusion process. There must be very much more PVC processed by calendering than any other material.

The process and the machinery were originally taken over from the rubber industry when PVC became available for non-electrical applications in the middle and late 1940s. On the whole, rubber calenders, being required to operate at temperatures never much above 100°C and incapable of attaining the temperatures required for PVC processing, proved inadequate, and much sheeting of poor quality was produced. As soon as the potential of plastics calendering became apparent machinery manufacturers began to develop machines specifically for the purpose, and the modern plastics calender is a highly sophisticated piece of machinery.

12.2 GENERAL PRINCIPLES

Strictly speaking, calendering is an extrusion process, in which moving roll-surfaces constitute both the material conveying means and the die. In its simplest form a calender is little more than a glorified mangle, and even in the most complicated machines the principle remains the same, namely the forcing of material through the gap or nip between two contra-rotating rolls which convey the material and form it to film or sheet by squeezing in the nip.

There is a tendency to think of calenders and calendering as embracing only the three- and more usually four-roll machines used to produce film or sheeting continuously from softened plastic material,[1] but the terms also include what is essentially the mangling action of two-roll machines used to compress and unify vinyl flooring, and the pressing cum polishing of already formed sheet, for example as a post-extrusion operation.

Except for the last-mentioned, the scale of operations of most plants is so large that the number of companies involved in calendering is relatively small and the techniques have become highly specialised. Consequently there has been a natural reticence in disclosing detailed information, and the number of technical papers on the subject has been very limited, having regard to its commercial importance. A few papers giving fairly detailed descriptions of the machinery and discussions of design problems have appeared,[1-7, 24-27] and there are also a few descriptions of procedure, usually very general and superficial.[8-14] A limited amount of work has also been published on the various aspects of fundamental behaviour of plastics materials during calendering[15-17] and of formulation/processing problems, e.g. 'plate-out'.[18]

The essence of all calendering is the production of film or sheet of uniform thickness and properties, and the main problems of calender design and operation are consequently concerned with control and distribution of temperature, control of the dimensions of the nips between the rolls, particularly the final pair, and haul-off.

12.3   PRODUCTION OF UNSUPPORTED FILM AND SHEETING

12.3.1   INTRODUCTION

Continuous flexible sheeting is produced by calendering plasticised PVC in the thickness range 0.04–0.8 mm and from unplasticised PVC up to about 0.18 mm. Thicker foil or sheeting of unplasticised PVC is too stiff to reel and is usually cut into appropriate lengths as produced. The thicker sheeting tends to have poor surface finish and is therefore either polished or produced by laminating thinner sheeting.

In order to ensure that the material is adequately and uniformly heated it is almost invariably the practice to feed calenders with pre-compounded and still 'plastic' material and to this end a calendering plant usually includes compounding equipment (Chapter 10). At one time, the most common installation comprised a simple mixing machine, such as a ribbon blender, an internal mixer, and one or two two-roll sheeting mills. Occasionally a two-roll mill is used for compounding with a second mill for sheeting, but this arrangement is quite impracticable for the highest calendering speeds in use today. Some plants use continuous compounding machinery, and this arrangement has the advantage that the compound production rate can be matched to the demands of the calender and present to it a continuous stream of softened material in the required state. This avoids the problem of variations of heat input into material being calendered, arising from the variable dwell-time of a charge in the feed nip. It is this problem which is responsible for the adage 'little and often' – charging material from the mill to the calender nip in small quantities at relatively high frequency, so as to minimise the difference in heat treatment received by material from a charge which is first and that which is last to pass through the nip.

In another common calendering installation a single-two-roll mill follows the internal mixer and discharges to an extruder-slabber feeding compounded material continuously to the calender nip.

In some small unsophisticated calendering plants material is fed manually from the sheeting mill to the calender, but automatic feeding by means of a conveyor belt or directly from a continuous compounding machine is more satisfactory and more common. Quite apart from its general convenience, cleanliness, and low-demand for labour, the latter method permits more simple permanent installation of the necessary metal detector before the calender nip, whereas with manual methods the detector can often be by-passed. The function of the detector is to prevent pieces of metal from entering the calender nips and damaging the rolls, and it should therefore be set as near the calender as possible, and should automatically stop the machine if metal is detected in the compound. It may be expensive to do this; but it could be very much more expensive to repair or replace damaged calender rolls.

Although there are technological advantages and disadvantages associated with each of the possible arrangements,[28] the reasons for the existence of a particular arrangement are often not obvious. A full-scale calendering plant represents a large capital investment, and the selection of the particular components and lay-out may involve historical and economic elements.

12.3.2   CALENDER CONSTRUCTION AND OPERATION

12.3.2.1   *General arrangement*

At least six different arrangements of rolls have been used for calenders, and with each of these, different paths of material through the machine are possible;[2] but the most popular arrangements today are the 'inverted-L' and 'inclined-Z' types (Figure 12.1), the latter having advantages which seem to be leading it to a position of pre-eminence for new calenders.

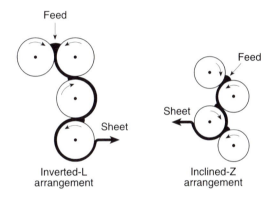

Figure 12.1. 'Inverted-L' and 'inclined-Z' calender bowl arrangements

In an inverted-L calender the central roll of the three in the vertical 'stack' is mounted in fixed bearings, while the bearings of the other rolls can be moved to adjust the nips. Thus the offset or 'breast' roll can be moved horizontally to adjust the nip between it and the top roll of the stack. This nip is usually (but not invariably) the point of feed. The top stack roll can be adjusted vertically to control the second nip, while the bottom roll can also be adjusted vertically to control the bottom nip. The forces developed between the nips in calendering are quite considerable and consequently adjustment of the top rolls of an inverted-L calender often interferes with the adjustment of the bottom roll, and vice versa.

This problem is avoided in the inclined-Z arrangement, which is one of the major advantages of calenders of this type, in which each nip can be adjusted independently of the others.

It may be noted here that the final roll is often given a 'frosted' surface to reduce sticking between layers of plasticised PVC when reeled. The 'frosting' can be produced by a form of sand-blasting using carborundum powder, by direct grinding with emery paper, or even by passing emery paper through the nip or calendering a PVC composition containing carborundum as filler.

12.3.2.2   *Thickness control*

Calendered sheeting is normally produced with a minimum of drawdown, generally not more than 10 per cent, so that the final nip of the calender must be close to the dimensions required in the sheet produced. The preceding nips will be set wider so that the material is squeezed gradually as it passes from the first (feed) nip to the last. Adjustment of nips an be a highly skilful operation. It does not involve merely setting the final nip to the gap necessary to deliver the required thickness; the other nips must be set so as to pass melt at the appropriate rate, thickness, and width, without building up large rolling banks of material on the feed sides of the nips, as the larger the banks are (1) the greater the chances of entrapping air in the melt, and (2) the less uniform will be the heat treatment of the mass passing through the nips.

On a modern production calender movement of the bearings is motorised and controlled by press-buttons on the control cabinet, or the whole operation may be regulated by a computerised control system.

Calender rolls are massive pieces of equipment in a large calender. Recommended width/diameter ratios range from 2.5:1 to 2.7:1 for plasticised PVC and 2:1 to 2.2:1 for unplasticised PVC.[1] Typical roll dimensions are 1.68 × 0.76 m and 2.3 × 0.91 m. Material for calender rolls is usually chilled cast iron or cast steel, the former being cheaper and less inclined to promote sticking of plasticised PVC, while the latter is stronger and has higher modulus and consequent resistance to deformation by bending forces. It also has the greater hardness, which is an advantage for the working surfaces and for the bearing shafts, where cast iron requires bearing bushes to be shrunk on.[1] Heat transfer is also better with cast steel than with chilled cast iron. Certainly for unplasticised PVC cast steel rolls are preferred.[1]

In spite of the massive construction of calender rolls they can still deform under the action of the forces developed between the nips, sufficiently to result in undesirable variation in thickness of sheet, particularly when producing thinner films. This deformation is more complex than might be imagined.[28] Three main devices have been introduced to overcome this problem. The least satisfactory of these, which however can be applied to an existing calender whereas the others cannot, is known as 'contouring'. This involves honing the surface of the delivery roll and possibly the other rolls so that the deformation during operation will produce a parallel nip. Roughly speaking this means that the roll will have a slightly greater diameter at the centre than at the edges, but the differences are so small that they can only be detected by accurate measurement. A disadvantage of this method is that a given contour can only match a limited range of materials and processing conditions.

A second method of compensating for the increased gap arising towards the centre of the nips is that known as 'cross-axes', in which the bearings of a roll are moved in opposite directions so as to incline the axis to that of the adjacent roll, thus effectively increasing the nip gap outwards from the centre.

A third method, commonly though inaccurately known as 'roll-bending', involves constructing the rolls with extensions outside the bearings and loading these extensions in such a way as to oppose and counterbalance the deforming forces in the nip.

In a modern calender line, mechanical sheet thickness gauges are often replaced by beta-ray scanning gauges. The information from these can be fed back electronically to adjust the nip settings automatically.

12.3.2.3   Temperature control

As already indicated, uniform heat treatment of material being calendered is essential in order to ensure uniform thickness, surface finish, and physical properties. Temperatures required for PVC calendering can be as high as 185°C, and even above 200°C, which places severe demands on the mechanics of the machines and on their lubrication, as well as on the performance of the heating medium. Heating is usually effected by circulating high pressure steam, high pressure hot water, or heat-exchange fluid through the rolls. Since it is usually necessary to have a positive temperature gradient passing through the machine, of the order of 2–8°C between each pair of rolls, it is necessary for the temperature of the heating medium for each roll to be controlled separately. With steam this is commonly achieved by means of independent pressure control valves, all fed from a common source of steam at higher pressure than the maximum required. The main purpose of the temperature differentials between the rolls is to ensure smooth transfer of the PVC from one roll to the next, and with some heavily filled compositions it is found that transfer takes place more readily from one roll to a cooler one, a negative temperature gradient being indicated in such cases. In all cases the actual level of the various temperatures needs to be adjusted so that material reaches the necessary plastic state in the final nip. Too low a level of temperatures will result in failure to attain the potential surface finish and possibly even mechanical properties, while too high a level of temperatures will result in sticking and possible 'plate-out'. Actual temperatures depend on the softening and melt flow, and hence on the formulation of the compositions, the thickness of film or sheeting to be produced, on the speed of calendering, and on the construction of the calender itself. As would be expected, other things being equal, optimum calendering temperatures decrease with increasing plasticiser content, increasing comonomer content in the polymer, and decreasing molecular weight in the polymer. It should be apparent that no precise guidance about optimum calender temperatures can be given, but the laboratory results given in Table 12.1 indicate the relative orders of temperatures for a limited range of plasticiser contents, and show the difference in behaviour between emulsion and suspension polymers of the same molecular weight.[19]

Temperatures generally need to be increased as thickness of sheet to be produced increases, the amount being anything from 3°C to over 20°C for an increase in thickness from 0.2 to 0.6 mm. With the advent of improved stabilisers calendering temperatures have tended to increase in order to achieve better fluxing of the material.

Increasing speeds of calendering, which can be as high as 135 m/min, clearly decrease the

Table 12.1 TYPICAL CALENDERING TEMPERATURES FOR A LIMITED RANGE OF PVC COMPOSITIONS

	A	*B*	*C*	*D*	*E*	*F*
Suspension homopolymer of ISO Viscosity No. 125	100	100	100	—	—	—
Emulsion homopolymer of ISO Viscosity No. 125	—	—	—	100	100	100
DOP	45	50	55	45	50	55
Breast roll temperature (°C)	162	158	156	150	149	148
Top roll temperature (°C)	166	162	159	153	151	150
Middle roll temperature (°C)	170	166	162	155	155	152
Bottom roll temperature (°C)	175	173	170	159	158	156

dwell-time of material on the rolls and in the nips, but at the same time increased frictional heat is developed and material temperatures may rise well above set-roll temperatures.

In modern calenders extra friction may be induced in the nips between the rolls by constructing them with friction ratios obtained by fixed gearing or by providing separate variable speed drives for each roll. This extra friction increases the amount of heat generated within the plastic mass and hence gives more uniform heating of the mass, but above a certain level it may necessitate roll cooling rather than heating.[1]

Obtaining sufficiently high temperatures for PVC processing is difficult enough, but an even greater difficulty is to ensure uniform temperatures across the faces of the calender rolls. Heating fluid has to be delivered to the rolls and removed from the rolls somewhere, and naturally it loses heat and drops in temperature as it passes. Even if circulating speeds are fast enough to render this heat loss negligible, heat transfer to the roll surfaces is not likely to be uniform, particularly with the internally bored arrangements of the earlier types of calender. This problem has been alleviated by drilling a number of relatively narrow channels for the heating fluid, passing as close as possible to and along the roll surfaces. Another source of temperature variation arises from heat loss to the bearings, which are usually 'flood oil' lubricated, the oil being cooled in an external heat-exchanger before re-circulation. The consequent heat loss can be compensated by locating radiant heaters at strategic positions towards the ends of the rolls.

12.3.2.4 *Formulation aspects*

Formulation of PVC compositions for calendering follows the broad principles described earlier (Chapters 5, 6, and 7), and although a good deal more than these broad principles is involved in production calendering, there is little that can be discussed in detail because the information is mainly held secret as part of the operating know-how of the companies which have been in production for many years.

The main problems in formulation are concerned with attaining the requisite degree of adhesion to each of the calender rolls, and with avoiding plate-out. It is necessary for the material to maintain close enough contact with the rolls to ensure efficient heat transfer and shearing, while transferring cleanly from one roll to the next. Adhesion to the final roll is particularly critical, since if the calendered material adheres too strongly haul-off forces will introduce longitudinal strains, whereas if adhesion is too slight the film or sheet will release too readily from the roll and will consequently crease or wrinkle. Optimum lubricant compositions vary with the nature and concentrations of the other ingredients of the composition, the characteristics of the calender, the thickness of film or sheeting being made, and the production rate. Production formulations are developed by trial and error based on past experience, and are so specialised that little if any really operatively useful information is published, and no further guidance can be given here.

Some indications of formulations for calendering are given in Section 8.9.4.

12.3.3   THICK SHEETING

As mentioned earlier thick calendered sheeting tends to have poor surface finish owing to inadequate and non-uniform heating, insufficient compaction, and air entrapment. For this reason, flexible sheeting is usually either produced slightly thicker than the required thickness and then polished, or produced by laminating an appropriate number of thinner sheets.

Both operations are commonly carried out in continuous laminating machines. In these machines the sheeting is heated while it is compressed between two continuous moving bands or between one continuous moving band and a large rotating drum.

Unplasticised thick sheeting is usually made by press-laminating and polishing in one operation between polished metal plates between the platens of a hydraulic press, in much the same way as PF, UF, and MF laminates are produced, except, of course, that unlike the latter process, cooling has to be applied before the pressed sheets can be removed and handled.

12.4   PRODUCTION OF SUPPORTED SHEETING OR COATED FABRICS

Coated fabrics, such as leathercloth, can be produced by feeding the fabric to be coated into the final nip, where is becomes laminated to the PVC sheeting. This is best accomplished by bringing the fabric to the final roll on the delivery side of the nip and allowing it to pass round with the roll.[2] In this way it becomes heated by the time it reaches the nip. The main problem is to maintain constant tension and to avoid wrinkles and creases in the fabric, but in other respects the procedure is similar to that for unsupported sheeting.

Cellular leathercloth and similar coated fabrics can also be produced by calendering, as well as by paste spreading (Section 14.4.8.1), using chemical blowing agents (Section 7.8). Particular care has to be taken in respect of compounding and calendering conditions in order to avoid premature decomposition of the blowing agent. The following are typical production procedures.[18] A premix is compounded from the following formulation:

Suspension polymer of ISO Viscosity No. 120–135	100 parts
DIOP or DOP	70 parts
Butyl benzyl phthalate	70 parts
Barium/zinc or potassium/zinc stabiliser/'kicker'	3 parts
Azodicarbonamide (blowing agent)	4 parts
Stearic acid	0.5 part

Such a formulation is suitable for lightweight cellular leathercloths, e.g. for clothing. For heavier materials as used for baggages, etc., the phthalates may be replaced completely by about 80 parts of TXP. The blowing agent and the stabiliser/kicker are best thoroughly dispersed in plasticiser by triple-roll milling before adding to the remaining plasticiser and the rest of the ingredients. The composition is compounded to 140°C, preferably in an internal mixer, sheeted on a two-roll mill at 145°C, and then fed to the calender. As usual in calendering, there should be a rising temperature gradient as the material passes through the machine, but the temperature should not exceed 145–150°C, or premature decomposition of the blowing agent will occur. Friction between rolls should not be used where the coating is applied, as excessive penetration into the fabric may result.

Complete gelation and expansion can be carried out as the coated fabric leaves the calender, or as a completely separate operation. Surface treatment at about 300°C for 0.5–1.5 min is used to form a very thin gelled surface layer, after which a temperature gradient down to 250°C expands and gels the bulk of the coating.

12.5 POST-CALENDER MACHINERY

Where the calendered sheeting is to be unembossed it is usually taken off the final or the penultimate roll into a train of highly polished cooling drums. Opinions vary as to the validity of chrome-plating these drums,[1] though they certainly can work satisfactorily. PTFE-coated rollers are said to be very satisfactory, and even asbestos paper has been used as a coating for take-off rolls.[1]

After cooling, plasticised sheet and thin unplasticised foil are wound up into reels, while thicker unplasticised sheet is cut into appropriate lengths, e.g. by guillotining.

Wrinkling and creasing of calendered film can be a problem, and specially designed 'wrinkle eliminators' can be incorporated into the calender line.[20]

Much plasticised sheeting, both supported and unsupported, is embossed. This can be done in line with the calendering process, but is often carried out as a completely separate operation. Most embossing machines have removable embossing rolls so that rolls bearing different patterns can be interchanged at will. It is important that the PVC should be in the correct state for embossing to be satisfactory, not too cold to receive the full pattern and not so hot as to stick excessively to the embossing roll; but this is largely a matter of trial and error, and experience (see Section 14.2.7.6).

12.6 THE 'LUVITHERM' PROCESS

The first 'Luvitherm' installation in the UK was in 1956, and it was therefore surprising to read[1] in 1964 that 'this processing method, known also as the sinter process, was employed almost exclusively up till some years ago'. It this was true of Germany, it was certainly not true of the UK.

In this process, usually confined to unplasticised PVC, the composition is agglomerated or gelled, generally in a continuous mixer or extruder-compounding machine, and is then only partially consolidated by passing through a calender at a temperature below that required to flux the composition to form homogeneous sheeting. There is some preference for L-type calenders because their use precludes the possibility of portions of material falling from the feed nip on to sheeting passing through the calender.

After leaving the calender, the sheeting is passed over a train of drums heated to well above the flow temperature of the composition, so that thorough, rapid, and brief fluxing results. Usually the homogenised sheeting is stretched longitudinally, and sometimes transversely as well, thus reducing the thickness and increasing tensile strength and other properties. Unstretched film can be produced in thicknesses ranging from 0.1 to 0.6 mm, and stretching can take the thickness down to 0.025 mm.

This process permits the use of polymers of relatively high molecular weight, and thus with somewhat better mechanical properties than those obtainable with unplasticised sheeting produced by conventional plastics calendering from polymers of relatively low molecular weight (ISO Viscosity No. about 85–90). Typically emulsion polymers having ISO Viscosity numbers in the region of 145–165 are used. Particular attention to stabilisation and lubrication is, as ever, necessary, but non-toxic compositions of adequate stability for the process can be formulated.

The high strength of the thinner stretched films makes them suitable for use as a base for the production of magnetic tapes. The thicker forms vacuum-form well, and are used to produce packages, particularly for fatty foodstuffs.

12.7 MELT-ROLL MACHINES

In about 1957 there was introduced a somewhat novel machine which is essentially a calender. Its main application is for coating thin layers of the thickness of about 0.025 to 0.1 mm on to substrates such as paper and fabrics, but it is claimed that machines of this type can produce satisfactory unsupported films. The novel features of these machines are that material is fed in granule form direct to melt-rolls, and that these melt-rolls are sufficiently hot to 'melt' the granules to form a small rolling bank.

A typical 'melt-roll' machine[21] consists of what is essentially a three-roll in-line calender, with the axes in the same horizontal plane. The 'melt' rolls are stainless steel rolls (hardened chrome-nickel steel), heated, usually electrically, to sufficiently high temperatures (maximum about 260°C) that the granules of plastics material fed into the nip are rapidly softened to a fairly fluid state. Thickness is controlled by varying the nip between the melt rolls by moving the first relative to the second, while maintaining parallelism between the two. Melt passing downwards through the nip follows the surface of the second roll to the nip between the latter and the third roll, which is covered with heat-resistant rubber. At the same time, substrate, preheated by passing around separated heated rolls (maximum about 300°C) is passed into the second nip against the surface of the rubber roll. The substrate and the plastics melt pass together through this nip and around the rubber roll, against which the plastics layer is embossed by an embossing roller, and the laminate is finally cooled and reeled.

The process is seen as an intermediate between calendering and extrusion melt-coating. Its advantages over the former lie in its low capital cost and the ease with which very thin coatings can be applied. Its advantage over extrusion melt-coating is the short dwell time, which makes it eminently suitable for thermally sensitive materials such as PVC, although coating rates are lower than those obtainable by extrusion of materials such as polyethylene where stability is not a major problem.

As the size of this type of machine is increased the plasticising efficiency of the melt rolls generally decreases, and with the larger production machines (e.g. about 1.25 m wide), compositions harder than a BS Softness of about 25 based on homopolymers cannot be processed successfully, and low molecular weight copolymers are required if unplasticised or lightly plasticised compositions are to be processed at all satisfactorily. Because of the short dwell-time the feed stock needs to be fully compounded granules rather than dry blends.

Table 12.2 shows typical roll temperatures for plasticised compounds of varying plasticiser content.

Substrates which can be coated with PVC using melt-roll machines include paper, cotton, and even hessian, depending on the application for which the laminate is intended. Applications include packaging laminates, which may be heat-sealable, bottle closures, interliners for clothing, and imitation leatherette for book-binding.

Table 12.2

BS Softness	Front roll	Rear roll
40	190°C	180°C
50	180°C	170°C
60	170°C	160°C

12.8   CALENDERED VINYL FLOORING

12.8.1   INTRODUCTION

There is a wide variety of forms of vinyl flooring, ranging from the soft and flexible continuous flooring and tiles to the hard and somewhat rigid vinyl tiles, and there is likewise a variety of processes and equipment used. While other processes, such as extrusion, compression moulding, and paste spreading, are used for certain types of flooring, the bulk is manufactured essentially by calendering processes of one kind or another. Even the thinnest forms are generally too thick for ordinary calendering to produce good quality surfaces. Considerable compaction of the surfaces is required if defects are to be avoided, and as thickness increases the more difficult does it become to impart this compaction in the nips of a normal plastics sheeting calender. Adequate compaction usually involves considerable compression, and this requires markedly increased speeds through successive nips. A moment's thought will reveal the fact that this is not possible in a multi-roll calender because each middle roll contributes to two nips. For these reasons special arrangements are employed.

The effects of variations in formulation on properties of vinyl flooring have been examined,[22] but assessment is often difficult, because the service conditions are often not reproducible in the laboratory.

12.8.2.   FLEXIBLE FLOORING

The required quality of surface finish of calendered flexible flooring is most commonly obtained by one or other of two main methods, either:

(1) separately producing on a conventional multi-roll plastics calender two or more layers, at least the top layer being sufficiently thin for it to be produced with an acceptable finish, and subsequently laminating the layers in a continuous laminator; or
(2) using a conventional multi-roll calender to produce sheeting rather thicker than required in the final product, and subsequently press-polishing in a continuous laminator.

In both cases pattern can be introduced either by printing or by adding to the melt in the feed nip small quantities of pigment or pigmented compound. The latter is, of course, only capable of yielding relatively limited and poorly controlled smeared, marble-like, decoration. In the first process, however, it is also possible to make the top layer from a clear composition and sandwich the pattern between it and the next lower layer, thus offering the possibility of using a wide variety of complicated printed patterns protected from wear by the top clear layer. The lamination procedure also permits a gradation in composition from the bottom to the top of the laminate. The bottom or base layer can be formulated from relatively cheap plasticisers and stabilisers, and contain very large proportions of cheap mineral fillers, while only the upper layers need incorporate the more expensive ingredients and have a relatively low filler content. A clear top layer naturally has to be formulated without filler and with relatively high quality stabilisers and plasticisers, but it is generally not thicker than 0.125 mm so that its contribution to cost is not out of proportion. Thus, referring to the formulation of Section 8.9.4.2, the bottom layer of the laminate might contain about 300 phr of a cheap mineral filler, while the uppermost filled layer might contain about 100 phr of a better quality filler (finer and whiter). Any intermediate layers might contain intermediate concentrations of filler. The number of layers in the laminate will be largely determined by the overall thickness, three or four being most common.

After laminating or press-polishing, both forms can be reeled up as continuous flooring, or cut into tiles.

Laminated flooring in which the PVC is supported by a fabric backing such as hessian can be produced in much the same way, the lamination usually being effected at the calender as in the production of supported films.

Some non-laminated flexible flooring, both continuous and tiles, is produced by very similar techniques to that used for vinyl tiles (Section 12.8.3). In this case the pattern is usually of the 'smeared pigment' or 'marbled' variety, achieved by adding pigments just before calendering. In some forms, however, compounded granules of various colours are dropped on to the sheeting before calendering, thus becoming embedded and producing a speckled effect. Over the last couple of decades there has been a considerable increase in the production of laminated vinyl flooring with one or more expanded layers providing a cushioning effect. These may include glass fibre to provide dimensional stability and indentation recovery, and the top layer may be polyurethane.[29]

12.8.3.   'RIGID' OR 'SEMI-RIGID' TILES

At one time these were mainly 'vinyl/asbestos' floor tiles, which are too stiff for reeling as continuous flooring, but the hazards associated with asbestos have led to attempts to replace it by other materials, such as talc, and to a decline in the market for these tiles in favour of alternatives, particularly the more sophisticated laminates. As seen earlier (Section 8.9.4.3), formulations for these tiles include very high proportions of fillers, and the actual 'vinyl chloride' content can be as low as 13–15 per cent of the total composition. Vinyl/asbestos tiles were developed from 'thermoplastic' or 'asphalt' tiles. The function of the asbestos has generally been regarded as increasing dimensional stability and heat-resistance, but there is some doubt as to whether there is any particular advantage in using asbestos when the concentration of other minerals is very high. In this kind of composition the polymer and plasticiser are seen as acting as a binder for the minerals, rather than the latter being fillers for a PVC composition. In fact, true 'melt flow' of the whole composition does not occur with these high filler contents, and it is 'wetting' of the filler particles by plasticised polymer that is required. It is in order to achieve this wetting that low molecular weight copolymers are commonly employed in vinyl tiles (e.g. ISO Viscosity No. 57, vinyl acetate content 15 per cent).

To obtain reasonable surface finish it is usual to compress sheeting by as much as tenfold. The composition is compounded in the usual way, in an internal mixer, a continuous compounding machine, or even on a two-roll mill. In the first two cases the compounded mass is then sheeted on a two-roll mill. After addition of pigment or pigmented compound to impart a pattern, the sheet is removed at up to as much as ten times the thickness required in the tiles, and is then fed through a vertical two-roll calender. In a simple plant this calender compresses the sheeting to the required final thickness in one pass, and it is cut into tiles by a punching or stamping machine after leaving the calender. It is, however, preferable to have a second and even a third two-roll calender after the first one, these assisting greatly in providing a good surface finish to the tiles.

As usual, the temperatures required for processing depend on the precise nature of the formulation and equipment being used, but it can be taken as a general rule that operating temperatures should decrease along the flow stream of the plant. Typical temperatures might be 150°C in the internal mixer, 130°C on the sheeting mill, 80–110°C at the first calender, and correspondingly lower temperatures at any subsequent calenders.

12.9   CALENDERING FAULTS

12.9.1   INTRODUCTION

A good deal of harm was done to the image of 'plastics' in the public eye in the first few years after the 1939–45 war, owing to the considerable amount of PVC sheeting which became available in various forms, particularly as 'plastic raincoats'. This inferior performance was due partly to formulation and partly to processing on rubber calenders which were incapable of giving the conditions, in particular the temperatures, necessary for the production of high quality PVC sheeting. The main shortcoming in formulation was concerned with plasticisers, which were often such that stiffening at low temperatures was excessive, or too volatile, so that embrittlement due to plasticiser loss occurred (*see* Section 6.3.4.4). With the plasticisers and other additives available, and the highly sophisticated plastics calenders used to produce sheeting, this kind of shortcoming should never arise nowadays. Nevertheless, formulation aimed at achieving a particular combination of desired properties is even now often of necessity a compromise, and considerable care and ingenuity is sometimes necessary if particular grades of sheeting are to meet all requirements specified by end-users. There is also the possibility, though it should be very rare, that calendering conditions might occasionally depart from the optimum, resulting in some deterioration of properties.

Particular attention must be paid to roll temperatures. To achieve the full potential of mechanical properties, these temperatures should be as high as possible without causing the PVC to stick too strongly or interfering with transfer of the material from each roll to the next.

12.9.2   VARIATIONS IN THICKNESS

Absolute uniformity of thickness is, of course, impossible to attain, but a modern calender is capable of producing sheeting to a tolerance of 0.0125 mm or better on a nominal thickness of 0.6 mm. In practice the variation attainable will depend on the design and construction of the particular calender, and on the composition being processed, the nominal thickness, and the rate of production. Variations from the best possible can arise from (1) incorrect setting of cross-axes and 'roll-bending' adjustments, (2) variations in temperature of the material charged to the calender, (3) fluctuations in calender roll temperatures, (4) fluctuations in calender roll speeds, and (5) fluctuations in haul-off speed.

12.9.3 INADEQUATE MECHANICAL PROPERTIES

Generally the most important mechanical properties of concern with calendered sheeting are tear strength, tensile strength, and elongation. Assuming that the nominal formulation has been properly designed, it is clear that formulation errors in preparing the compound could result in deviations from the required mechanical properties. (This is, of course, true of compounding for any kind of subsequent processing; but it can be more important but less obvious in calendering than with most other forming operations where the precise mechanical properties are often not critical.)

Apart from some applications where anisotropic distribution of mechanical properties is required (e.g. in tapes), it is usually desirable that they should be as near uniform as possible in both transverse and longitudinal directions. This objective will rarely, if ever, be achieved in practice, but any major deviations are likely to be due to running with too high a tension in the sheeting as it is taken off the calender. This can be an indirect result of insufficient lubrication in the formulation or to unduly high temperatures, both of which cause excessive adhesion to the calender rolls and hence require relatively high tension to remove the sheeting. In this sort of situation, not only will the mechanical properties be distributed non-uniformly, but the sheeting will also be under strain. This is liable to result in dimensional instability, raising problems particularly in subsequent sealing or welding.

12.9.4 'PLATE-OUT'

This phenomenon has been referred to in a number of previous places (Sections 7.2.1, 7.7.3, 11.2.5.12, 12.3.2.3, 12.3.2.4) and little need be added to what has already been stated. Its occurrence may be observed in the formation of deposits of pigment and/or filler, or even lubricant alone, on the rolls of the calender, greasiness of the surface of the calendered sheeting, or discoloration of the sheeting arising from contamination of the surface by pigment plated-out on to the rolls by a previous run of a material of different colour.

It is particularly prone to occur as calendering speeds are increased, with consequent increase in shear conditions and in melt temperatures. Its avoidance lies in somewhat delicate balancing of operating conditions and formulation, particularly with respect to lubricants and stabilisers.[18]

12.9.5 SURFACE DEFECTS

12.9.5.1 *Mattness*
In what should be glossy sheeting, general mattness or surface is usually due to calendering at insufficiently high temperatures.

12.9.5.2 *Grain and 'pimples'*
These are defects due to inadequate dispersion of ingredients of the composition being calendered, particularly polymer or pigment; or to contamination by non-dispersible materials;

or sometimes to variable adhesion of the sheeting arising from local hot spots on the calender rolls or uneven deposition of lubricant on the roll surfaces.

12.9.5.3   'Dimples'

These are small depressions in the surface of the sheeting, usually due to variable contraction resulting from partial release of strain induced during haul-off (*see* Section 12.9.7).

12.9.5.4   Rough or dull patches and spots

Areas of dullness on an otherwise glossy surface are usually the result of variations in temperature of material in the first nip of the calender. This can be due to feeding material at too low a temperature, or in too large doses which lead to chilling of the later portions passing through the nip. The dullness is frequently in the form of streaks, when it is commonly known as 'cold streak'.

When very small particles of relatively cold material pass through the nips, they frequently result in small surface markings resembling the shape of birds' footprints, and hence are commonly known as 'crow's feet'. Very similar markings are sometimes produced by spot sticking and consequent plucking of the surface at the final nip.

12.9.5.5   'Blisters' and 'windows'[23]

Both these defects are caused by entrapment of air in the 'rolling banks'. The nature of 'blisters' is self-evident. 'Windows' arise in the same way, but appear as surface depressions due to bursting of the very thin skin of material enclosing the air.

This entrapment of air usually results from inadequate fluxing of the rolling banks, and is most likely to be eliminated by raising temperatures and reducing the sizes of the banks.

Depressions somewhat like 'windows' can also result from particles of pigment, filler, or contamination, or even decomposed PVC, which are extracted or fall from the surface, leaving spaces behind them.

12.9.5.6   Greasy streaks

These can arise from irregular plate-out of lubricant or from contamination by lubricating oil.

12.9.5.7   Surface tackiness

Intrinsic tackiness or otherwise of sheeting is very dependent on formulation. It increases with plasticiser content and decreases with increase in filler content. There appears to be some variation between different plasticisers compared at concentrations designed to yield equal flexibility, but there has been little or no fundamental examination of this aspect of plasticiser behaviour.

Quite apart from the general mild nuisance of slight tackiness in some finished products, e.g. unholstery, it can lead to blocking of adjacent layers of sheeting, particularly in reel form where there may be some pressure increasing the tendency for blocking to occur. As indicated earlier (Section 12.3.3.2), calender rolls are frequently frosted, to avoid blocking of the sheeting, but this does result in slight translucence of otherwise transparent material. Some fillers whose refractive indices are close to those of the plasticised PVC can be incorporated to reduce tackiness without appreciably affecting clarity (Section 7.3.2.2).

Another somewhat messy and generally unsatisfactory procedure is to dust talc on to the surface of the sheeting before it is reeled.

12.9.6 'PIN-HOLES'

'Pin-holes' are small holes right through the sheeting, and can arise from a number of causes. One form is really a gross case of 'blisters' (Section 12.9.5.5), in which air entrapped within the melt forms a bubble which burst through both sides of the sheeting. It is unusual, but not unknown, for such bubbles to remain entrapped to form what would appear to be voids in the sheeting.

Holes in sheeting can also result from particles of pigment, filler, contaminating matter, or decomposed PVC which are plucked or fall out of the sheeting, leaving holes behind them. Another possible cause is plucking of small pieces from the material at hot spots on the rolls or in parts of the material which have been starved of lubricant as a result of inadequate dispersion.

12.9.7 CREASES AND WRINKLES

Creases arise mainly from folds in the longitudinal direction which occur as a result of faulty wind-up equipment or procedure, and which become set in the material during cooling. Careful selection of design or haul-off equipment is the best way of avoiding this trouble.[20]

Wrinkles arise from differential shrinkage as a result of partial strain (*see* Section 12.9.3).

REFERENCES

1. E. MIEBACH, *Int. Plast. Engng*, **4**, 174 (1964); **4**, 7, 215 (1964)
2. J. BROWN, *Plast. Prog., Lond.*, 167 (1951)
3. R.S. COLBORNE, *Br. Plast.*, **28**, 134 (1955)
4. K.J. GOOCH, *Br. Plast.*, **30**, 105 (1957); *Plast. Technol.*, **3**, 3, 187 (1957)
5. P. SIEGEL, *Kunststoffe*, **47**, 242 (1957)
6. W. HAZEL, *Kunststoffe*, **52**, 230 (1962)
7. G. ARDICHVILI, *Kunststoffe*, **54**, 520 (1964)
8. F.B. MAKIN, *Br. Plast.*, **28**, 12, 500 (1955)
9. E.C. BROWN, *Plast. Technol.*, **2**, 4, 226 (1956)
10. R.C. RUMBERGER, *S.P.E. Jl*, **16**, 1, 87 (1960)
11. R.A. MANSFIELD, *S.P.E. Jl*, **16**, 1, 89 (1960)
12. J.F. SALHOFER, *Kunstoff-Rdsch.*, **7**, 12, 571 (1960)
13. P. CHAIGNET, *Industrie Plast. mod.*, **14**, 1, 41 (1962)
14. G. DOST, *Plastverarbeiter*, **13**, 6, 268 (1962)
15. A.A. BERLIN, G.S. PETROV and V.F. PROSVIRKINA, *Zh. fiz. Khim.*, **32**, 2565 (1958)
16. W.E. WOLSTENHOLME and P.E. ROGGI, *Mod. Plast.*, **37**, 7, 131 (1960)
17. D.I. MARSHALL, Chapter 6 in *Processing of Thermoplastic Materials* (E.C. Bernhardt, ed.), Reinhold, New York (1959)
18. F.R. HANSEN and S.F. DENNIS, *Rubb. Plast. Age*, **76**, 5, 715 (1955)
19. *'Corvic' Vinyl Chloride Polymers and Copolymers*, I.C.I. Ltd. (1962)
20. W.M. GALLAHUE, *Mod. Plast.*, **42**, 2, 123 (1964)
21. Anon., *Plastics*, **24**, 260, 89 (1959)
22. A.W. MCKEE, *S.P.E. Jl*, **18**, 186 (1962)

23. M. FINSTON, *J. appl. Mech*, **18**, 12 (1951)
24. R.A. ELDEN and A.D. SWAN, *Calendering of Plastics*, Iliffe, London (1971)
25. R.A. ELDEN, *Developments in PVC Production and Processing – 1* (A. Whelan and J.L. Craft, eds), Applied Science Publishers, London (1977), Chapter 10
26. N. STACKHOUSE, 'PVC Processing', *PRI Intl. Conf.*, Royal Holloway College, Egham Hill, Surrey, England (6–7 April 1978) Paper C1
27. E. SALO, 'PVC Processing II', *PRI Intl. Conf.*, Brighton, England (26–28 April 1983) Paper 19
28. R.A. ELDEN and A.D. SWAN, *Plast. Engg.*, **32**(6), 50–63 (1976)
29. R. BLASS, 'PVC Processing II', *PRI Intl. Conf.*, Brighton, England, (26–28 April 1983) Paper 27

13
Moulding PVC

13.1 INTRODUCTION

A PVC moulding, like a moulding of any other thermoplastic material, requires to be cooled in its mould before it can be removed, consequently extending moulding cycles considerably if the mould is also used to heat the material to make it flow. For this reason compression moulding is not a widely used technique for PVC, although it is probably used more for PVC than other thermoplastics materials, because of a few rather important applications where other methods of production are not suitable.

Until the 1960s injection moulding of PVC was also somewhat limited. With unplasticised PVC this was largely due to the difficulty of thermally plasticising the material sufficiently for it to negotiate the relatively complicated flow paths involved without degrading too much. Increasing demand for injection moulded products in unplasticised PVC, such as pipe fittings, coupled with improvements in polymers, stabilisers, general formulation know-how, and developments in injection moulding machinery, resulted in a remarkable expansion in injection moulding of unplasticised PVC during the 1960s and 1970s.

Because of its relatively low softening and flow temperatures plasticised PVC has always generally been relatively easy to injection mould, provided that it was reasonably stabilised, but appropriate applications were limited until the late 1960s. The rapid expansion of the use of the material in footwear, for sandals, wellingtons, and shoe soles, all produced by injection moulding in one form or another, led to an expansion in the injection moulding of plasticised PVC comparable with, but less publicised than, that which took place with the unplasticised material.

13.2 COMPRESSION MOULDING

13.2.1 PRESS-POLISHING AND LAMINATING

Under the heading of compression moulding might be included conventional compression moulding, and also press-polishing and laminating of sheet material. The usual limitation of long cycle times in compression moulding thermoplastics materials can sometimes be reduced by arranging the heating and cooling sources to be as near the cavities as possible, for example by having the actual mould parts cored for heating by steam and cooling by water.

Press-polishing and laminating is extensively used to produce thick sheets from both unplasticised and plasticised PVC. The sheets to be laminated and/or polished are assembled

227

between highly polished metal plates and placed between the platens of a hydraulic press, possibly with several units stacked together to increase productivity. Apart from the cooling part of the cycle the process resembles superficially high pressure laminating with phenolic and amino resins.

13.2.2   PRODUCTION OF GRAMOPHONE RECORDS

13.2.2.1   Introduction

The major use of compression moulding of PVC at the present time is undoubtedly in the production of vinyl gramophone records, which in the UK accounted for 10 000 t of copolymer in 1968, and 17 000 t in 1975. Usage of PVC in gramophone discs has fallen in recent years, following the advent of compact discs with superior sound reproduction qualities. Compression rather than injection moulding is used because it is extremely difficult to produce by the latter process records of the usual dimensions sufficiently free from moulded-in strains, which make the records very prone to deformation. This is particularly the case with the larger (12 in) records, but although some of the smaller (7 in) records are injection moulded[1] the problem is by no means absent from their production. Equipment and procedure for the production of records have been described,[2, 3] the stress conditions encountered by a record during playing have been investigated,[4] and the influence of formulation on moulding and playing behaviour has been studied.[2, 5, 6]

13.2.2.2   Equipment and procedures

Vinyl compositions were first introduced as replacements for the heavily filled shellac 78 rev/min records in common usage up to that time, and were therefore required to process in the same equipment in similar moulding times. This, together with the property requirements for sound reproduction, were the main reasons for the selection of low molecular weight (ISO Viscosity No. *c.* 60) copolymers (e.g. 15–20 per cent vinyl acetate) for this application.[5]

Early procedure consisted in compounding the composition (Section 10.4) on a two-roll mill or in an internal mixer, sheeting the mass to around 3.2 mm thick, and cutting the sheet into rectangular biscuits of size appropriate to the size of record to be produced. To mould a record a biscuit is heated, usually by infra-red radiation, until sufficiently soft, rolled into a 'dolly', and placed in the open heated mould, usually with the labels already in position. The mould is closed, heating stopped, and cooling applied until the mould can be opened and the moulded record extracted safely. In moulding records in the tilt-head presses specially designed for the job, closing of the tilt-head actuates the pressure system, and at the same time switches the heating to cooling, so that during moulding the mould is actually cooling down. Opening the tilt-head cuts out the pressure and switches on the heating, so that ejection occurs while the press is actually heating up. Extremely high production rates are thus obtained. Typical pressing cycles are 40 s for a 12 in and 20 s for a 7 in record.[2] Slight flash is removed by rotating the record on a mandrel in association with a suitably located cutter. Apart from control of each pressing cycle, namely heating and cooling, and mould opening and closing, the whole procedure was originally manually operated. With one process operator working one or at best two presses, the labour content in the production is seen to be very high, and the main developments have been directed towards making the process more and more automatic.

A record mould comprises two stampers, one for each side of the record, one mounted on the lower and one on the upper platen of the press. The stampers are produced by a multi-stage electrodeposition sequence from wax masters cut during the original recording or from tapes.

Developments from this procedure have been mainly concerned with automating the pressing equipment, and in presenting material to the mould in a more suitable and more uniformly softened state. The first of these consisted of dispensing with the biscuits and using compounded material in granule form. This latter is produced by 'cube-cutting' sheeted material compounded in an internal mixer (Section 10.4.2.3), or by using a continuous compounding machine (Section 10.4.4) to extrude laces which are cut by a suitable granulator. The appropriate weight of granules is then heated in similar fashion to a biscuit and moulded in the usual way. More recently extruder-compounding machines specially designed for gramophone record production were introduced. These usually consist of a vertical single- or twin-screw extruder, accepting dry blend, and extruding rod vertically downwards. Screw rotation is automatically controlled so as to deliver thermally plasticised charges of the required size at the required frequency, and on the latest machines the charge can be delivered automatically to the mould.

13.2.2.3   *Formulation*

Performance requirements for gramophone record compositions are very demanding, for, in addition to the peculiar flow behaviour required accurately to fill a most difficult mould cavity, the record has to meet some stringent service demands, and these two factors are frequently opposed.

The choice of polymer is largely determined by the flow behaviour required and is limited to some extent by the desire to keep moulding cycles as short as possible. It is possible to produce satisfactory records from homopolymers of reasonably high molecular weight, and, indeed, in playing performance in terms of wear and fidelity such records can be superior to those currently produced; but the moulding cycles involved are unacceptedly long. It is largely for this last reason that low molecular weight copolymers are commonly employed.[5] Typical copolymers have ISO Viscosity Numbers between 50 and 75, and vinyl acetate contents between 10 and 20 per cent, the most common combination of values being around 58 and 15 per cent, respectively. Where granules or biscuits are being used, compounded in an internal mixer, ordinary suspension copolymers are generally used. They process satisfactorily and are the cheapest form. Where extruder-compounders are being employed, either to produce granulated compound or preplasticised charges, copolymers of higher bulk and packing density give faster outputs, and polymers having the non-porous type of particle are preferred.

Stabilisation is not difficult, provided that any stabilisers selected do not interfere with the lubrication of the composition; but selection of lubricants is critical. The lubricant system not only aids flow of the melt and assists release from the stampers (and, of course, other processing equipment) but also dramatically affects wear during playing. From these points of view there are lower limits below which behaviour is unsatisfactory. On the other hand, as the concentration of lubricant increases so also does the tendency for 'plate-out' or 'staining' to occur, depositing lubricant on the stamper surface, particularly in regions where there are relatively abrupt changes in configuration. The deposit builds up in successive pressings and prevents complete filling of the cavity, thus producing 'unfills' in the pressed records which become evident as 'clicks' or 'crackle', according to the precise distribution and severity of the plate-out. Because of the somewhat abrupt change in general configuration at the beginning and the end of a playing groove, there is a tendency for this problem to arise more commonly in those regions, and for this reason it is often known as 'end-line staining'. With many lubricants the concentration range in which satisfactory lubrication without plate-out can occur is very limited and, indeed, often non-existent, and achievement of lubricant balance is very difficult and largely determined by trial and error.

Most gramophone records are made black by the incorporation of 0.5–1.5 per cent of carbon black. Whenever there are large areas of plate-out this shows up rather clearly as dull greyish against the black background. In addition to colouring and aiding inspection, the black has some effect on record properties. Increase in carbon black content slightly increases wear life, presumably by a lubricating effect, but usually greatly increases background noise, possibly owing to the presence of agglomerates.

Static has long been regarded as a bugbear with gramophone records, and vinyl compositions are little if at all better than other materials in this respect. A few antistatic agents, usually quaternary ammonium salts,[7] are satisfactory in vinyl gramophone records. Although they generally discolour during process, they retain their antistatic behaviour, and the discoloration is masked by the black. Most antistatic agents, however, have some lubrication effect, and their introduction into record compositions does tend to aggravate the plate-out problem.

13.2.2.4   Faults in vinyl gramophone records

When one considers the dimensions of the sound track of a gramophone record,[8] and the loads imposed by the stylus on the modulated groove,[4] one should not be less than amazed at the incredible fidelity attainable. Thus there may be as many as 14 lines per millimetre,[8] and this includes the level regions separating the grooves. The stylus tip is usually of a mere 50 μm radius. In a 33⅓ rev/min record, a roughness corresponding to a granule size of as little as 30 μm will produce a continuous audible background noise.

Coupled with these aspects of dimensions, one should also consider the complicated sequence of recording, preparation of stampers, compounding, and pressing, and particularly the shortness of the pressing cycles employed, which are typically of the order of 40 s for a 12 in record and 20 s for a 7 in record.[2]

In the face of these considerations it may seem churlish even to mention the subject of faults in gramophone records!

Some possible faults will be evident from the discussion of processing and formulation problems in the two preceding sections, and there is no point in covering the same ground again. Much of the assessment of quality is highly subjective and individual, and it is extremely difficult to assess the quality of a particular record in numerical terms. Naturally, the quality of the sound reproduced by a record cannot be better than that of the original recording or that possible within the limitations of the playing equipment.

The most common audio defects are background noise in the form of 'crackle', 'hiss', and 'clicks'. In addition to the possible causes of these defects discussed in the preceding sections, they can also result from inadequate thermal plasticisation of the polymer, inadequate dispersion of carbon black and possibly other particulate ingredients, and contamination by extraneous materials such as other polymers and grit. Contamination can be effective both by its presence on the stamper surface preventing complete filling of the cavity, and by providing hard bumps in the playing groove walls.

Warping is often encountered. It is almost if not completely impossible to produce a vinyl gramophone record quite free from strain, and heating to quite moderate temperatures is liable to release strain and produce warping. Loading records unevenly can also cause warping. It is for these reasons that records should always be stored on edge in places where excessive warmth (e.g. against hot radiators) will not be met.

13.2.3   MOULDED CELLULAR PVC

13.2.3.1   *Introduction*
Two different methods are available for producing microcellular mouldings in plasticised PVC by what are essentially compression moulding techniques. One involves moulding of granulated compounds containing blowing agent,[9, 10] similar to those described earlier (Section 12.4), and the other involves gelling and moulding a plastisol containing blowing agent.[11]

13.2.3.2   *Compression moulding microcellular PVC from granulated compound*
This process is particularly applicable to the production of light-weight material for shoe soles.[9, 10] It produces flexible sheets with fine closed cell structure of densities around 0.35.

The moulding process employs a semi-positive mould. First the cavity is filled with granules, and the mould is then closed completely at a pressure of about 1540 MN/m^2. The mould is heated to 165°C and the pressure generated within the cavity is then allowed to raise the top force of the mould about 13 mm. After cooling to 30°C the mould can be opened and the microcellular sheet removed.

13.2.3.3   *Moulded microcellular PVC from plastisols*
Although this process involves plastisols (Chapter 14), the process is essentially a form of compression moulding and is conveniently described here.

In general a low viscosity paste is preferred, but this is not critical. A chemical blowing agent (Section 7.8) such as 1,1′-azobisisobutyronitrile is incorporated into the paste and thoroughly dispersed by triple-roll milling at least three times with a gap of about 0.1 mm. The concentration of blowing agent depends on the density required but is typically in the range 5–15 per cent of the total weight. Other suitable blowing agents include other azobisnitriles; benzenesulphonic acid hydrazide; *p,p′*-oxybis(benzenesulphonyl hydrazide); aminoguanidine dicarbonate; and monobenzenesulphonhydrazine.

The plastisol is then poured into a gastight mould which is placed in a press and heated to 160–175°C for 30 min under a pressure of not less than 15.5 MN/m^2. The mould is then cooled under pressure to 25°C or lower and opened. The gelled product is removed and expanded by heating to 80–100°C for 15–20 min in a water-bath or oven.

Direct blowing in one moulding operation is possible but requires very careful formulation and control. Larger cells are obtained than with the two-stage process.

13.3   INJECTION MOULDING

13.3.1   INTRODUCTION

Except for very small mouldings screw-preplasticising injection machines (particularly of the screw reciprocating type) are preferred to other types of injection machine for all materials, but this is particularly so for PVC. Indeed, for compositions containing less than about 40 phr of plasticiser injection moulding by ram machines is scarcely conceivable. This is due to the combination of low thermal conductivity, melt flow behaviour, and thermal instability.

The major use for injection moulding of plasticised PVC is in the production of footwear and footwear components, and although conventional screw-injection machines such as are used for unplasticised PVC are also used for plasticised materials, special injection machines have been developed with footwear applications mainly in mind.

13.3.2   INJECTION MOULDING PLASTICISED PVC

On the whole the lower softening points and melt viscosities of plasticised PVC make the injection moulding of such material a much easier proposition than for unplasticised PVC.

The machines employed fall broadly into two types, namely the conventional 'high pressure' reciprocating screw machine commonly employed for injection moulding generally,[12, 13, 22] and the 'low pressure' type in which injection is effected by screw rotation as in a conventional screw extruder.[12–14] The latter type was designed specifically to produce footwear such as sandals, shoes, and bootees, but there is no technical reason why it should not be used to mould other items from plasticised PVC of sufficient softness.

A 'low-pressure' screw injection machine typically resembles a short extruder ($L/D = c.$ 8/1–9/1) with a short-pitched smear-headed screw rotated at relatively high speeds (85–200 rev/min) to plasticise and inject simultaneously, and is held stationary between successive injections. The melt is at relatively low viscosity during injection, and pressures reach values only a little over 13.8 MN/m². Consequently moulds can be constructed of aluminium or light alloys, and can be held closed by relatively simple and cheap clamping mechanisms. This makes it possible to mount up to about a dozen moulds on a turntable which presents each one in turn to the extruder nozzle for injection and then removes it for successive cooling, opening, and closing, while other moulds are being filled. Machines of this kind can be used to produce solid mouldings such as sandals and unit soles in what are really moulds of conventional shape, to produce lined footwear such as bootees using moulds incorporating lasts which carry the linings, and to mould soles on to welts and uppers, also using moulds incorporating lasts. In moulding on to fabric sufficient bonding is usually obtained by penetration of the PVC melt into the weave, but with some materials (e.g. leather uppers) adhesives have to be applied before moulding. For leather several isocyanate adhesives are available. Where the material to be bonded is a PVC laminate, treatment with a solvent such as cyclohexanone is adequate.

'High pressure' reciprocating screw injection machines for plasticised PVC are of conventional type. They are not generally used for moulding of PVC sandals or lined footwear, but are used for moulding of unit soles or direct moulding of soles on to welts and uppers.

The general design and procedure requirements for processing plasticised PVC (Section

9.3), particularly those relating to degradation, are applicable to its injection moulding, and no special comments are necessary. Both types of machine can produce satisfactory mouldings from self-coloured compounded granules, natural granules blended with colour masterbatch, and even from dry blends. However, dry blends can be troublesome in conventional machines,[23] and special attention to machine design is necessary if they are to be used for injection moulding.

13.3.3   INJECTION MOULDING UNPLASTICISED PVC

There can be no doubt that at the present stage of development reciprocating screw-preplasticising machines are the most satisfactory for injection moulding of unplasticised PVC. Single- and twin-screw machines are available but there are difficulties in designing twin-screw machines for PVC. Tapered screws are most efficient but are not suitable for reciprocating screw-type injection units, while the cross-sectional areas of parallel screws are so large that effective injection pressure may be reduced unacceptably.[23] The principles of screw injection machines are by now too well known to merit detailed description in a book dealing with one class of material only. Construction, control systems, and procedures with single-screw injection machines have been fully described in the literature.[15–18, 21, 23–25] This type of machine succeeds with unplasticised PVC where others fail by virtue of (1) the more uniform heating of the melt resulting from frictional working of the screw, (2) the relatively short dwell times (usually not more than a few minutes), and (3) the avoidance of stagnation. These factors permit operation at what are relatively high temperatures for PVC (e.g. 190–210°C), so that melt viscosities are relatively low and thus require relatively low injection pressures. The good mixing action of the screw permits moulding of PVC in any of the possible forms, such as granulated compound or dry blend, but the latter require careful and detailed analysis of material and machine parameters for consistently good performance.[25]

A typical screw has an *L/D* ratio of 17/1 and a compression ratio of 2:1–3:1 achieved by decreasing flight depth. Preferably, the tip of the screw should be tapered and shaped to fit fairly closely in the head flow channel (e.g. 2 mm gap), which should be correspondingly conical in shape. The angle of the taper of the cone should be about 30°.

Set temperatures on the cylinder or barrel are normally from about 130°C at the feed end to 175°C at the discharge end of the cylinder, with the nozzle temperature at about 180°C.

Screw speeds are generally slower than those used for other materials, but overall cycle times are not increased because the plasticising time is generally shorter than the time required for cooling. With smaller machines, up to 0.85 kg shot size, speeds of 30–45 rev/min are suitable, and for larger machines speeds should be lower, e.g. 15–30 rev/min. Back pressure on the screw during plasticisation is preferably between about 0.35 and 0.7 MN/m², in order to pack and compress the material in the screw and prevent the passage of air down the screw and into the melt. Injection pressures need to be fairly high compared with other materials. A pressure of no more than 96.5 MN/N² will generally be adequate, but pressures as high as 140 MN/N² may be desirable for moulding shapes with relatively thin sections. Injection speed should be as high as is possible without producing decomposition or plateout due to excessive frictional working. Shot size should be adjusted to the minimum compatible with full moulding so as to keep the clearance between the tip of the screw and the head to a minimum at the end of the injection stroke.

A high proportion of mouldings in unplasticised PVC are relatively large. They are often required to meet quite demanding mechanical criteria, and are of such a shape (e.g. T- and cross-shaped pipe fittings) that the mould cavities are complex, with retractable cores. These factors tend to place restrictions on the size of components that can be moulded. A number of variations on 'conventional' injection moulding are designed to overcome or reduce these problems. In one of these the screw rotates but does not move axially during the earlier stages of mould filling, and then moves to complete the injection in the normal way. In another system the screw rotates at the same time as it injects material into the mould. Both these systems are deficient in melt viscosity control.[26] The 'flow moulding' process also offers the possibility of increasing shot size, but with improved control of viscosity. This process is essentially one of 'intrusion', injection being achieved by a rotating extruder screw which does not move axially, and rotation of the screw being interrupted during cooling and ejection of the moulding. This is made possible by adjustment of the axial gap between the screw tip and the barrel, such that the shear energy applied to the melt eliminates variations in melt viscosity of material from the transport zone of the screw, and the melt viscosity is reduced to its optimum condition for injection. Material in the transport zone of the screw is maintained at relatively low temperature, the temperature being raised and viscosity reduced very rapidly at the shear gap, resulting in short residence times.[26] The process requires well controlled and effective temperature control systems.

As with any PVC processing equipment, moulds should be designed without any sharp corners or pockets where stagnation can occur. Depending on the particular resin and formulation used, the melt flow behaviour of unplasticised PVC under permissible processing conditions limits the thinness of mouldings that can be produced, and a thickness of 1 mm is about the thinnest that can be achieved. However, at this figure the length of flow-path is limited to a few inches, and 2 mm is generally a more reasonable minimum thickness of section at which to aim. Thick sections up to 16 mm can be satisfactorily moulded without particular difficulty, preferably by 'intrusion', that is direct injection into the mould cavity. Where runners are employed, as in multi-cavity moulds, they should be relatively short and of large cross-section in order to keep pressure-drop between nozzle and cavities to a minimum, 8 mm being about the minimum acceptable diameter. A cold slug well is advantageous. Gates should not be too narrow, the minimum possible diameter varying from 1.6 to 2 mm for small mouldings up to about 16 mm for a large moulding such as a pipe fitting weighing 1.8 kg.

From other points of view mould construction follows similar principles to those for other thermoplastics materials. The cavities are preferably hard chromium plated and highly polished. Venting can be valuable in avoiding scorching.

Mould temperatures near 20°C usually yield the optimum combination of surface finish and moulding cycle, but temperatures up to a maximum of 60°C can be used if there is a tendency to produce short mouldings. Higher mould temperatures are liable to result in distortion of mouldings on extraction.

Mould shrinkage is between 0.3 per cent and 0.8 per cent, usually between 0.5 per cent and 0.7 per cent.

13.3.4 INJECTION MOULDING FAULTS

PVC is liable to faults similar to those encountered in injection moulding of other thermosplastics materials, and is also prone to a number of faults more or less peculiar to itself and resulting from its thermal instability and the complex nature of additives necessary to its processing.[19]

13.3.4.1   Decomposition

In unpigmented or lightly coloured compositions decomposition is usually detected by the usual type of discoloration or by 'silver streaking' at a fairly early stage. In dark compositions, decomposition can often be masked, but it can be revealed by viewing under filtered ultra-violet radiation, when degraded regions show up as greenish-yellow fluorescence[20] quite distinct from that exhibited by oil. The distribution of degraded material in a moulding varies with the cause.

General decomposition is only likely to occur (1) if the composition being processed is inadequately stabilised, (2) if the temperatures get too high due to incorrect setting or faulty operation, or (3) if the machine stands idle for an extended period while full of material. These situations should rarely if ever arise. Should they do so, the cylinder heaters should be switched off, the carriage removed so that the nozzle is retracted from the screw bushing, the feed stopped, and the cylinder emptied as far as possible by ejection into the air. Meanwhile the temperatures must be watched closely to ensure that they do not fall so far that the PVC overloads and damages the screw. When no more material can be ejected the screw should be removed from the cylinder, and both cleaned, taking care to avoid damage to the surfaces in any way.

Where mouldings contain a line or streaks of decomposed material through the centre of the sprue and running out into the moulding (i.e. 'black-heart'), this is usually due to excessively high temperatures at the forward end of the cylinder. If the fault persists after reducing the temperatures and emptying the cylinder a few times, it will probably be necessary to remove the nozzle and head and clean the screw tip. The same fault can result if the tip of the screw is not properly designed in relation to the flow path in the head.

Randomly scattered dark specks or spots are usually material which has degraded in the cylinder and is subsequently being displaced. It can usually only be cleared by removing the screw and cleaning it, as well as the cylinder, the head, and the nozzle.

Streaks of decomposition originating in the region of a gate are due to excessive frictional heating there. If the material and mould have previously operated satisfactorily reduction in injection speed and/or increase in melt temperature may alleviate matters, but the former will increase cycle times and may yield inferior surface finish. Where possible an increase in cross-section of the gate will help, and is a preferable solution.

Where decomposition occurs under what appears to be normally correct conditions, particularly after a period of satisfactory moulding, the cause is usually too large a shot size which allows material to accumulate on the nozzle wall. Flashing of the mouldings often occurs at the same time. On the other hand decomposition at the commencement of moulding, particularly with silver streaking, can be due to material remaining in the nozzle from a previous run, and thus will usually disperse after a few injections if operating conditions, screw tip design, and nozzle design are correct.

Decomposition which appears at the same point or points in successive mouldings, very often at weld lines, is usually due to entrapment of air. This can be avoided by reducing the injection speed or increasing the venting facilities of the mould. Somewhat similar recurring decomposition can arise as a result of turbulent flow of the melt, to which the only real solution is redesign of the mould.

13.3.4.2   Surface defects

PVC is prone to surface defects similar in appearance and identical in cause to those obtained with other thermoplastics materials, and in general the cures are also the same. Thus, overall dullness or mattness is due to the material being too cold, and can be alleviated by increasing cylinder temperatures and injection speed. Patches of dullness or mattness are due to inhomogeneity, particularly in respect of melt temperature, and generally require an increase in cylinder temperatures. Where an initially glossy surface fades to dullness or mattness, this is usually due to premature removal of the moulding, and can usually be avoided by slight increase in cooling time.

Dull patches can sometimes occur as a result of the 'cold slug' being too cold, so that some material entering the cavity is cooled to too low a temperature for gloss to result. The cure is usually to raise the nozzle temperature. If possible it also helps to separate the nozzle from the mould during the cooling and plasticising periods of the moulding cycle.

Silver streaks are not always due to decomposition, but can also be caused by moisture. It should usually be possible to differentiate between the two causes by examination under ultra-violet light. Moisture is most likely to be picked up by compounded granules as a result of condensation from the atmosphere when a container is brought into the processing shop from a cold store and opened without allowing sufficient time for the material to reach ambient temperature. Dry blends are particularly prone to pick up moisture because of their relatively high surface area/volume ration. Drying is the only obvious solution, and a minimum of one hour at 50–100°C is usually desirable.

Sink marks, which are surface depressions due to contraction of the cooling moulded material, arise from the usual causes of inadequate pressure, inadequate time of holding pressure, or an excessively cooled sprue region.

13.3.4.3   Internal defects

Weld lines, where two separate flow paths meet, are perhaps more liable to occur with unplasticised PV than with most other materials. This is due to its relatively high melt viscosity under processing conditions and to the presence of lubricants which are necessary for satisfactory processing. Wherever it is possible weld lines are preferably avoided by designing the mould so that no welding of separate fronts of melt is necessary. Mould lubricants should certainly be eschewed if possible. Higher temperatures should help, although this can increase lubricant separation. Higher mould temperatures up to 55–60°C may also help. Injection speed can also have an important effect on weld line formation, but unfortunately there are conflicting requirements. Up to the point where welding commences injection should be as rapid as possible to avoid chilling, but thereafter injection should be slow and smooth so as to eliminate air from the interface between the two fronts.

Delamination of PVC mouldings usually occurs in the region of the sprue and is most likely to be due to incompatibility of one kind or another. This may result from contamination by different polymers, such as polyolefins or polystyrene, which may have been used to purge PVC from the machine following the previous run, or from separation of ingredients such as lubricants from the composition. In the latter case increasing the temperature gradient along the cylinder or increasing the shot size may help. Increasing the cross-section of the gate, if possible, may also be of value.

Voids can arise from a variety of causes, including the obvious one of contraction of the core in a thick section during cooling from the melt after the outside has already set. Other causes include (1) entrapment of air bubbles, (2) moisture, (3) plasticiser vapour, and even (4) advanced degradation of the material, in severe cases of which a honeycombed interior may result.[19]

Embedded lumps may be due to granules of material, or the 'cold slug', which have not been thermally plasticised sufficiently, or to contamination by other materials. Smaller specks of material which are not decomposed PVC are due to contamination which may or may not be polymeric in nature.

Black or grey discoloration is likely to occur where PVC containing lead compounds comes into contact with material containing sulphur compounds, resulting in the formation of lead sulphide. Similar but less marked staining can occur with PVC containing cadmium, though a small amount of zinc octoate or similar compound is usually sufficient to depress this to an acceptable level.

Haze in what should be basically a clear moulding usually arises from the material being insufficiently hot as it is moulded. Possible remedies include increasing cylinder, nozzle, and mould temperatures, and decreasing the cross-sections of the gates and the nozzle. A dull mould surface may also produce a hazy appearance in what is essentially a clear material.

Warped or distorted mouldings can result from non-uniform shrinkage or from strains in the mouldings. The former situation can be alleviated by increasing the length of the cooling part of the moulding cycle. Possible causes are the usual ones for thermoplastics injection moulding, e.g. (1) material too cold, (2) injection pressure too high, (3) 'holding' pressure too high or time too long, or (4) incorrect feed setting. Similarly the causes and remedies of short mouldings, oversized mouldings, and flashings are the same as for other thermoplastic materials, and no special discussion is appropriate here.

REFERENCES

1. G.C. PORTINGELL, *Br. Plast.*, **35**, 3, 122 (1962)
2. C.J. MARTIN, *S.P.E. Jl*, **18**, 392 (1962)
3. D.H. CAMBDEN, 'Compression Moulding of Thermoplastic Materials', Chapter 6 in *Injection Moulding of Plastics* (J.S. Walker and E.R. Martin, eds), Iliffe, London (1966) P.I. Monograph
4. D.A. BARLOW, *Wireless Wld*, **70**, 4, 160 (1964)
5. B.S. DYER. *Trans. J. Plast. Inst.*, **27**, 69, 84 (1959)
6. P. RANGNES, *Kunststoffe*, **51**, 8, 428 (1961)
7. J.A. RHYS, 'PVC Stabilisation', Chapter 3 in *Advances in PVC Compounding and Processing*, (M. Kaufman, ed.), Maclaren, London (1962)
8. R. TREGAN, P. GOURIOU, A. BONNEMAYRE and R. TERTIAN, *Industrie Plast. mod.*, **16**, 1, 57 (1964)
9. K.I. BLOOM, I.C.I. Ltd, Plastics Division, Tech. Data W. TD 201 (1967)
10. *Brit. Pat.* 1 150 803
11. C.A. REDFARN and J. BEDFORD, *Experimental Plastics*, Iliffe, London, 85 (1960)
12. Anon., *Shoe Leath. News*, 1 (March 1965)
13. R.W. GOULD, *Br. Plast.*, **35**, 11, 572 (1962)
14. Anon., *Industrie Plast. mod.*, **7**, 8, 12 (1955)
15. Anon., *Br. Plast.*, **29**, 12, 442 (1956)
16. Anon., *Plastics*, **23**, 248, 176 (1958)
17. H. GOLLER, *Plastics*, **26**, 280, 89 (1961)
18. H. FRIMBERGER and J.G. FULLER, *Plastics Technology*, **7**, 5, 53 (1961)
19. B. GRIEFF and G.C. PORTINGELL, *Br. Plast.*, **36**, 6, 319 (1963)
20. Anon., *Plastics*, **31**, 344, 732 (1966)

21. I.I. Rubin, *Plastics Engineering Handbook*, Soc. Plast. Ind. (J. Frados, ed.), Van Nostrand Reinhold, New York (1976) Chapter 4
22. A.R. Carter, 'PVC Processing', *PRI Intl. Conf.*, Royal Holloway College, Egham Hill, Surrey, England (6–7 April 1978) Paper A6
23. F. Cooke, *Developments in PVC Technology*, (J.H.L. Henson and A. Whelan, eds), Applied Science Publishers, London (1973) Chapter 10
24. V.T. Gardner, 'PVC Processing II', *PRI Intl. Conf.*, Brighton, England, (26–28 April 1983) Paper 29
25. A.J. Ganzeman, *ibid*, Paper 30
26. D. Nash, 'PVC Processing', *PRI Intl. Conf.*, Royal Holloway College, Egham Hill, Surrey, England (6–7 April 1978) Paper A5

14

PVC Pastes

14.1 INTRODUCTION

The various possible responses of different types of PVC resin to admixture with plasticiser have already been discussed (Sections 4.4, 10.4.1.2, 10.4.2.3, and 11.2.4.2). Paste resins are those which form more or less stable dispersions of polymer particles in liquid plasticiser at temperatures below those required for gelation. On heating to temperatures in the region of 110°C and above, PVC pastes gel to the solid state and cannot be recovered to true paste form. The properties of the gelled product are determined by formulation in accordance with the same principles as apply to compounded PVC, and also on the conditions of the gelation, particularly temperature. The viscosity of a paste, which may be important in its processing, also depends on formulation as well as on shear conditions and temperature.

There has been some confusion relating to the terminology of PVC pastes. At one time it was suggested[14] that the word 'paste' was being replaced by the word 'plastisol', but the former is still in common usage. The latter term was originally limited to pastes containing no diluent liquids, but nowadays many plastisols are diluted with diluent prior to application as coatings, and the term 'organosol' is now used only for pastes containing relatively high proportions of diluent. Some pastes can be formulated to produce hard mouldings on gelation by using a polymerisable plasticiser, and these are known as 'rigisols'. This is not a common procedure, and it is more usual to produce hard mouldings by using high proportions of 'filler' polymers with 'low viscosity' plasticisers plus diluents. It is also possible to produce pastes in thick putty-like form, and these are known as 'plastigels'.

14.2 FORMULATION AND PROPERTIES OF PVC PASTES

14.2.1 PROPERTIES OF GELLED PASTES

At first it might seem odd to consider gelled pastes before the formulation and properties of the ungelled pastes, but the properties required in the finished product constitute an overriding factor which delineates within fairly narrow limits what formulation variations are permissible.

The properties of fully gelled material follows closely the principles discussed earlier (Chapters 6, 7, 8, and 9), and the properties required will largely determine the nature and content of the plasticiser, whether or not filler can be included, and if so how much.

Paste resins are now available from a wide selection of molecular weights, generally with ISO Viscosity Nos. in the range 110 to 170, though some resins with considerably higher molecular weights are also available for speciality applications. As would be expected, with a given formulation mechanical properties such as tensile strength increase with increase in molecular weight of resin, provided that the full potential is obtained by adequate gelation. Although the use of polymers of lower molecular weight allows lower processing temperatures or faster processing, which is useful in some applications, interaction between polymer and plasticiser may be excessively rapid, and thus result in rapid viscosity increase and premature partial gelation.

14.2.2   RHEOLOGICAL PROPERTIES OF PASTES

14.2.2.1   General

Having delineated at least the polymer, plasticiser, and filler contents for the formulation of a PVC paste in terms of mechanical properties required in the end product, it is possible to vary its viscosity under specified shear conditions by attention to other aspects of the formulation. This is necessary because different processes require different rheological behaviour which may be within narrow limits.

PVC pastes are generally non-Newtonian, and complete rheological characterisation requires measurements over a wide range of shear conditions, e.g. shear rates of from 1 to $5000 \, s^{-1}$ (see Chapter 18). This is especially so because different pastes exhibit different behaviour with varying shear rates, some being dilatant and some shear-thinning or pseudoplastic, while some exhibit different behaviour at different shear rates.

Other things being constant, viscosity decreases with increasing plasticiser content and increases with increasing filler content, but the actual viscosity and viscosity/shear relationships obtained will also depend markedly on the nature of the resin and plasticiser constituents. However, generalisations about the performance of specific plasticisers is unjustified since most can exhibit different behaviour depending on the other constituents and on the test conditions.[3]

Figure 14.1 illustrates how viscosity varies with concentration of plasticiser, and how actual values may differ for two different paste resins. Viscosities can be varied by other means, as discussed later. Explanations for the different behaviour with different resins, and indeed for the very formation of pastes, have been somewhat confused. Clearly the size and size distribution of polymer particles in the paste must be such that in the important range of concentrations sufficient plasticiser is available to disperse the particles as well as merely fill the voids between them. Absorption and solvation at room temperature have to be kept sufficiently low to avoid premature gelation, and this requires[4] that the surfaces of the resin particles should resist interaction with plasticiser until gelation is deliberately induced by heating to 150–200°C. Since these requirements are opposed and cannot be completely separated, actual behaviour falls short of the ideal.

There has been some disagreement about the mechanisms involved in room temperature interactions between polymer and plasticiser. It has been suggested[4] that at room temperature the smallest particles tend to dissolve in plasticiser, decreasing the total volume of particles and thus tending to reduce viscosity from that point of view but also causing a slight increase in viscosity of the plasticiser phase. On the other hand it has been argued that there is no reason why the smallest resin particles should dissolve, as they contain no less crys-

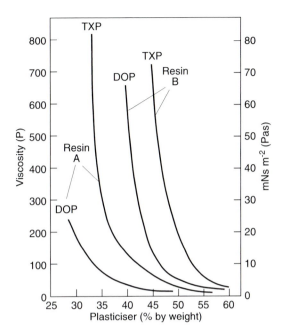

Figure 14.1. Variation in plastisol viscosity with plasticiser content for two different resins[1]

tallinity than larger particles. However, doubts have been expressed that crystallinity has any effect, partly on thermodynamic grounds,[41] and also because the proportion of crystalline regions is small. It has also been stated that solubilised polymer is normally of low molecular weight and will rapidly raise the viscosity of the paste.[41] If solubilisation of polymer does occur it seems likely that the process would occur at particle surfaces. Molecules of relatively low molecular weight there would tend to dissolve first, and if the process continues for a sufficiently long period the smaller particles would tend to lose their identity faster than larger particles. At the same time some absorption of plasticiser causes swelling of the larger particles and reduces the total volume of free plasticiser, thus tending to increase viscosity.

The ageing of pastes is not merely dependent on particle size but is also affected by the nature and amount of residual emulsifying agent. With this complex behaviour it is not surprising that irreproducible and erratic results are sometimes obtained from viscosity measurements.[4]

With a given formulation the viscous behaviour is dependent on the nature of the resin. As mentioned earlier, two different resins may yield appreciably different viscous behaviour i.e. different viscosities as measured at the same shear rate and different variations of viscosity with varying shear rate.

The different viscosities obtained with different resins have been explained in terms of the voidage between particles.[5, 6] Resins that yield relatively lower viscosities than others at the same plasticiser levels are seen to be those whose particles pack relatively close together, so requiring less plasticiser to fill the voidage and leaving more for dilution. This generally requires a fairly wide particle size distribution, the smaller particles fitting into the voids between the larger particles. As has been pointed out,[7] many published rheological comparisons have been made at equal weight fractions of resin and plasticiser, and it is instructive to convert these to volume fraction comparisons. If this is done, for example, with data from Fig. 14.1, the two plasticisers are seen to be closer in behaviour, although there are still appreciable differences. This approach is, however, academic, since the choice of plasticiser and its concentration are determined largely by property and performance requirements of end-products, so that in practice it is of more interest to be able to account for the different behaviour for different resins.

14.2.2.2   *Effects of types of resins*

As indicated earlier (Section 3.3.2.3) most paste resins are derived from emulsion and microsuspension polymers. These are usually isolated by spray-drying. For paste production the resins may be the polymers as dried, or may be produced therefrom by grinding to produce the required particle ('grain') characteristics.

If the drying process is relatively mild, so that the primary particles (i.e. those formed in the latex) are only loosely bound together to form secondary particles, dispersion of the latter in plasticiser may break them down again more or less completely to the original primary particles. In this case there will be a tendency for the resin to form pastes consisting substantially of primary particles suspended in the plasticiser. If these particles are very small (much less than 0.1 µm) the resultant pastes are of relatively high viscosity, and, moreover, increase in viscosity with time even at room temperature owing to the relative ease with which plasticiser can penetrate to the centre of the particles. For this reason it will be found that the powder particles of most commercial resins as supplied for paste production (i.e. 'grains of Allsopp's and Geil's nomenclature[42, 43] – see Chapter 4) are appreciably larger than 0.1 µm, generally averaging around 10 µm. However, paste resins having mean particle sizes as low as ⅓ µm have useful properties provided that the size distribution is narrow and finer particles are essentially absent.

Where the drying conditions have been severe, sintering of primary particles may be sufficient to retard or inhibit breakdown, and the particles suspended in the plasticiser will tend to be secondary particles (or 'grains') obtained by drying or grinding. The amount of breakdown in such cases may depend on the severity of the shear conditions during mixing. It may be assumed that any commercial resin will lie somewhere between the two extremes, depending on the behaviour required of it. Thus to be able to predict the behaviour of a polymer in plasticiser, it is not sufficient to study merely the size and size distribution of primary particles as formed in the latex, though it is accepted that these can be important.[7, 8] Nor is it sufficient to study the size and size distribution, and shape of the secondary particles formed during the drying process with or without subsequent grinding. It is, in particular, necessary to know how strongly the secondary particles resist breakdown by interaction with plasticiser. Rate of breakdown is likely to vary with the activity of the plasticiser, as also is rate of penetration of particles, so it is not really surprising that different kinds of behaviour are observed with different plasticisers, and the implication[7] that volume concentrations of polymer and plasticiser are of overriding importance is probably not justified.

As indicated above, the instantaneous viscosity of a paste will depend on the particle size, size distribution, and shape of the particles suspended in the plasticiser. Particles that can pack closely together yield pastes of relatively low viscosity, while particles that cannot will yield pastes of relatively high viscosities. In spite of what has been stated in the preceding paragraph, this is probably still true, but it is important to note that it is the particles as they exist in the paste which determine the behaviour. From a study of twelve commercial paste resins[34] it was concluded that 'high viscosity' resins have a narrow size distribution of primary particles produced from the latex, and that the nature of secondary particles produced by drying and grinding are of little significance. Mean size of primary particles covered the range from 0.23 to 0.46 μm, the smaller size yielding the higher viscosity. 'Medium viscosity' resins had a broad primary particle size distribution ranging from 0.8 to 1.5 μm. 'Low viscosity' resins also had broad primary particle size distributions, with a relatively high proportion of secondary particles, which lead to increased dilatancy. The accuracy of these conclusions has been questioned on the ground that the twelve resins were probably produced using different emulsifiers, and, since these can affect viscous behaviour to particle sizes may be irrelevant.[41] However it is generally agreed that grain particle size distributions that optimize packing normally produce relatively low viscosities in a given formulation.[44]

If secondary particles suspended in the paste are slowly broken down, or if primary or secondary particles are slowly penetrated by the plasticiser, the viscosity will change with time, generally increasing, possibly to a point where something like gelation occurs.

As well as being dependent on size and size distribution of primary and secondary particles, paste viscosity and rheological behaviour are also dependent on the amount and nature of surfactant used in the polymerisation, and on the molecular weight of the polymer.[8, 34, 44–47]

It has been suggested[5, 6] that where shearing disturbs the particles suspended in the paste to locations where they pack together less closely dilatancy will be observed. Thus it is postulated that irregularly shaped particles will tend to pack closely together at rest, tending towards the most favourable close-packing position. On application of shear the particles will be disturbed into positions that correspond to less close packing, thus increasing the total space between them and requiring an increased volume of plasticiser to fill this space. This is seen as reducing the excess of plasticiser available for dilution, and thus increasing viscosity. It has not been explained why the particles when at rest should pack into a position of minimum voidage, or indeed why the implicit excess of plasticiser from regions of close-packed particles should reduce overall viscosity. Nevertheless dilatancy is observed in pastes containing irregularly shaped particles. A more likely explanation is that dilatant behaviour is due to increasing interaction between particles resulting in increased mutual interference with their movement with increasing shear rate. This would also explain why addition of smaller particles can produce a change from dilatant to pseudoplastic behaviour while maintaining a constant total plasticiser content.[7]

Explanations advanced for 'shear-thinning'[5, 6] are even less satisfactory. A paste exhibiting this behaviour is seen to contain spherical particles of rather regular size. It has been suggested that these develop a structure which 'is one of cubic close packing resulting from the interaction of uniform reticulate forces', and that 'on shearing the particles assume a hexagonal closely packed structure resulting in reduced voidage and hence more free plasticiser resulting in enhanced diminution of viscosity'.[5] This explanation overlooks the fact that continued flow would require the particles to alternate between the proposed cubic and hexagonal packing, and, moreover, does not explain why 'free' plasticiser should produce a reduction in viscosity. Perhaps an even more serious objection is that photomicrographs of particles of shear–thinning polymers,[5, 6] presumably at rest, show the particles lined up in order much more nearly approaching an hexagonal than a cubic type of packing.

An alternative explanation of shear–thinning is that weak structures due to particle–particle interactions form in a paste at rest, and that these break down under shear.[44, 48]

Other objections to the theories that have been advanced to explain viscous behaviour of PVC pastes are that many exhibit dilatancy over one range of shear rates and shear–thinning over another, which may be higher or lower (i.e. a minimum or maximum may be found in viscosity versus shear rate curves), and that a single resin can exhibit dilatancy in combination with some plasticisers and shear–thinning with others.[2, 3]

By selecting appropriate production procedures the various types of viscous behaviour obtainable with emulsion polymerised paste polymers can be reasonably well matched with microsuspension polymers, except that it has not so far been possible to produce forms of the latter that are highly pseudoplastic. This precludes their use in applications where such behaviour is required, such as for vehicle undersealing applied by spraying. On the other hand, emulsion polymerised paste polymers are generally not suitable for drum gelling and release processes, though this is more related to the latex particle size distribution than the process.

As would be expected the range of rheological behaviour of pastes formed by individual resins can be extended and filled in by using selected combinations of different resins. Sometimes the effect is merely one of providing a range of particle sizes in the paste suspension, and pastes of lower viscosities than those obtained from the individual resins result. This is illustrated in Fig. 14.2, which shows the way viscosity can be varied by com-

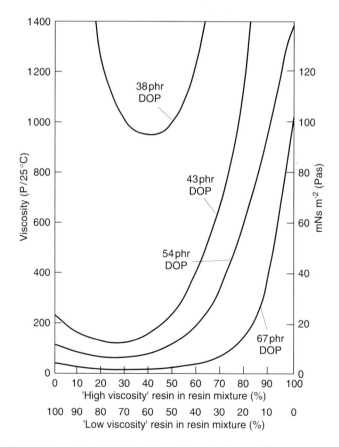

Figure 14.2. Variation of plastisol viscosities with ratio of two different paste resins[1]

bining in various proportions two resins, one of which tends to give relatively low viscosities, and the other relatively high viscosities in specific paste formulations. Of course this kind of behaviour is only found for pairs of resins whose particle characteristics are suitably matched, and will not be observed for other pairs.

Reduction in viscosity for a given polymer/plasticiser ratio can also be achieved by incorporating a fine particle suspension polymer as part of the total polymer content[1, 4, 9] (Fig. 14.3). Here the particles of suspension polymer are probably scarcely penetrated by plasticiser at room temperature, and thus behave more like the particles of a mineral filler, and hence are known as 'filler' polymers. On gelation at high temperatures, however, the particles of filler polymer and those of the paste resin both gel, and contribute to the overall polymer content and thus to the mechanical properties of the fully gelled material.

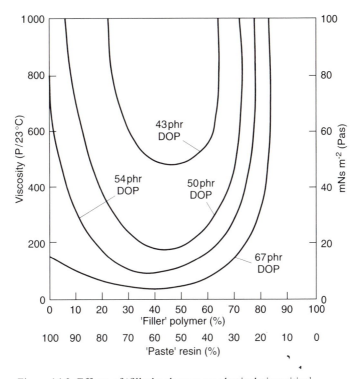

Figure 14.3. Effects of 'filler' polymers on plastisol viscosities[1]

14.2.2.3   Plasticisers in PVC pastes

It was pointed out earlier (Section 14.2.2.1) that most published rheological comparisons (and also, indeed, comparisons of mechanical properties: see Section 6.3.4.2) have been made at equal weight fractions of resin and plasticiser, and that it has been suggested that comparisons at equal volume fractions might be more meaningful.[7] In commercial practice, however, it is the properties required in the end-product which largely determine the nature and concentration of plasticiser to be used, and it would be of much more practical value if comparisons were made at concentrations required to yield equivalent mechanical properties. To extract such comparisons from existing published literature would be tedious and is possible only to a very limited extent; what follows can only be taken as a guide to some general trends. Tables 14.1[1] and 14.2,[3, 10] indicate some trade literature values of viscosities determined on pastes 'as made' and after ageing for a number of different plasticisers at equal concentration.

To the approximate ranking that might be deduced from Tables 14.1 and 14.2 may be added the fact that butyl benzyl phthalate tends to yield pastes of much higher viscosity than does dioctyl phthalate. Particular attention is drawn to the variations in rate of increase of viscosity with time. For example (Table 14.1), dibutyl phthalate with a 'low viscosity' resin produces pastes of lower initial viscosity than other listed plasticisers, but the resultant pastes thicken rapidly on standing. This is presumably due to the relatively high activity of the interaction of dibutyl phthalate with PVC resins. As stressed in the first edition of this book, dibutyl phthalate is really too volatile for most applications, and in practice nowadays this plasticiser is rarely encountered.

The viscosity of the plasticiser itself, as also its molecular size and polarity, are also likely

Table 14.1 TYPICAL VISCOSITIES (IN POISES) MEASURED[1] ON PASTES MADE WITH DIFFERENT PLASTICISERS AT 85 PHR MEASURED AT 25°C

Plasticiser	'Low viscosity' paste resin		'High viscosity' paste resin	
	Initial	7 days	After	7 days
Dibutyl phthalate	3	26	600	*
Dioctyl phthalate	9	14	190	800
Dinonyl phthalate	10	16.5	63	30†
Tritolyl phosphate	22	34	600	450†
Trixylyl phosphate	27	41	500	700
Dioctyl phthalate + di-'alphanyl' sebacate (65:20)	6	8	41	190
Dioctyl phthalate + chlorinated hydrocarbon (2:1)	31	35	300	150
Polyester	82	105	600	1500†

* Too high to measure by same method.
† Sedimentation.

Table 14.2 TYPICAL VISCOSITIES (IN POISES) MEASURED[10] ON PASTES MADE WITH DIFFERENT PLASTICISERS AT 70 PHR MEASURED AT 23°C

	Viscosity in poises at indicated times after mixing					
	3 h	1 day	3 days	6 days	10 days	20 days
Dibutyl phthalate	46	80	250	650	1275	3280
Di-isobutyl phthalate	43	65	108	175	265	490
Di-n-heptyl phthalate	26	36	54	74	92	112
Dioctyl phthalate	48	55	67	80	94	110
Di-iso-octyl phthalate	44	49	58	68	79	101
Dinonyl phthalate	60	65	74	86	97	114
Di-isodecyl phthalate	58	61	67	76	86	102
Ditridecyl phthalate	160	181	210	230	248	255
Dioctyl adipate	20	31	52	76	98	116
Triethylene glycol dicaprylate	10	11	12	13	14	16

to contribute to viscosity effects. These and other variations in behaviour greatly reduce the value which single point determinations common in the literature may have. In addition, as mentioned earlier (Sections 14.3.4.2 and 14.3.4.3), one is in practice usually concerned with comparing plasticisers, not at equal concentrations, but at concentrations designed to give identical properties in the end product. Some trade literature[11] attempts to do this, as illustrated in Table 14.3, which compares viscosities of pastes made with a number of different plasticisers at concentrations designed to give equal moduli.

Table 14.3 TYPICAL VISCOSITIES OF PASTES MADE WITH DIFFERENT PLASTICISERS AT CONCENTRATIONS DESIGNED TO YIELD MODULI, AT 100% ELONGATION, OF MN/m²

	phr	Viscosity in poises at indicated times after mixing		
		2 days	16 days	30 days
Dibutyl phthalate	65	100	380	525
Di-isobutyl phthalate	75	22.6	39	63
Dioctyl phthalate	75.5	36	52	57
Di-iso-octyl phthalate	77	30	39	43
Dinonyl phthalate	83.5	20.6	28.6	33
Di-isodecyl phthalate	80	26	35.3	37.2
Ditridecyl phthalate	89.5	33	39	43
Di-iso-octyl adipate	63	12.6	19	21.5
Dibutyl sebacate	53	72	155	204
Di-iso-octyl azelate	67.5	8	9.8	10.5
Tritolyl phosphate	85	25	28.6	30
Trixylyl phosphate	90	43	46	46
n-Butyl epoxystearate	66	12.2	16.4	18.8
Iso-octyl epoxystearate	67.5	7.6	19.7	26
Epoxydised soya bean oil	80	152	177	190

The data of Tables 14.2 and 14.3 were determined using the same grade of resin, so the reasons for the apparent discrepancies between the two tables must be looked for elsewhere, possibly in the methods of mixing and measurements employed.

Allusion has already been made (Section 14.2.2) to the dependence of viscosity/shear rate behaviour on the nature and concentration of the plasticiser.[2, 3] Effects of shear rate on apparent viscosity appear to increase markedly with decrease in plasticiser concentration,[2] and the indications are that close approach to Newtonian behaviour is observed at concentrations of around 50 per cent or more, dependent on the particular plasticiser. This is not particularly surprising since decrease in concentration of polymer corresponds to an approach towards pure plasticiser, which would be expected to exhibit near-Newtonian behaviour when tested by methods employed to evaluate PVC pastes.

It was indicated in Table 14.1 that some pastes exhibit sedimentation, with apparent decrease in viscosity due to drop in polymer content. In addition to sedimentation, separation of polymer from paste may take the form of surface scum. This kind of behaviour tends to occur most commonly in pastes formed from 'high viscosity' resins, but of relatively low viscosity due to the plasticiser content, and pastes of this kind should be stored in cool places and stirred from time to time.

14.2.2.4   Fillers in PVC pastes

Fillers are used in pastes mainly for the same reasons as in other forms of plasticised PVC (Section 7.3.2), namely cost reduction. Effects on mechanical properties are similar to those already described, but a few special features require separate consideration. Thus, although mechanical properties suffer in the expected way, incorporation of fillers into highly plasticised compositions can be advantageous in reducing surface tack.

Other things being equal, incorporation of fillers increases viscosities of pastes, and can limit the maximum loading, which generally should not exceed 30 to 60 phr, though much higher loadings are occasionally used. This increase in viscosity is due to the usual effect of suspending particulate solids in liquids, but plasticiser absorption by filler particles will lead to a reduction in effective plasticiser concentration thus also increasing viscosity. The absorptive capacity of a filler is therefore of rather more importance in considering its possible use in a paste formulation than it is in other forms of PVC. However, although a good deal has been published about the importance of oil absorption values of fillers,[4, 9–12] both in respect of viscosity of pastes and of mechanical properties of processed pastes and compounds, calculations from values published in trade literature show that these are not of great practical value in predicting paste viscosities. This is illustrated in Table 14.4, based on viscosities and oil absorption values in parts absorbed per 100 parts of filler, taken from trade literature.[4, 9] The viscosities were measured after one day's ageing, on a range of pastes containing equal volumes of filler and based on a formulation of 100 parts paste resin to 67 parts DOP.

Table 14.4 RELATIONSHIPS BETWEEN VISCOSITY OF PASTES AND OIL ABSORPTION OF A NUMBER OF FILLERS COMPARED AT EQUAL VOLUME

Filler	*phr*	*Oil absorption*	*Oil absorption × phr*	*Viscosity (P)*
None	0	0	0	100
Barytes	57.3	16	917	70
Precipitated calcium carbonate	34.8	36	1252	180
Calcium carbonate	34	45	1530	400
Blanc fixe	51.5	38	1957	150
Lithopone	55.2	37	2042	400
Calcined clay	33.3	66	2198	170
Diatomaceous earth	29.5	148	4366	400

If oil absorption determines the effect a filler will have on paste viscosity, the viscosities quoted in Table 14.4 should correlate with the oil absorption values or with the products of these with the concentrations, but neither correlation occurs.

The fillers most commonly employed in PVC pastes are the various forms of calcium carbonate, both mineral and precipitated. Some additional possible fillers are listed in Table 14.5, together with some typical oil absorption values.[14]

14.2.2.5   Organosols[15–17]

Where for a required formulation the viscosity of a paste is too high for convenient processing even after using the most favourable resin combination (Section 14.2.2.1), viscosity can often be appreciably reduced by incorporation of appropriate organic solvents. At

Table 14.5 TYPICAL OIL ABSORPTION VALUES FOR SOME
POSSIBLE FILLERS

Filler	Oil absorption (phr)
Mineral calcium carbonates	15–36
Whiting	20–36
Precipitated calcium carbonate	25–45
Calcium silicate	45
Dolomite	33
China clay – medium fine	35
China clay – very fine	55
Silica	42

one time the term 'organosol' was applied to any paste containing such a diluent, but many plastisols are now produced containing small proportions of solvent and the term is now commonly restricted to pastes containing relatively high proportions.

Solvents with a strong solvating effect on PVC, such as ketones, are preferably avoided since they tend to yield increased rather than decreased viscosities, and their use usually results in rather rapid increases in viscosity on ageing. Nevertheless some solvents of this type, such as methyl ethyl ketone and methyl isobutyl ketone, are sometimes used in admixture with hydrocarbons. More commonly non-solvents for PVC, such as white spirit or solvent naphtha, are used. After gelation the diluent is removed by volatilisation, and this in itself is not always simple.

14.2.2.6　*Other viscosity depressants*
In addition to organic diluents the viscosity of a given paste formulation can also be reduced by small amounts of some other materials, mainly specific surface–active agents, particularly derivatives of polyethylene glycol and nonyl phenyl ethoxylates.

14.2.2.7　*Thickening agents*
It will be apparent that one is commonly concerned to reduce viscosity, but there are occasions when it is desired to increase this, which can be done by incorporation of small amounts of metal soaps, particularly alumunium stearate or finely divided silica. If the quantities are sufficiently high (e.g. 5–15 phr), viscosity increases to such an extent that putty-like 'plastigels' results, the amount of thickening agent required to achieve this increasing with plasticiser content, as would be expected.

14.2.2.8　*Other ingredients of pastes*
Since pastes are usually processed at temperatures of the same order as those used for processing PVC compounds, stabilisation is generally necessary. The principles of stabiliser selection discussed previously (Chapter 5) apply generally, but additionally attention has to be paid to the fact that many stabilisers, particularly metal soaps, increase paste viscosities and rates of increase of viscosity on ageing. Metal soaps are also liable to produce 'blooming' and print adhesion problems. Most manufacturers now offer stabiliser systems specially designed for use in pastes.

Lubricants (Section 7.2.1) are also necessary where gelation of a paste occurs in contact with a surface from which the gelled product is to be removed, but are often omitted where lamination is required. As already indicated soaps tend to increase viscosity, and this factor must be taken into account when selecting a lubricant. Typical lubricants include butyl stearate and ethyl palmitate at about 1–2 phr concentration.

Colorants are selected in accordance with the general principles discussed previously (Section 7.7), with the added complication, already considered for fillers, that plasticiser absorption increases viscosity. However, although it is clearly desirable to employ pigments with low plasticiser absorption, the proportions used in most cases are not sufficient to make a really appreciable difference to viscous behaviour. Where high proportions of pigment are used, as in some wall coverings which contain quite high proportions of titanium dioxide, there is an appreciable effect on viscosity.

Chemical blowing agents (Section 7.8) are commonly incorporated into PVC pastes in the manufacture of cellular products, such as cushioned floor coverings and wall coverings.

14.2.2.9   Effects of temperature on viscosity

In addition to the normal dependence of viscosity on temperature, the rate of interaction between polymer and plasticiser is also temperature dependent, so somewhat complex behaviour is experienced. The effect of plasticiser interaction will depend to some extent on the nature of the resin particles in the paste, but in general viscosity decreases with increase in temperature up to about 45°C, and from there upwards increases more or less steeply until gelation occurs. In fact replotting some published data[4] over the temperature range 14–28°C, in the form of log viscosity versus reciprocal K, yields an essentially linear relationship, suggesting that plasticiser/polymer interaction is too slow at those temperatures to affect viscosity within the time scale involved in the viscosity testing. However, this clearly depends on the nature of the plasticiser and the time at which measurements are made.

Rate of increase of viscosity with time increases with temperature, but there is considerable variation from one paste to another, some becoming almost unusable within a matter of months, while others remain fluid with relatively little change for several years. The following figures indicate the effect of temperature on viscosity increase of a typical paste:[4]

Temperature	Viscosity after 7 days
23°C	18 Pas
35°C	54 Pas
50°C	400 Pas
70°C	gelled

14.2.3   GELATION OF PVC PASTES

As already mentioned, on heating a PVC paste to temperatures of about 160°C, it passes from the viscous fluid state to a solid, usually rubbery, state, the transformation generally being known as 'gelation'.

Rate of gelation is dependent on temperature and formulation. At any particular plasticiser concentration the rate depends on the activity of the plasticiser.

The mechanisms of gelation have not been elucidated precisely, and although evidence has been offered to suggest that critical temperatures are involved, the process is almost certainly a sharply temperature dependent rate process (Section 6.3.3). Progress of gelation can be assessed experimentally by measurement[4] of tensile strengths of specimens at different time intervals (Fig. 14.4). At relatively low temperatures (e.g. around 135–145°C) the shape of the tensile strength/time curves is very dependent on the nature of the plasticisers and resins employed, but at higher temperatures (e.g. 165°C and above) differences are largely confined to the first few minutes, and therefore often pass unnoticed. There is considerable variation and possibly discrepancy amongst published information. For example Fig. 14.4 suggests that gelation times of at least one hour are required for maximum tensile strength to be developed, whereas it has been suggested elsewhere[1] that at temperatures of 170°C–180°C there is little if any gain after 10 min. Thickness of specimen does have an influence because of poor heat conduction, but it is doubtful if this would account for the discrepancies. The maximum strength developed does, however, depend not only on the formulation but also markedly on gelation temperature, and temperatures of at least 170°C–180°C are clearly indicated for best results. There are indications that extended heating times can result in decrease from the maximum strengths, even at lower temperatures, but it is not clear whether this is due to degradation of polymer or loss of plasticiser by volatilisation. It is therefore more practical to ensure achievement of optimum properties by rapid gelation at relatively high temperatures rather than by prolonged gelation at relatively low temperatures. Because of poor heat conduction the thickness of paste layers or sections markedly affects heating times required to achieve full gelation and optimum possible mechanical properties. Thus in a typical case[18] of a paste gelled at 150°C a minimum time of about 9 min was required for a thin layer, increasing thickness requiring increasing time of gelation in an almost linear relationship, so that a layer of 6.5 mm thick required about 20 min. Thick products are perhaps better built up in stages by partial gelation of successive layers at 110–130°C, followed by full gelation of the laminate at a higher temperature. Full gelation of the successive layers as they are applied severely reduces interlaminar bonding, and the layers can then be easily separated.

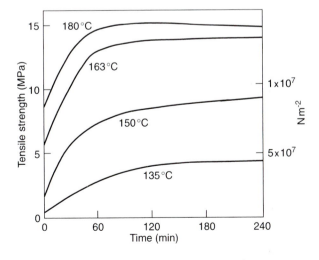

Figure 14.4. Gelation of PVC pastes[1]

14.3   PREPARATION AND STORAGE OF PVC PASTES

14.3.1   Mixing

Procedures for the preparation of PVC pastes appear deceptively simple. Almost any type of mixer (Sections 10.3.2, 10.3.3, and 10.3.4) can be used,[53] though triple-roll milling (Section 10.3.10) may be necessary to ensure thorough dispersion of some powdery additives such as pigments and blowing agents. Sigma-blade, dough, pony, paddle, planetary, and 'high-speed' mixers have all been used. Even ball-milling can be used where difficulty of dispersion in paste of low viscosity is encountered. The tendency nowadays is to use 'high-speed' impeller/dissolver type mixers.[54, 55]

   With mixers other than 'high-speed' types the amount of plasticiser mixed with the polymer initially may be limited so that the shear developed is sufficient to break down the resin particles and disperse all solids adequately. Any remaining plasticiser is added after a smooth paste has been formed. Solids such as pigments, lubricants, stabilisers, and fillers are preferably added in the form of previously prepared smooth dispersions in plasticiser. The quantity of plasticiser used in the first stage of mixing depends on the nature of the resin and plasticiser and on the degree of breakdown of the resin particles required. Depending on these and on the nature of the mixing machine there will usually be an appreciable rise in temperature, which must not be allowed to become excessive so as to result in premature thickening or partial gelation, and generally temperatures should not be allowed to exceed, for example, 38°C for DOP and DIOP, or 35°C for aryl phosphates. Cooling facilities may therefore be necessary. Typical proportions suitable for the first, thick, stage of mixing range from 40 to 60 phr for DOP and DIOP, and 45 to 85 for aryl phosphates. Total mixing times rarely exceed 30 min.

   With 'high-speed' type mixers all liquids other than diluent should be charged to the mixer first, the solids being added in discrete doses during the mixing cycle. Cycle times depend on the volume of the charge, but typically would not be longer than 30 min for a 500 kg batch.

14.3.2   Dearation of pastes

Unless the viscosity of a paste is low, air entrapped within it is usually very slow to escape unaided and, if still present in the paste when it is gelled, will result in voids or porosity. Mixers suitable for paste mixing are available with facilities for application of vacuum during the mixing process, thus making it possible to produce pastes with very low air contents. Otherwise subsequent dearation may be necessary. This can be carried out with the paste spread in a relatively thin layer in a vacuum chamber, or by drawing the paste in a stream into a container enclosed in a vacuum chamber.[4]

   Even after dearation, air is readily entrapped in a paste during handling and processing, and care is needed to prevent this happening.

14.3.3   Storage

The tendency of PVC paste viscosities to increase with time even at room temperature has already been mentioned (Section 14.2.2). Rates of increase in viscosity vary considerably for different pastes, but viscosity decreases only if separation of polymer occurs. Generally the higher the initial viscosity and the higher the temperature the faster the rate of thickening. Pastes should therefore be stored in fairly cool conditions.

14.4   PROCESSING OF PVC PASTES

14.4.1   GENERAL CONSIDERATIONS

Paste processing in general consists in distributing a paste of the required composition in the required form and gelling it. The flow and gelation behaviour for maximum convenience will be determined by the nature of the product and the process.

What is essentially compression moulding of pastes to form microcellular products has already been described (Section 13.2.3.3). Most other processes and equipment for pastes are almost specific to such materials, although some are essentially similar to those used for solutions and powders.

14.4.2   CASTING

The most obvious possible way of processing a PVC paste is to gel it in a mould. Though feasible, the technique is rarely used other than to produce specimens for test and experimental purposes.

Casting is used to produce unsupported films by spreading a thin layer on a suitable substrate to which the PVC will not adhere, gelling the paste, cooling, and stripping. The substrate may be a continuous moving stainless steel band or specially treated paper supported on a moving metal band. The lay-out is broadly similar to that used for laminating by spreading techniques (Section 14.4.6). Film cast from PVC paste is usually printed, commonly by the transfer technique in which the treated paper on which the casting is effected is first printed with inks which are easily taken up from the paper by the paste.[4]

The main problem in open casting is to avoid entrapment of air bubbles. Flow behaviour of the paste is not very critical except in that respect. Viscosities in the range 90–500 P are commonly used.

14.4.3   INJECTION MOULDING

What is essentially a form of injection moulding can be used to mould PVC pastes, but the equipment and technique are specific to pastes. The technique has been mainly employed to mould soles and heels directly on to conventional footwear uppers, but seems to have been largely displaced by other procedures.

The equipment is similar to that used for direct vulcanisation of sponge-rubber soles.[19] The upper to be soled may be of normal fabric, leather, or PVC-coated fabric. Sometimes, as with PVC-coated uppers, sufficient adhesion is obtained without any special attention, but priming with appropriate adhesives is often necessary. Each upper is carried on a last which is supported in an associated hollow mould, usually consisting of two side parts that swing into place on hinges, and a sole/heel plate. Paste is injected into the closed mould through a 'gun' fed with paste from a reservoir pressurised at about 0.3 MNm^{-2}. The last is preheated at about 140–150°C and the mould is maintained at 140–150°C, so that the injected paste gels in 3.5–5 min, depending on the thickness. To keep gelation times to a minimum thick heels usually require 'fillers' to be inserted in the mould to reduce thickness of paste layer to be gelled. These are conveniently blocks of compounded plasticised PVC held in position by pins attached to the heel base plate or to the shoe upper.

For optimum wear performance a BS Softness of around 75 is required, and this largely determines the plasticiser content of the paste, which is usually in the region of 70–80 phr for the common phthalates. To compete with rubber processing cycles have to be short, which requires relatively low viscosities, usually around 20–40 P, rapid gelation, and high strength when hot.

14.4.4   SLUSH MOULDING (HOLLOW CASTING OR 'FLOW MOULDING')

Like casting and injection moulding of pastes, this process is rarely used for production purposes. There are a number of variations on the process, but the essential feature is that one or more layers of paste are gelled on the inside of a hollow mould which is open at one end.

One variation involves filling the mould with paste, heating in an oven until a sufficiently thick layer has partially gelled, pouring out ungelled paste, completing gelation of the residue, cooling, and finally stripping the casting from the mould. More commonly the mould is preheated to about 110°C before being filled so that partial gelation is more rapid. Thicker mouldings can be produced by repeating the heating and filling operations two or three times before completing gelation at a higher temperature.

Moulds can be quite cheap compared with those necessary for most other processes, usually being pressed from aluminium or produced by electrodeposition. Undercut shapes are eminently feasible because of the flexible nature of the gelled product, although care is necessary to ensure that this does not lead to air entrapment. Flow behaviour is not critical provided the paste can be poured without trapping air bubbles. Viscosities in the range 10–150 P are common, 10–20 P being preferred for thin mouldings and 90–150 P for thick mouldings.

14.4.5   ROTATIONAL CASTING OR MOULDING

14.4.5.1   Introduction

This process is a development from slush moulding and is now the process most commonly used for the production of hollow articles from PVC pastes.[19] It is now also commonly used with other thermoplastics in powder form (e.g. polyolefins and nylons). Fairly sophisticated and automated machines are available for the process, and production rates of over 115 kgh^{-1} of moulded PVC per machine can be achieved. In rotational moulding wastage of material is reduced to a minimum and flash is almost entirely absent. Very fine detail can be reproduced faithfully, and yet mould costs are generally relatively low.

14.4.5.2   Process and equipment

The principles of the process involves a closed hollow mould or moulds, usually openable to two halves or fitted with a base plate, into which is placed a predetermined amount of material. The mould is rotated at different speeds about two axes at right-angles to each other while mounted in an oven at sufficiently high temperature to gel the paste (or other thermoplastic). Speeds of rotation are typically in the 5–20 rev/min range for the higher of the two speeds and down to as low as one-fifth of the faster speed for the lower speed of rotation. At such low speeds there is negligible centrifugal action, and a layer of paste is picked up by the whole surface of the mould as a result of slight adhesion and the viscous nature of the paste. Where incomplete filling of the mould might occur, as in a complicated shape, spinning in one plane before heating can sometimes help. Heating must be such as to give reasonably rapid gelation but not so fast as to prevent the formation of a gelled coating of uniform thickness.[20] After gelation the mould is transferred to a cooling station and the moulding removed by splitting the mould or by removing the base plate, and withdrawing the moulding with the aid of vacuum.

Simpler machines have a carriage for one mould only but more complicated automatic machines have a number of carriages capable of carrying moulds in clusters of up to twenty. Moulds are of cast or machined non-porous aluminium, or electro-formed or nickel-plated copper, and are relatively cheap as they do not have to withstand high pressures.

Oven temperatures are typically up to 240°C, at which gelation is usually complete in 10 min, except for pastes of low plasticiser content which require somewhat longer.

Since its introduction the rotational moulding process has become more and more complex in respect of machine design and moulded products, and is now applicable to polymers other than PVC pastes, including polyethylene, polypropylene, polycarbonate, and PVC powder blends. A problem has been the difficulty, arising from the complex movements of the moulds and the machine arrangements necessary to achieve them, of obtaining information about the progress of material changes within the moulds. Hence process control has been somewhat arbitrary. However, Crawford and co-workers have developed an ingenious system that overcomes these problems and offers the possibility of precision control, as well as obtaining a better understanding of the process.[36-40] The essential feature of this system is a radio transmitter, attached to an arm of the machine, which takes readings from thermocouples located inside and outside the moulds and transmits signals to a receiver outside the oven. The signals are processed by a computer.

14.4.5.3 Formulation
Rotational casting generally requires pastes of relatively low viscosity, e.g. not greater than 60 P unless there is a rapid drop in viscosity during heating to gelation. With softer compositions, e.g. in the BS Softness range of 60 upwards, DOP or DIOP are usually adequate plasticisers, though DNP is advantageous in the 60–70 BS Softness range because it yields lower viscosities. In harder compositions some proprietary mixed ester plasticisers are available which yield sufficiently low viscosities to permit mouldings as hard as BS Softness 10 to be produced, provided that fairly high proportions (i.e. up to 50%) of filler polymer are used, together with a viscosity depressant. Organic phosphates are used only where flame-resistance is required. They tend to yield pastes of high viscosity for a given softness. Polymeric plasticisers are rarely used, for obvious reasons.

Owing to the use of high temperatures to obtain short cycles, processing conditions are rather severe, and good stabilisation is required. Combinations of epoxydised oils with liquid calcium/zinc or barium/zinc complexes with organic costabilisers are commonly employed.

Fillers are only incorporated up to about 10 phr to produce 'drier' surface finish than is obtained in their absence.

Table 14.6 TYPICAL FORMULATION FOR ROTATIONAL CASTING

	Soft mouldings e.g. toys	Medium soft mouldings e.g. play balls	Hard mouldings
'Low' viscosity paste resin	100	100	100
Filler polymer	—	—	50
DOP or DIP	85	—	—
DNP	—	65	—
Mixed ester plasticiser	—	–	25
Viscosity depressant	—	2	2
Epoxydised oil	2	2	2
Liquid calcium/zinc or barium/zinc with organic costabiliser	2	2	2
Viscosity (poises at 25°C)	10–20	10–20	40–60
BS Softness	95	60	10

14.4.5.4   *Problems*

If extraction of mouldings is laborious, owing to sticking, lubricant can be used, but since sticking is usually due to temperatures being too low, increased processing temperature may help. On the other hand, difficult extraction can also result from distortion due to excessively high temperature. The latter can also cause 'burning', though this is usually due to poor heat distribution. Blisters can arise from air entrapment, moisture, volatiles, or porous moulds. Incomplete mould filling (Section 14.4.5.2) can sometimes be cured by wetting the mould surface with silicone oil or plasticiser, but if a lower viscosity paste can be formulated this is a more convenient solution.

14.4.6   DIPPING PROCESSES

These processes include dip-coating and dip-forming. They were adapted directly from the rubber industry, where dipping into latex with subsequent drying and curing have been used for many years to produce such articles as lined and unlined rubber gloves. Essentially the same process can be used to produce PVC-coated articles such as fabric gloves and wire goods, and other laminated products, or unsupported articles of appropriate shape such as disposable gloves. The process consists in dipping a former or an article to be coated into a suitable paste, which is allowed to drain and is then gelled in the usual way. Equipment is available having continuous conveying systems with associated dipping stations, gelation ovens, cooling stations, and stripping stations, so that production rates can be quite high and handling low.[21] The formers or article to be coated may be introduced into the paste at ambient temperatures or pre-heated to 100–160°C. Immersion must not be too rapid or bubbles are likely to be formed by air entrapment. The amount of paste 'pick-up' depends on:

(1) The viscosity of the paste.
(2) The temperature and heat content of the dipped article.
(3) The shape of the dipped article.
(4) Time of immersion.
(5) Speed of withdrawal from the paste.

Too rapid removal is liable to produce flow marks. Shear rates are low, about $1-5\,\text{s}^{-1}$, but a measure of dilatancy is often advantageous, as thixotropy or shear–thinning cause 'tear-drops'.

Generally viscosities in the region of 10–150 P are preferred, but in coating fabrics (e.g. in lined gloves) viscosity must be matched to the degree of penetration required. Relatively thick coatings or castings can be produced by judicious increase in viscosity where possible, or by building up a number of partially gelled layers followed by full gelation of the whole.

In metal coating the article is preferably previously treated with a primer such as a solution of a vinyl copolymer in a ketone, although the mechanical key is often enough, depending on the shape of the article. Inclusion of a cross-linking epoxy resin system in the paste can also yield strong PVC-metal bonding.[22] Hot dipping is also preferable in metal coating by the paste-dipping technique. A disadvantage of hot-dipping is that in long runs the viscosity of the paste increases as a result of the repeated heating, and blobs of partially gelled material may be formed. These tendencies can be kept to a minimum by careful control of temperature and by continuous slow stirring of the paste.

14.4.7   SPREADING

14.4.7.1   General

Unsupported film can be produced (Section 14.4.2) from PVC pastes by spreading on a suitable substrate, gelling, and stripping off the gelled film. In principle the same technique and arrangements of equipment are used to produce laminates,[23] though there are points of difference in detail. Several different types of laminate are produced by this technique including leathercloth for upholstery, protective clothing, and tarpaulins, floor coverings, coated felt, backed carpet, wall coverings, conveyor belting, and coated metal plate and strip.[24]

Most spreading is carried out continuously and a variety of machines is available to meet different requirements.

Viscosity and flow behaviour of pastes for spreading depend on the thickness required, the nature of the substrate, the degree of adhesion required, design of the equipment, particularly the means used for the spreading, and processing speeds. Thicknesses can be varied from below 0.01 up to about 0.5 mm or even thicker, corresponding to coatings of approximately $14-680\,\text{gm}^{-2}$.

Many different substrates are used to produce laminates with PVC, including fabrics such as cotton cloths; twills; ducks; drills and cambrics; synthetic fibres including nylons, polyesters, polypropylene, and glass fibre; and various forms of paper. The nature of the substrate, in particular its weave or fibrous form, contributes to the bond formed between it and the PVC. Paper must not be so smooth that inadequate adhesion is obtained. With synthetic fibres an adhesion promoter (e.g. an isocyanate compound) is usually desirable. 'Natural' fabrics may need to be scoured to remove seeds, etc., before use, and light calendering is desirable to press down longer surface fibres if the PVC coating is going to be thin. Moisture in the substrate will cause blisters or reduce adhesion, and drying before applying the paste may be necessary.

14.4.7.2 Knife coating

Probably the most common method for spreading PVC pastes onto fabrics is to pass the substrate between a 'doctor' knife and a backing roll, or a 'doctor' knife and a backing blanket of oil-resistant rubber running over a pair of rolls, the paste being applied to the substrate just before it passes under the knife and thus being spread to a layer the thickness of which is determined mainly by the clearance between the knife and its backing roller. The coated substrate then passes through a gelation oven, whose length depends on the gelation time required, thence to an embossing roller, and then reeled when the coating is sufficiently cold. The principles of these two arrangements are illustrated in Fig. 14.5.

Several layers of paste coating can be applied provided that the undercoatings are only partially gelled before application of the top coating, the general principle being the same as in building up thick mouldings. Also fabric coated on one side can be subsequently re-processed to provide a coating on both sides.

As well as producing leathercloth, coated paper, and similar products, the method can be used to apply PVC backing to woven, needle-tufted, or needle loom felt for carpeting, but where the pile might be crushed the 'lick' roller technique (Section 14.4.7.3) is preferable.

In spreading machines it is important that the doctor knife carrier should be adjustable in angle, position, and height above the backing roll or belt, so that coating thickness, penetration of the substrate, and surface quality can be controlled. Generally the thicker the blade of the doctor knife, and in particular the length of the flat parallel to the substrate,

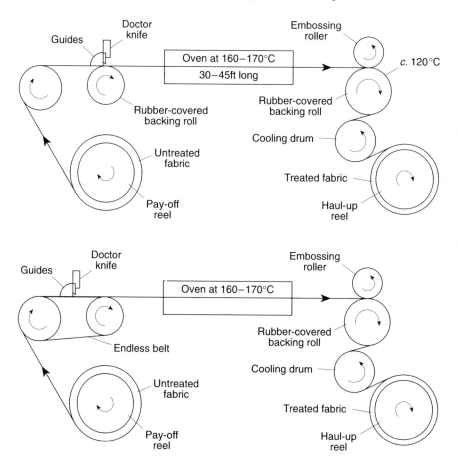

Figure 14.5. Typical knife-coating arrangements[14]

which is usually about 2.5–15 mm, the greater the penetration or 'strike-through' of lighter fabric substrates or adhesion to heavy fabric substrates. However, pastes sometimes tend to work up the back of the knife and then fall on to the surface of the coating to give 'weeping', 'creeping', 'blobbing', or 'tear-drops', a tendency reduced or eliminated by increasing the length of the flat parallel at the tip of the doctor blade. To permit rapid and convenient exchange of doctor blades it is usual to mount two or more on a revolving carriage with appropriate adjustments in relation to the substrate.

An alternative method of making PVC leathercloth is to spread paste on embossed silicone-coated release paper, and, after gelling and while still hot, rolling on the backing fabric, finally peeling off the release paper to give an embossed finish. The release paper can be re-used as long as it is not damaged.

Shear rates under the doctor blade in a paste spreading operation tend to be relatively high, in the range of 1000–7000 s^{-1}, though rarely as high as the latter figure, and 'shear–thinning' pastes are consequently preferred. 20 m min^{-1} is a typical knife coater line speed, and this is equivalent to about 3000 s^{-1}. Most knife coaters are run slower and thicker.[41] Dilatancy tends to encourage blobbing. The viscosity required depends on the nature of the fabric to be coated and the amount of penetration required, the lower the viscosity the greater the penetration, and vice versa. Obviously, open-weave fabrics tend to be penetrated by paste more than close-woven fabrics. Too low a penetration may result in unacceptably low adhesion between fabric and coating, but too deep a penetration will generally result in marked stiffening of the laminate, which spoils its handling behaviour. Viscosities should generally be not less than about 120 P, and are more usually in the range 350–700 P.

Formulation follows the principles already enunciated, the only special points to note being concerned with the preference for pastes which are shear–thinning, and with the need for careful consideration if there is any desire to include coarse particles, e.g. of resin or filler. Pastes containing coarse particles, say of filler polymer, tend to exhibit blobbing more readily than pastes without them, particularly as the thickness of coating is reduced. For this reason it is generally inadvisable to include filler polymers in coatings less than 0.1 mm thick, and coatings of 0.05 mm or less thickness require the paste to be triple-roll milled to remove small lumps.

Wall coverings account for about 15 per cent of the W. European and 25 per cent of the UK market, but only a small proportion is now produced by knife coating. Also much higher

Table 14.7 SOME TYPICAL PASTE FORMULATIONS FOR SPREADING

Applications	Upholstery		Wall covering	Clear top coats	
	Soft	General purpose		Soft	Hard
Paste resin (100 parts)	'High' viscosity	'Medium' viscosity	'Low' viscosity	'Medium' viscosity	'Low' viscosity
DOP or DIOP	85	65	45	70	30
Whiting or similar	15	15	15	—	0
Epoxydised oil	3	3	3	3	3
Calcium/zinc or barium/zinc with organic costabiliser	2	2	2	2	2
Pigment	1–5	1–4	1–4	—	—

filler concentrations with correspondingly higher concentrations of plasticiser are now common.[41] Tin maleates are also used in compact floor coverings of this type. Much wall covering now has expanded PVC, where foam structure, density, and colour are key considerations. For this reason potassium/zinc stabiliser/kickers have been developed. Foam density and colour can be controlled by variation of the potassium/zinc ration and reaction with appropriate acids and costabilisers.[49] As with other applications involving chemical blowing of PVC pastes the viscoelastic properties of the pastes are critical in determining the foam cell structure.[50]

Table 14.8 TYPICAL PASTE FORMULATIONS FOR SPREADING APPLICATIONS

Applications	Protective clothing	Tarpaulins	Conveyor belting	Carpet backing
Paste resin (100 parts)	'Low' viscosity	'Medium' viscosity	'Medium' viscosity	'Low' viscosity
Filler polymer	—	—	—	40
DOP or DIOP	70	30	15	30
Triaryl phosphate	—	—	75	35
Whiting or similar	—	15	15	55
Epoxydised oil	3	3	3	3
Calcium/zinc or barium/zinc with organic costabiliser or organotin	2	2	2	2
Calcium stearate	1.5	1	1.5	—
Pigment	1–4	1–4	1–4	—

A form of floor covering rather like linoleum can be produced by paste spreading on felt or other cheap base materials such as paper. These are usually made up in at least three layers, consisting of a cheap heavily filled base layer, an intermediate opaque layer which may be pigmented or printed, and a good quality clear top wear layer.

Nowadays most spread flooring incorporates one or two foamed layers (Section 14.4.8). In its simplest form this involves merely chemically blowing the intermediate layer, but it is common practice to introduce a layer of glass fibre between the foamed and compact base PVC layers. This provides good lay flat behaviour and dimensional stability. The base layer may also be foamed, by chemical blowing or mechanical foaming; and this layer may be substituted by a polyurethane foam with an integral skin.[51]

14.4.7.3 *Spreading by dip-coating*

Fabric can be impregnated or coated on both sides by passing it through paste and then through squeeze rolls or scrapers to remove excess before passing to a gelation oven. The supporting fabric may be heated to 100–110°C before passing through the reservoir of paste. In this way laminates such as conveyor belting can be produced in one pass with coatings up to about 1.25 mm thick on both sides and around the edges.

14.4.7.4 *Roller coating*

Single, reverse, and 'lick' roller techniques can be used for paste spreading on paper, fabrics, glass fibre mat and scrim, hardboard, fibre-board, and steel sheet or strip.[33] More accurate adjustment of thickness of coating is possible than by doctor knife techniques, but as a result of the hydrostatic pressures set up by the paste there is a tendency for coatings to be thicker towards the centre. Also, turbulence created by rotation of the 'pick-up' roller in the paste reservoir may result in entrapped air bubbles. The most common arrangements are reverse roller coating, the principle of which is illustrated in Fig. 14.6, or a combination of 'lick' roller and reverse roller coating, illustrated in Fig. 14.7. The second of these was particularly suitable for 'low' viscosity pastes, but these are now usually processed by reverse roller coating using 3-roll machines with 'nip-feeding'. This is the most common arrangement for the production of wall coverings, where viscosities are generally less than 50 poises.[41]

Straightforward 'lick' roller coating can be useful where the substrate being coated has a pile which might be crushed by doctor knives or rollers. Using the 'lick' roller technique to pick up paste from a reservoir requires viscosities generally between 100 and 400 P.

Formulation principles for roller coating processes are the same as for pastes generally. Where the substrate is fibrous in nature, such as fabric and fibre-board, adhesion is usually adequate without special treatment. With relatively smooth surfaces additional steps are necessary to achieve adhesion. These include priming the surface with heat-activated adhesive before coating or incorporating a cross-linking epoxy resin system in the paste formulation.[22]

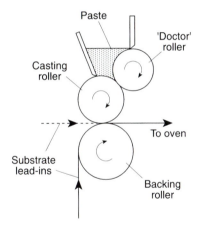

Figure 14.6. 'Reverse' roller coating[4]

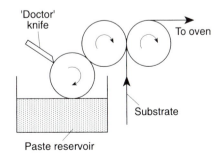

Figure 14.7. 'Lick' and 'reverse' roller coating[4]

14.4.7.5   Curtain coating

Higher speeds of coating, for example of steel sheet and strip, can be achieved by 'curtain' coating, which in principle involves dropping a thin 'curtain' of paste on to the substrate as it passes below the feed point on its way to the gelation oven. Viscosities need to be relatively low for this technique and organosols are usually necessary rather than undiluted plastisols.

14.4.7.6   Spraying

Normal paint-spraying techniques, including airless and electrostatic, can be used to apply PVC pastes as coatings, particularly to metal surfaces. Spray paints usually have viscosities as low as 4–5 P, so it is usually necessary to employ organosols rather than undiluted plastisols. Suitable viscosities are in the region of 10–30 P, but the largest air gap and fluid tip for the spray gun are usually necessary. Pressurised systems can cope with thicker pastes and thus permit thicker coatings. Diluents such as branch-chain dodecylbenzene are used at concentrations of 5–10%. White spirit is sometimes used, but yields thixotropic creams in some formulations. Ethylene glycol monobutyl ether is a very effective diluent, and is also useful in adjusting conductivity for electrostatic spraying.

When spraying on to vertical surfaces running can be minimised by using thixotropic pastes. An appropriate degree of thixotropy can be achieved by incorporating, for example, 0.5–2% of fine silica, 5–10% of a pigment dispersion, such as titanium dioxide in plasticiser, 10–30% of fine china clay, or a small concentration of carbon black.

After spraying, diluents are flashed off in an oven at 150–170°C, or even lower, for 5–20 min. Primers can be used to obtain adhesion.

Coatings as thick as 0.75 mm can be applied to a vertical surface in one operation.

The spraying process is widely used for car upholstery and under-body sealants where copolymers and pseudoplastic paste polymers are required with formulations having low fusion temperatures.

14.4.7.7   Finishing

Coatings made from PVC pastes on flat surfaces are often embossed, partly for decorative purposes but also to some extent to reduce surface tackiness. As indicated in Fig. 14.5 embossing equipment can be used in conjunction with spreading equipment, but there are distinct advantages in carrying out embossing as a separate operation. Where two or more layers of paste are being applied the embossing unit is only required for the top layer, and is therefore idle for a large proportion of production time. A separate embossing unit can serve two or more spreading units. Moreover, although coatings leaving the gelation oven are already soft ready for embossing, they are soft right through and consequently tend to be too deeply embossed. Separate heating of cold coated material softens the surface first, and can be more readily controlled to yield the required depth of emboss (Section 12.5).

Where the surface of a coating is undesirably tacky, so that it tends to block, soil, or be unpleasant to the touch, finishing lacquers are usually applied. These can be also used for decorative purposes by pigmenting them and applying them differentially in the valleys or on the peaks and ridges of the 'embossment. Most finishing lacquers comprise solutions of mixtures of vinyl and acrylic resins in suitable solvents. Where dull or matt finishes are required these may be obtained by including 0.2–0.5% colloidal silica in the lacquer.

14.4.8 PRODUCTION OF CELLULAR PRODUCTS FROM PVC PASTES

14.4.8.1 Using chemical blowing agents (Section 7.8)

The production of cellular mouldings from pastes containing chemical blowing agents has already been described (Section 13.2.3.3), since the methods employed are essentially compression moulding processes. It is possible to produce cellular 'moulded' products by free-blowing at atmospheric pressure, the resultant cell structure being intercommunicating, but control and reproducibility are poor. Usually the blowing agent should decompose below the paste gelation temperature, a suitable compound being azodicarbonamide, usually with a 'kicker'. Other possible blowing agents include 1,1′-azobisisobutyronitrile and monobenzenesulphonylhydrazine. As is usual with chemical blowing the agent must be thoroughly dispersed in the paste, usually by triple-roll milling. Also the paste must have a consistency which allows the liberated gas to expand and intercommunicate without escaping, at least up to gelation temperature. To this end a small proportion of a high viscosity plasticiser such as a polyester is beneficial.

In the formation of cellular PVC using chemical blowing agents an open cell structure results if decomposition of the blowing agent precedes gelation, whereas closed cell structures tend to result if decomposition occurs after or at the same time as gelation. Cell structure is largely determined by the melt viscosity of the fused plastisol and the rate of decomposition of the blowing agent. Relatively low melt viscosity combined with rapid decomposition of blowing agent favours formation of large cells, while higher melt viscosity and sustained slow decomposition of blowing agent favours development of small cells.[52]

Increase in blowing agent concentration up to about 2 per cent reduces the density of the cellular product, but beyond that cell structure generally becomes irregular without decreasing density. If it is necessary to reduce density still further a 'kicker' system must be used. 'Kickers' are nearly always required to obtain decomposition temperatures of 160–200°C.[41]

Open chemical blowing can also be used to produce cellular leathercloth and similar products like those produced by calendering, as well as unsupported cellular sheeting. It is not always necessary to incorporate the 'kicker' into the paste. In the production of coated embossed wall coverings, for example, the blowing can be varied to match the pattern by coating the 'kicker' onto the pattern layout.

The usual practice in the direct spreading process is first to spread an expandable paste containing chemical blowing agent directly on the fabric by the normal methods, to gel the coating partially, by heating at 120–140°C, then to apply a top coat of ordinary paste without blowing agent, and finally to gel the layers and expand the first layer in one operation. The latter preferably involves an initial rapid heating to about 190–250°C to form a thin gelled outer layer, followed by a decreasing temperature to complete expansion and gelation. Infra-red heating is a convenient means of achieving the rapid heating required, but the colour of the paste greatly affects heating time. For example, black pastes may require less than 0.5 min compared with 1.5 min for white pastes.[25] Infra-red is used for pre-heating/pre-gelling for leathercloth and for some floor coverings, the main fusion being accomplished with heated drums.

Formulation of expandable pastes is quite critical, and almost every ingredient has an effect on cellular structure and gelling behaviour. As mentioned earlier the blowing agent in particular must be very thoroughly dispersed, either by triple-roll milling into plasticiser or by triple-roll milling of the whole paste. Formulation of unexpandable top layers is not very critical provided they are sufficiently flexible. About 80 phr of DOP or DIOP constitutes an appropriate plasticiser concentration.

Indirect coating, also known as transfer or counter coating, is now the most common method used for coating fabrics. It involves forming a cellular sheet on a backing or carrier from which it can be stripped, and then sealing the sheet to the fabric by heat or by means of an adhesive. Preference for this type of process, particularly for lightweight products and with knitted fabric substrates, is due to the fact that penetration into the fabric is kept to a minimum, so that flexibility and drape of the laminate is at a maximum. Usually a thin layer of unexpandable paste is first applied to the carrier and partially gelled at 120–140°C. The carrier may be a fabric or paper with a coating (e.g. of silicone rubber) which limits adhesion of the PVC, steel belting, or even wire mesh. Maximum flexibility of the final laminate is achieved by keeping the thickness of the unexpanded layer down to 0.05 mm or even less. A layer of expandable paste is spread to the required thickness (usually 0.125–0.25 mm) on to the partially gelled compact layer and the laminate is then fully gelled with simultaneous expansion as in the direct coating process.

Where unsupported cellular sheeting is being produced the expandable layer can be first partially gelled at 120–140°C, and a thin layer of unexpandable paste applied before the final expansion process, so that a three-layer laminate results. This can be stripped off the substrate after cooling.

Where cellular leathercloth is being produced the layers of PVC can be laminated to the fabric by applying the latter to the expandable paste layer before it is gelled and expanded, or by means of adhesives. In the latter case the expanded/gelled product is cooled to below 70°C, coated with adhesive, stripped from the carrier, and laminated to the fabric by passing through nip rolls. Proprietary adhesives are used, and these are effective at room temperature, provided that sufficient time is allowed for them to act fully. This can take several hours, but setting time can be reduced considerably by heating. For example, activation at 60°C can reduce the time to as little as 5 min.

The various types of cellular leathercloth can be treated by the usual leathercloth finishing operations, such as embossing and lacquering. Care has to be taken, of course, that heating required for embossing is not sufficient to cause undue collapse of the cellular structure. Direct heating of the top layer to about 120–130°C is usually satisfactory.

14.4.8.2   Mechanical foaming

The essential feature of mechanical foaming is the use of a gas or volatile liquid to convert a paste into foamed form, followed by gelation in a mould or some other convenient forming device.[26, 37]

In the 'Elastomer' process a plastisol and inert gas such as carbon dioxide are pumped into and pressurised in a stirrer unit which mixes paste and gas vigorously. On discharge the mixture is ejected in a foamed state into moulds or on to a moving belt. Gelation is effected by the usual means, but HF heating is preferable where the thickness is more than 25 mm. Paste properties are rather critical for satisfactory performance, and a viscosity less than 150 P is indicated, corresponding to plasticiser concentrations of 70 phr and upwards with a 'medium' viscosity paste resin.

The 'Dennis' process[29] uses a countercurrent technique[30], in which a gas such as carbon dioxide at low pressure passes up a packed absorption tower down which the plastisol flows.

The resultant gassed plastisol is pumped under pressure from the base of the tower to moulds for gelation.

The 'Vanderbilt' process[31] is somewhat similar in principle to the 'Elastomer' process, but utilises air as the foaming agent. The air is pumped into the plastisol containing foam stabiliser[32] in a special mixer at about 0.85 Mnm^{-2} pressure. On release of the pressure the resultant foamed paste can be gelled in the usual way. The process is very specific in the nature of the resin used, and to a lesser extent on other ingredients.

The most common process involves mechanical frothing with air using 'foam stabilisers' and specially designed foaming equipment.[41]

All these processes yield open-cell structures.

14.4.8.3 'Physical' methods

A method which, though perhaps seeming rather crude, has been used effectively to produce micro-porous cellular products, consists in including in the composition a filler which can be subsequently extracted by a suitable solvent. Though the process can be used with sheeting made by calendering or extrusion it has been used more often with pastes of rather high viscosity. These are deposited and gelled in the required form and then treated with the solvent. The filler used depends on the pore size required but can be a simple substance like common salt which can be extracted with water.

So-called 'breathing' leathercloth has been produced by a process of this type.[32, 33]

REFERENCES

1. *'Corvic' Paste-making Polymers and Latices*, Tech. Service Note C104, I.C.I. Ltd (1968)
2. C. CAWTHRA, G.P. PEARSON and W.R. MOORE, *Trans. J. Plast. Inst.*, **33**, 104, 39 (1965)
3. 'Vestolit' polyvinyl chloride, Chemische Werke Hus AG (1966)
4. *'Geon' 121 paste resin*, Tech. Manual No. 2, British Geon Ltd. (1957)
5. R. HAMMOND, *Trans. J. Plast. Inst.*, **26**, 49 (1958)
6. B.S. DYER, *Trans. J. Plast. Inst.*, **27**, 84 (1959)
7. O. BJERKE, *S.C.I. Monogr.* No. 26, 370 (1967)
8. E. PEGGION, F. TESTA and G. GATTA, *Chimica Ind., Milano*, **46**, 9 (1964)
9. M.S. WELLING, *Plastics*, **21**, 225, 121 (1956)
10. *Plasticisers for PVC*, B.P. Chemicals (U.K.) Ltd. (1968)
11. *Plasticisers for PVC*, A. Boake, Roberts & Co., Ltd. (1953)
12. H.M. MACTURK and I. PHILLIPS, *Br. Plast.*, **28**, 10, 463 (1955)
13. I. PHILLIPS and P.G. YOUDE, *Br. Plast.*, **30**, 7, 297 (1957)
14. W.S. PENN, *PVC Technology*, Maclaren, London (1966)
15. E.R. NIELSON, *Mod. Plast.*, **27**, 9, 97 (1950)
16. G.M. POWELL, R.W. QUARLES, C.I. SPESSARD, W.H. McKNIGHT and T.E. MULLEN, *Mod. Plast.*, **28**, 10; 129 (1951)
17. A.C. WERNER, *Mod. Plast.*, **36**, 11, 126 (1959)
18. *'Welvic' Pastes*, I.C.I. Ltd. (1953)
19. R.W. GOULD, *Br. Plast.*, **35**, 11, 572 (1962)
20. C.C. MEAZEY, *Br. Plast.*, **32**, 2, 55 (1959)
21. Anon., *Br. Plast.*, **26**, 58 (1961)
22. P.P. HOPF and B.D. SULLY, *J. Polym. Sci.*, **48**, 367 (1960)
23. G.L. BOOTH, *Mod. Plast.*, **36**, 1, 91 (1958); **36**, 2, 90 (1958)
24. Anon., *Br. Plast.*, **34**, 5, 218 (1961); **37**, 10, 542 (1964)
25. *The Manufacture of Cellular PVC Leathercloth*, Tech. Service Note C105, I.C.I. Ltd. (1966)
26. I.D. MAXTON, *Trans. J. Plast. Inst.*, **30**, 86, 113 (1962)

27. A. COOPER, *Trans. J. Plast. Inst.*, **29**, 80, 39 (1961)
28. *Brit. Pat.* 737 367; 767 465; 770 237; 819 163; 819 164; *U.S. Pat.* 2 666 036
29. *U.S. Pat.* 2 763 475
30. G.J. CROWDES, *Mod. Plast.*, **34** 11, 117 (1957)
31. Anon., *Vanderbilt News,* **29**, 1, 2 (1963)
32. *Deckor Cellset 5 Vinyl Foam Stabiliser*, Scott Bader & Co. Ltd.
33. L.I. BLOOM and M. CLASPER, *Brit. Pat.* 1 150 803; Tech. Data Note W. TD201 I.C.I. Ltd. (1967)
34. L. UNDERDAL, S. LANGE, O. PALMGREN and N.P. THORHAUG, 'PVC Processing', *PRI Intl. Conf.*, Royal Holloway College, Egham Hill, Surrey, England (6–7 April 1978) Paper C4
35. N. STACKHOUSE, *ibid,* Paper C1
36. J. DE GASPARI, *Plastics Technology*, 23 (December 1990)
37. R.J. CRAWFORD, J.P. NUGENT and W. XIN, *Intl. Polym. Processing VI*, **1**, 56 (1991)
38. R.J. CRAWFORD and P.J. NUGENT, *Plast. Rubb. Comp. Proc. Appn.*, **17**, 23 (1992)
39. D-W. SUN and R.J. CRAWFORD, *Plast. Rubb. Comp. Proc. Appn.*, **19**, 47 (1993)
40. R.J. CRAWFORD, ed,, *Rotational Moulding of Plastics*, John Wiley & Sons, Inc., (1996)
41. M.J. HITCH and colleagues, private communication (1995)
42. M.W. ALLSOPP, *Manufacture and Processing of PVC*, M.W. Burgess ed. Applied Science Publishers Ltd., (1982) Chapter 7
43. P.H. GEIL, Macromol. *Sci.-Phys*, **B14** (1), 171 (1977)
44. M.J. BUNTEN, *Encyclopedia of Polymer Science and Engineering*, John Wiley & Sons, Inc., Vol. 17 (1989) pp 366 *et seq.*
45. W. SCHUBERT and H. STROLLER, *Plast. Kautsche.*, **14** (8), 552 (1967); **15** (4) 252 (1968)
46. O. BERGER and W. BRAUN, *Plast. Kausche*, **14** (9), 663 (1967)
47. G. GATTA, G. VIANELLO and G. BENETTA, *Chimica Ind.*, Milano **51** (11), 1234 (1969)
48. T. GILLESPIE, *J. Colloid Interface Sci.*, **22** (6), 554 (1966)
49. S.G. PATRICK and R.A. METCALFE, 'PVC 93 The Future', *IOM Intl. Conf.*, Brighton (27–29 April 1993) Paper 37
50. D.A. BUSCOTT and S.A. COULSON, 'PVC 93 The Future', *IOM Intl. Conf.*, Brighton (27–29 April 1993) Paper 36
51. R. BLASS, 'PVC Processing II', *PRI Intl. Conf.* Brighton (26–28 April 1983) Paper 27
52. H.A. SARVETNICK, *Plastisols and Organosols*, Van Nostrand Reinhold Coy. (1972)
53. J.K.L. BAJAS, *Plast. Tech.*, **22** (10), 34 (1976)
54. Anon, *Adhesive Age*, **20** (2), 26 (1977)
55. G. MATTHEWS, *Polymer Mixing Technology*, Applied Science Publishers (1982)

15
Miscellaneous processes

15.1 SINTERING

15.1.1 General considerations

Sintering processes using PVC are almost entirely based on powders rather than granules, but the powders may be resins, more or less as formed, usually in admixture with fillers, plasticised powders, or powders formed by grinding fully compounded material. Broadly speaking the processes may be divided into two distinct types, depending on the method used to distribute the powder before sintering. This may be either a spreading of some sort, spraying, or dipping into a fluidised bed of powder. Products manufactured by these sintering processes include unsupported porous sheets, mainly used as battery separators, protective coatings for metal plate and strip, heat-sensitive coatings for fabric interliners for clothing, etc., and protective coatings for metal articles other than plate and strip. All processes require the powder to be sufficiently thermoplastic to bond to a high enough strength for the purposes envisaged, but some require the material to 'flow-out' to form a more or less homogeneous melt, whereas others require little or no flow-out.

15.1.2 Production of porous sheeting

This process, mainly used to produce battery separators,[1-4] is superficially quite simple, but is not nearly so simple to operate in practice. The basic principle is to deposit polymer as a layer of powder on a continuous moving belt, which conducts the powder under a forming device which introduces ridges in the layer, and then conveys the formed powder through a sintering oven from which sintered sheet issues as a somewhat brittle, stiff sheet which can be stamped into appropriate shapes and sizes.

The sintering process and the pore size requirements of the finished product are most readily met by emulsion polymers, particularly spray-dried cenospherical types, either as such, or in the form of paste resins. Even with these shrinkage during sintering is liable to cause cracking of the sheet, but this can usually be avoided by incorporating fillers such as woodflour or silica with the polymer. In spite of the high temperatures of sintering, which are usually in the region of 250–270°C, and the relative instability of emulsion polymers, stabilisers are not required, the yellow or brown colour which results from degradation being immaterial in the applications for which the sintered products are used.

Requirements for the spreading operation are rather demanding and critical. Various different designs of feed hopper are in use, and each requires its own kind of powder flow behaviour. In general, smooth regular flow under gravity is required. Emulsion and paste polymers flow rather poorly and irregularly as a rule, and in this respect spray-dried polymers with cenospherical particles are advantageous. The types of behaviour required for the formation and retention of ridge structures in the spread powder are somewhat opposed to each other and to the requirements for satisfactory delivery from the feeding device. The ridges are introduced by passing the powder under a grooved roller, or under a forming 'comb', which is merely a fixed bar with slots corresponding to the ridges. It is obvious that the powder flow behaviour required for satisfactory shaping can be extremely critical, for not only does the powder have to take up the required shape under the action of the forming roll or bar, but it has to retain the shape after leaving the former, in spite of the movement of the conveyor belt. It is difficult to determine the behaviour of a particular polymer without actual production trials. Some help can be gained by laboratory evaluation of powder flow behaviour using a manually operated bar with slots of different dimensions or measuring the angle of repose of the conical heap of powder formed by running a standard quantity from a standard funnel at a standard height above a horizontal surface.

The conveyor must be such that corrosion by hydrogen chloride is absent and adhesive between it and the sintered product is minimal. For these reasons stainless steel or resin-treated fabrics are used.

An essentially similar technique can be used to apply plasticised powders to fabrics for the production of interlining material, though the temperatures used are very much lower. The powders may be either dry blends made from porous particle polymers with high proportions of plasticiser (so that softening temperatures are low) or powders made by freeze-grinding of compounded material of similar formulation.

15.1.3   COATING METAL PLATE AND STRIP

This process was developed by the British Iron and Steel Research Association as their 'Pacplate' process for the coating of steel strip for packaging applications. In basic principle it is similar to the sintering process previously described (Section 15.1.2), but the raw materials used are different and the products bear little if any resemblance, partly because here the objective is to provide the steel substrate with a strongly adhesive continuous film of PVC.

The steel strip is first degreased and coated with a phenolic resin/nitrile rubber adhesive in solvent, which latter is removed by hot air. The strip is next passed through a curing oven and heated to a temperature sufficient to melt the polymer. It then passes under the powder applicator, a double-roll feeder device which accurately controls the thickness of the applied layer of powder. The softened powder is compacted and smoothed by passing the strip through the nip between two rollers coated with polytetrafluoroethylene which form what is essentially a vertical two-roll calender, the coating preventing adhesion of the melt to the rolls. The coating is finally smoothed by pressing it by means of a silicone rubber roller against a highly polished steel cooling roll. When sufficiently cold the coated strip can be coiled up into reels.

As with the process described in Section 15.1.2, smooth flow of the powder feed is required; but it is also required to flow out under heat to form a continuous film. In the particular application for which the process and product were developed it is also required that the laminate should be formable by cold drawing and pressing without fracture of the PVC film. Processing conditions are sufficiently severe to volatilise some of the dioctyl type of phthalate plasticiser, and polymeric plasticisers are preferred, though there seems no reason why higher phthalates should not be satisfactory. For these reasons formulations of the following type were developed:[6]

Suspension resin of ISO Viscosity No. 80–90	100	parts
'Low' molecular weight polyester plasticiser	20	parts
Epoxydised soya bean oil	10	parts
Tribasic lead sulphate	6	parts
Titanium dioxide	3	parts
Pigment	2	parts
Antistatic agent	0.5	part

The powder is prepared from such a formulation by forming a plasticised dry blend in the usual way, so a porous particle polymer is positively indicated, although the plasticiser content is low.

The relatively low molecular weight of the polymer is required to achieve adequate flow at the low plasticiser content required. The antistatic agent assists the flow of the powder during application to the substrate.

15.1.4 FLUIDISED BED COATING

The general principles of fluidised beds in general,[7–11] and of their use in applying coatings of plastics materials[12, 13] including vinyl compositions,[10–16] have been described in detail, and are too well known to merit description here. Suffice it to say that the fluidised bed coating technique can be used with PVC compositions, but for a number of reasons has not reached a position of great importance in PVC technology generally. The reasons are mainly technical and are concerned with conflict between the requirements for (1) particular mechanical properties in the coating, (2) powder particle size and structure which will fluidise smoothly, and (3) formulation and powder structure which will result in easy flow-out to form a smooth coating. Attainment of the first of these is made difficult by the requirements of the other two.

For fluidisation the particle size and shape should be as uniform as possible and an average particle size not greater than about 300 μm with not more than a very small proportion larger than 500–600 μm, seems to be indicated. Powders can be made by freeze-grinding,[11, 14, 17] or by dry blending.[10, 11, 14, 15] The former process is rather expensive, and it is difficult to produce a uniform powder with it. Particle sizes of dry blends are determined mainly by the resins from which they are made, and they therefore tend to be of relatively narrow size range, averaging about 100–150 μm. While for fluidisation dry blends appear to be superior to ground compounded material, and they are certainly cheaper, they have not made much headway, at least in the UK, probably because their relative inhomogeneity tends to give poor 'flow-out'; during the melt stage, resulting in less smooth surfaces.

Formulation for fluidised bed coating is difficult, and while the broad principles are suggested in the literature,[14, 15] formulation of a satisfactory specific composition is another matter. Obviously increasing plasticiser content lowers the softening temperature and makes it easier to obtain adequate melt flow, but, on the other hand, high plasticiser contents can interfere with the fluidisation. Completely unplasticised formulations require low molecular weight copolymers to obtain adequate melt flow, and the resultant coatings tend to be brittle. It is preferable to include a small proportion of plasticiser, say not more than 15 phr, for harder coatings, to permit the use of a copolymer of reasonable molecular weight, but even then the resultant coatings are none too tough. Even with appreciable proportions of plasticiser it is not possible to get good results with high molecular weight homopolymers, and an ISO Viscosity No. of about 85 is probably the maximum that will give satisfactory melt flow.[11]

While octyl and higher phthalates can be used, there seems to be some advantage in adipates or sebacates such as DOA and DAS. Good thermal stability is required, but selection of stabilisers follows the usual principles (Chapters 5 and 8). Fillers interfere with melt flow at any appreciable concentration, and, since mechanical properties are marginal, it is probably best to omit them altogether.

The usual technique for fluidised bed coating is to heat the object before dipping it in the fluidised powder. After withdrawal the article is reheated to homogenise and smooth out the coating. The amount of powder picked up by the object depends on its temperature and its specific heat, the softening behaviour of the powder, and on the time of immersion. Temperatures are usually in the range 200–320°C,[16] and times of immersion up to 20–30 s, depending on the thickness of coating required. There is usually little advantage in extending the time of immersion beyond the indicated figure, as the low thermal conductivity of the coating considerably slows down further deposition. Adhesion is rarely a problem, as most coatings cling to the coated article by surrounding it, and grip tightly, owing to the much higher coefficient of expansion of the PVC than that of metals. In fact, this is liable to result in stresses in the coatings, which may account in part for their tendency to exhibit brittle behaviour. If specific adhesion is required, the article to be coated can be primed with appropriate adhesive prior to dipping.

15.1.5   POWDER SPRAYING

PVC powders of the kind described in the previous section can be distributed on a substrate by spray guns adapted to handle powders, but adhesion to the sprayed article is a problem. Electrostatic spray guns attack this problem by charging the powders so that they are held on the substrate electrostatically. The basic principle[19] involves applying a high voltage to a powder spray gun. Powder is passed by low pressure air along a flexible conveying tube to the nozzle. Here the powder passes through a metal chamber and receives a high negative charge from a sharp-edged electrode system. The sprayed powder is deposited on the surface to be coated and is held there by electrostatic attraction until it is fused, either by heat from the object being coated or by external heat (e.g. infra-red).

15.2   SOLUTIONS

Vinyl chloride homopolymers are rarely used in solution, because of their low solubility and the high viscosities of solutions of any appreciable concentration. Copolymers, however, particularly those with 13 per cent or more of vinyl acetate, are readily soluble in a number of solvents, such as ketones, esters, and some aliphatic chlorine-containing compounds. Viscosities naturally depend on the precise nature of the copolymer and plasticiser, and on concentrations, but range from about 200 to 5000 cP at a solids content of 25 per cent. Variation of viscosity with concentration typically follows a linear relationship when plotted on a logarithmic scale against concentration.[22] As with organosols (Section 14.2.2.5), viscosity can sometimes be depressed by addition of non-solvent organic diluents, such as toluene or xylene, but the concentrations required are much higher (e.g. up to 40 per cent replacement of solvent).

Preparation of solutions is relatively simple, although careful attention to cleanliness is necessary if high clarity is required. Mixers which induce turbulence, e.g. high speed propeller and turbine-type mixers, are preferable to those that do not to any extent, such as paddle stirrers. The copolymer should be carefully added to the stirred solvent, sufficiently slowly to avoid any build-up of undissolved polymer which might produce agglomerates. Where a solvent/diluent mixture is being used it can be advantageous first to wet the polymer thoroughly with the diluent, and then to add the solvent slowly while stirring. Haze in solutions can arise from incompletely dissolved polymer, particularly contamination by homopolymer, or from impurities such as dust. Discoloration may arise if steel equipment is used, but this can be inhibited by incorporation of about 1 per cent of propylene oxide.[22]

The main applications of vinyl copolymer solutions are as surface coating lacquers, where adhesion may be important. On porous surfaces, such as paper, cloth, wallboards, plaster, and concrete, surface coating films deposited from these solutions are usually satisfactory without special treatment, but on impervious surfaces, such as metals and glass, thorough cleaning and special treatment such as shot-blasting, priming, and baking at 180–190°C to expel all solvent and flux the polymer is usually necessary. This latter will usually require incorporation of heat stabiliser (Chapter 5), particularly when coating on iron, brass, zinc, and tinplate.[22] Intrinsically good adhesion is obtained, even with room temperature drying in air, with terpolymers containing about 1 per cent of maleic acid or anhydride. Terpolymers containing small proportions of hydroxyl groups are also available, but these usually require special two- or three-stage coating procedures to achieve the required adhesion.

Plasticisers may be included in copolymer solutions, if increased flexibility is required. They also lead to more rapid loss of residual solvent than occurs with unplasticised coatings.

Vinyl copolymer solutions can be applied by the usual lacquer techniques, including brushing, roller methods, and spraying.

15.3   LATICES

Emulsion polymerisation produces more or less stable latices (Section 3.3.2). These can be given increased stability by addition of appropriate surfactant emulsion stabilisers, and concentrated up to about 55 per cent solids content. At room temperature latices are generally stable almost indefinitely, although occasional small additions of ammonia may be necessary to maintain the pH above 7.5, below which coagulation may occur.[24]

If homopolymer latex is allowed to dry by evaporation of the aqueous phase, the polymer is deposited in a discontinuous powdery form. Although deposition in this way has been used to prime some fabric surfaces before application of a plastisol, it is much more common to plasticise the latex first, so that a continuous film can be produced later. Principles of plasticisation for end properties are as previously described (Chapter 6), and a minimum of about 25 phr is required to produce a useful continuous film after drying and fusing. For stability to be maintained during and after plasticiser addition it is usually necessary to add previously additional surface active agent to the latex. A suitable material is ammonium oleate, which can be conveniently formed *in situ* by adding aqueous ammonia, followed by the equivalent weight of oleic acid dispersed in the plasticiser.[24] Optimum concentrations of ammonium oleate are between about 2 and 3 per cent by weight of the latex, decreasing with increase in the proportion of plasticiser to be added. The plasticiser addition step requires thorough stirring.

Plasticised latices can be applied by padding, dipping, and spraying methods, for which the latices are satisfactory as they stand, or by spreading, for which the viscosity generally requires to be increased. This can be done by the addition of thickening agents, such as sodium alginate, or carboxymethyl cellulose, at concentrations of from 0.5 to 3 per cent based on the weight of polymer in the latex.

Pigments and fillers are best added as aqueous dispersions, which can be ball-milled to break down pigment agglomerates.

After application, the water should be evaporated off at 80–90°C. Coatings greater than 0.08 mm thick should be built up in successive layers not thicker than that, each layer being dried before application of the next. When dry the deposit has not appreciable strength and requires fusing at 150°C or higher if a continuous film is to be produced, but only a few minutes at this temperature are required. Where a latex is being used as a primer for a paste coating, fusion takes place simultaneously with the paste gelation. This heat treatment requires some stabilisation of the latex, but as little as 1 phr of a relatively inefficient stabiliser added to the latex is usually sufficient.[24]

15.4   THERMOFORMING PROCESSES

15.4.1   GENERAL CONSIDERATIONS

Much PVC is thermoformed by various methods (Section 9.3.5), including straight forming of thermally softened sheet, vacuum-forming, the various methods used to produce packages such as blister-packs, bending of pipe, shrink wrapping and fitting, and, of course, embossing (Sections 12.5 and 14.4.7.4). In any of these processes some molecular orientation is bound to occur depending on formulation, temperature, and amount of stretching. Changes in structure and mechanical properties resulting from orientation have been studied in some detail, and, although the studies have mainly been directed towards deliberate orientation, the results are relevant to all thermoforming processes.[37] With both unplasticised PVC[38] and plasticised PVC,[39, 40] monoaxial orientation increases tensile yield stress in the direction of draw and decreases it at right angles to the direction of draw. These changes are accompanied by development of a two-dimensional order. They could be increased by annealing in the stretched state; at 110°C for unplasticised and 120°C for plasticised PVC. The degree of order is greater than the maximum predicted from the level of syndiotacticity.[37] Biaxial orientation also increases ultimate tensile strength, but the level of two-dimensional order is lower than that obtained by monoaxial drawing.[37]

15.4.2 FORMING PVC SHEETS

Thermoforming obviously has to be carried out above the softening temperature but below the temperature range at which any melt flow occurs. Clearly stress-strain relationships, and, in particular, their variation with temperature, are of fundamental importance, but only brief analyses of the behaviour of PVC sheet in thermoforming have been made[25-27] (Section 9.3.5).

While plasticised PVC sheet can be and is formed, it is difficult to reproduce deep embossed patterns, though excellent results can be obtained with shallow shapings using heated tools.[26]

Most thermoformed vinyl sheeting is unplasticised homopolymer or copolymer. For many years the bulk of sheeting was produced by calendering, and three broad classes of unplasticised sheeting can be distinguished. These are (1) sheeting based on fairly low molecular weight homopolymer (ISO Viscosity No. usually around 85–90), (2) sheeting based on fairly low molecular weight copolymers (ISO Viscosity Nos. in the region of 65–85, and vinyl acetate contents of 8–15 per cent), and (3) sheeting produced by the 'Luvitherm' process (Section 12.6) from relatively high molecular weight homopolymers (ISO Viscosity Nos. in the region of 135–170). More recently extrusion of unplasticised PVC sheeting has increased steadily, and mainly produces material similar to the first two categories just mentioned.

Clearly the molecular weight, and the proportion and nature of comonomer, if any, are important factors controlling the thermoforming behaviour of sheet material, but no studies of the relationships appear to have been published. This may be because to some extent the molecular weight is often limited to a narrow range by processing requirements on the one hand and physical property requirements on the other. From first principles it would be expected that increasing molecular weight would lead to associated increase in force required to produce a given deformation during forming (i.e. increase in modulus), but that this would be associated with increase in the possible deformation before tearing occurred, though clearly forming temperature also constitutes another critical variable. One advantage of copolymers as compared with homopolymers is the fact that they offer the possibility of reduced forming temperatures without reduction in molecular weight. While sheeting based on low molecular weight homopolymer can be thermoformed and vacuum-formed to a limited extent, copolymers are usually necessary for deep drawings and for close reproduction of fine detail.

Temperature affects not only the thermoforming behaviour of a particular sheet material, but also the properties of the formed product. As might be expected, dimensional stability, as measured by shrinkage at 74°C, increases with increasing forming temperature.[25, 26] Warping, however, has been found to be at a maximum at intermediate forming temperatures. Unplasticised sheeting based on homopolymer was found to shrink less than plasticised sheeting,[25, 26] but although the materials were not characterised in detail, this is almost certainly a reflection of softening temperatures. It would be expected that best results would generally be obtained by forming at temperatures at which elongation at break of the sheet passes through a maximum.[28] However, while forming temperatures are usually in the range 90 to 140°C, with a preference for temperatures around 120°C, tensile studies at elevated temperatures indicate that elongation at break maxima are obtained in the region of 90 to 95°C for unplasticised PVC, and 80 to 90°C for plasticised PVC, depending on specific formulations.[41] Forming should be rapid, taking, for example, less than 60 s.[29]

15.4.3   FORMING TUBING AND PIPE

Unplasticised PVC tube and pipe can be formed by rather similar techniques to those used for some metals, by inserting a forming spring or filling with dry sand, heating to forming temperature, bending, cooling, and removing the spring or sand. Hollow profiles such as window frame sections can be bent by means of outer formers in hot oil.

Flange-like and expanded ends, sockets, and chamfered ends, can also be produced by heating to forming temperature and shaping with an appropriate tool.

Shrink-fitting of tubing around cylindrical objects can be carried out by first stretching, usually by air pressure, while heated to the low end of the forming temperature range (120–130°C) and cooling in the stretched state. The object to be fitted can then be inserted in the tubing which is re-heated to about 140°C so that it shrinks on to the object by recovery of the strain induced during stretching.

15.4.4   ORIENTED FILM AND SHEETING

PVC film and sheeting, both unplasticised and plasticised, can be oriented by control of 'blow-up' ratio and haul-off speeds during blow extrusion (Section 11.3.4) or as a separate operation. Indeed it is practically impossible to produce film or sheet completely unoriented, and it may be necessary to relieve stresses by controlled heating if the orientation is likely to lead to dimensional instability. Thus calendered sheeting is sometimes heated in reels or stacks by HF heating, in order to relieve stresses induced during haul-off and reeling. Monoaxial orientation is used to develop high strengths in the linear direction, for applications such as magnetic recording tapes and cable lapping. Biaxially oriented sheet is used for shrink-wrapping.

15.4.5   WELDING

15.4.5.1   General considerations
Being thermoplastic, PVC can be welded to itself by heating to sufficiently high temperatures. As usual the liability to degrade requires that some care be exercised to ensure that the material is not over-heated. Decomposed material between abutting surfaces reduces the strength of bond obtained. Because of the polar nature of the polymer, it can be heated by means of high frequency fields, thus making it possible to ensure that the heat is developed uniformly through the mass of material and at the joint, thus avoiding the excessive heating of the surfaces which is usually necessary with ordinary heat-sealing. Welding processes have been rather fully described and discussed elsewhere,[27, 30, 31] and it is not intended to go into much detail here, since they are not specific to PVC.

15.4.5.2   Film and thin sheeting
PVC films can be welded together by heat-sealing in the same way and with the same equipment as that used for polyethylene. For production purposes, however, high frequency welding is usually employed as it is much more satisfactory and rapid, because the material is heated uniformly throughout its mass and not by conduction from hot tools applied to the surfaces. Many types of machines and electrodes are available,[31] and, in addition to the usual types of seals for joining separate sheets and forming bags, wallets and sacks, the

method is used to produce a variety of speciality goods such as quilted and inflatable products.

15.4.5.3 Thick sheeting

For the welding of thicker forms of PVC some form of direct heat is usually employed. This may take the form of heated tools or jets of hot gas. Thus, in one method the surfaces to be joined are heated by passing a hot tool along them and then pressing them together. Alternatively the surfaces to be joined are passed over the heated tool before being pressed together to form the bond. Hot gas applied to the surfaces through a specially designed heating torch can be used to soften the surfaces for bonding in much the same way. Unplasticised sheeting of thickness greater than about 3 mm is conveniently welded by first chamfering the edges to be joined so as to form a V-shaped groove, and then softening the sloping sides by means of a hot-gas torch and filling the groove with softened welding rod of similar material to the sheeting.[29] Tools are available which feed the welding rod at a controlled rate as the torch traverses the groove. Similar devices and techniques can be used to weld joints between floor tiles or continuous sheeting when laid, so as to form an impervious floor covering, and tools which can be made to traverse the joints while welding are produced specially for this purpose.

15.4.5.4 Profiles

An important use of welding, for which special apparatus has been developed, is for the assembly of window frames from mitred extruded profiles. The technique, known as 'mirror-plate' welding, involves plates coated with PTFE which are located between adjacent mitred profile ends. They are heated to 260°C for about 43 seconds, followed by a similar 'off' period. The plates are removed and the profiles abutted together under controlled pressure to make the joints with a minimum of 'flash'. It is only necessary to machine off about 8.5 μm to produce a very neat joint. This may be achieved with knives attached to the assembly machine or as an independent operation.

15.4.5.5 'Non-flat' products

Joints between shapes other than flat sheets can also be made by means of hot-gas torches, usually using welding rod as previously described.

Where the shapes are appropriate joints can be made in unplasticised or lightly plasticised PVC by friction welding, that is by generating the heat required for softening by friction developed by movement between the two surfaces when moved rapidly relative to each other.

15.4.6   CEMENTING

Bonding plasticised PVC to itself by means of cements is rarely satisfactory, as wrinkling in the region of the bond usually occurs. Cementing, however, is often a convenient means of uniting unplasticised PVC products. A number of proprietary cements are available, but good bonds can be obtained by the use of fairly simple compositions of low molecular weight copolymer in solvent mixtures, e.g. 5 parts of a 15 per cent vinyl acetate copolymer of ISO Viscosity No. 60, in 25 parts of cyclohexanone and 70 parts of tetrahydrofuran.[29] Cements can also be based on other polymers, such as some acrylics, polyvinyl ethers, chlorinated PVC, and modified rubbers, and on other solvents, such as ethylene dichloride.

This technique is particularly useful in making joints between two pipes. When doing this it is usual to expand the end of one pipe and chamfer the end of the other pipe (Section 15.4.3), before applying cement and inserting one end in the other. Gap-filling cements are available for this purpose.

Cementing is also used to stick calendered sheeting to metal plate, and sometimes to produce laminates of PVC itself.

When using solvent-based cements caution is always necessary to avoid inhalation, and in the case of flammable solvents, such as toluene, to ensure that there is no possibility of sparks being generated in the vicinity. Proper protective gloves, masks, etc must always be worn.

15.4.7   MACHINING

Thin material, whether plasticised or unplasticised, can be cut by scissors, shears, or guillotines.

Unplasticised PVC can be machined by conventional wood and metal working techniques, but, as with other thermoplastics, special tools and operating speeds are desirable.[29]

15.4.8   SURFACE DECORATION

PVC can be printed,[32–36] painted, and decorated by such techniques as embossing and hot foil stamping, but metallisation does not appear to have met with much success.

Most methods of printing can be used on PVC, depending on the shape of the product to be printed and the type of print. Thus film and foil can be printed by rotogravure, aniline, letterpress, screen, and offset methods, though the first of these seems to be generally preferred.[34, 36] For textured and foamed vinyl floor and wall coverings the rotary screen system is preferred, but chemical embossing processes are also available.[35]

In printing processes the choice of solvent-base for the ink depends on the particular printing process used, and, in particular, the contact times involved. Thus ketones such as methyl ethyl ketone and cyclohexanone, or mixtures of these with hydrocarbons, are usual for rotogravure and aniline printing, since these give good adhesion to the PVC substrate. In letterpress printing, however, such solvents evaporate too quickly, and the consistency of the ink consequently changes, so that oil-based inks are necessary. Common ink pigments include titanium dioxide, chrome yellow, phthalocyanine pigments, and 'gas black' (Section 7.7.6). It is not clear that the EC Council Directive and the corresponding UK Statutory Instrument on the use of cadmium specifically ban cadmium pigments in printing inks, but they are in any case generally to be avoided. In addition to the pigment the composition usually includes some dissolved vinyl chloride/vinyl acetate copolymer.

Generally speaking it is easier to print on PVC than on polyethylene, and pretreatment, as required for polyethylene, is usually unnecessary. Poor print adhesion can sometimes occur as a result of a surface film of lubricant over the area to be printed. The only real solution to this problem lies in reformulation to avoid excessive migration of lubricant to the surface.

Unplasticised PVC can if desired be painted with normal paints, though some care needs to be taken to ensure that the solvent medium is not one which attacks PVC excessively. Moreover gloss paint finishes are commonly more brittle than PVC, and damage to the paint may initiate cracks in the PVC. For this reason two-part polyurethane paints are desirable, to obtain flexibility and so avoid the problem.

REFERENCES

1. *Brit. Pat.* 841 092
2. Anon., *Aust. Plast.*, **11**, 122, 19 (1955)
3. B. JEVELOT, *Industrie Plast. mod.*, **8**, 3, 6 (1956)
4. Anon., *Br. Plast.*, **36**, 2, 66 (1963)
5. *Brit. Pat.* 906 414
6. W. BULLOUGH and T.A. CANNING, *Metal Finish J.*, **9**, 108, 475 (1963)
7. J. GAYNOR, *S.P.E. Jl*, **15**, 12, 1059 (1959)
8. D.W.A. FARMER, *Br. Plast.*, **37**, 10, 558 (1964)
9. P.N. ROWE, *Sci. J.*, **1**, 4, 59 (1965)
10. R.J. BORSCH, *Mod. Plast.*, **37**, 11, 124 (1960)
11. W. ADAMS, 'PVC Sintering Processes', Chapter 7 in *Advances in PVC Compounding and Processing* (M. Kaufman, ed.), Maclaren, London (1962)
12. *Brit. Pat.* 643 691; 759 214; 852 638; *U.S. Pat.* 2 974 059
13. A. VAN DER HOEVEN, *Plastica*, **14**, 12, 1130 (1961)
14. T.H. MACEWAN, *Plastics*, **29**, 316, 109 (1964)
15. A.J. RENKIS, *Plast. Technol.*, **8**, 10, 33 (1962)
16. H.H. REINSCH, *Kunststoff-Rdsch.*, **10**, 5, 239 (1963)
17. A. BRACKEN and L.J. BRITTAINE, *Mfg Chem.*, **27**, 12, 497 (1956)
18. *French Pat.* 1 261 473
19. A.W. BRIGHT, *Product Finish*, **15**, 3, 54 (1962)
20. Anon., *Int. Plast. Engng*, **2**, 5, 223 (1962)
21. Anon., *Chem. Engng News*, **40**, 23, 42 (1962)
22. *'Corvic' S46/70 Vinyl Chloride/Vinyl Acetate Copolymer for Solution Coating*, I.S. Note No. 638, I.C.I. Ltd. (1960)
23. Anon., *Appl. Plast.*, **4**, 10, 54 (1961)
24. *'Corvic' Vinyl Chloride Polymers and Copolymers*, Tech. Service Note C104, I.C.I. Ltd. (1968)
25. N. PLATZER, *Mod. Plast.*, **31**, 3, 144 (1954)
26. N. PLATZER, Chapter 8 in *Processing of Thermoplastic Materials* (edited by E.C. Bernhardt), Reinhold, New York (1959)
27. J.M.J. ESTEVEZ and D.C. POWELL, *Manipulation of Thermoplastic Sheet, Rod and Tube*, Iliffe, London (1960), P.I. Monograph
28. S. MOTTRAM and D.A. LEVER, *Ind. Chemist*, May (1957)
29. W.S. PENN, *PVC Technology*, Maclaren, London (1966)
30. H.P. ZADE, *Plastics*, **24**, 259, 136 (1959); *Heat Sealing and High Frequency Welding of Plastics*, Temple Press, London (1959)
31. E.C. STANLEY, *Trans. J. Plast. Inst.*, **25**, 59, 15 (1957)
32. T.B. PYE, *Rubb. J. int. Plast.*, **137**, 10, 356 (1959)
33. P.P. BIRNBAUM, *Trans. J. Plast. Inst.*, **29**, 80, 60 (1961)
34. F.R. CLAYTON, 'PVC Processing', *PRI Intl. Conf.*, Royal Holloway College, Egham Hill, Surrey, England (6–7 April 1978) Paper C6
35. W.G. NIVEN, *ibid*, Paper C5
36. P. BAKER, 'PVC Processing II', *PRI Intl. Conf.*, Brighton, England (26–28 April 1983) Paper 21
37. M. GILBERT and Z. LUI, *Plast. Rubb. Proc. Appn.*, **9**, 67 (1988)
38. J.W. SUMMERS, *J. Vinyl. Tech.*, **3**, 107 (1981)
39. M. HERNER and G. HATZMANN, *Kunststoffe*, **62**, 81 (1972)
40. Z. LUI and M. GILBERT, *Polymer*, **28**, 1303 (1987)
41. D.J. HITT and M. GILBERT, *Mat. Sci. Tech.*, **8**, 739 (1992)

16

Properties and applications of PVC

16.1 INTRODUCTION

16.1.1 GENERAL

Variations in properties of PVC with variations in formulation have been discussed previously in some detail (Chapters 4, 6, 7, and 8), but some aspects of these properties are discussed more conveniently here, because they have such an important bearing on suitability or otherwise for particular applications.

The applications of PVC are now so diverse and so many that a mere catalogue would occupy several pages, and would perhaps serve little useful purpose. On the other hand, the diversity of the formulations used and of the applications into which they go, makes it difficult to consider the subject from the standpoint of basic principles, for one is really concerned with a whole range of different materials. Unplasticised PVC clearly differs in behaviour from plasticised PVC, and many different types of the latter exist. Even amongst unplasticised PVC there are some quite marked variations in properties and behaviour.

As a compromise, applications of PVC are here considered within the framework of a number of non-exclusive fields of applications or industries, but it must be recognised that the coverage cannot be comprehensive, or exhaustive. In any case the list of applications is changing almost daily, and there is little merit in compiling a full list which will be out of date within a short time.

For the above reasons it is not easy to distinguish those properties of the material which lead to its selection for so many applications.

16.1.2 COSTS

Clearly cost must be a major factor in determining whether or not PVC will be selected for a given application, and indeed it might be imagined that 'cost per unit of service or performance' might be the only major factor. Vinyl chloride homopolymers, and even copolymers, are among the cheapest polymers per unit weight in most if not all countries of the world. Density, however, at 1.4 for homopolymer, makes it more expensive than polyethylene at equal volumes.

The raw materials cost of a PVC composition, however, is usually markedly affected by the nature and proportions of ingredients other than the polymer, and these may tend to increase or decrease the overall cost. In addition costs of blending or compounding may also be involved. Consequently, the cost of feedstock to converting machinery such as extruders, moulding machines and calenders, varies widely, from a figure not much different from the price of polymer itself to figures four or five times as high. In the earlier discussion of the calculation of volume costs (Section 8.8.1) an example was given in which the raw materials cost of a composition worked out to be less than that of the polymer on which it was based, on both weight and volume bases, and this is not particularly exceptional.

16.2   PROPERTIES

16.2.1   GENERAL

PVC compositions fall into two fairly clearly distinguished classes, namely plasticised and unplasticised. The relatively rigid forms containing small proportions of plasticiser (up to about 20 phr) to reduce softening temperature and/or melt viscosity, are nowadays of comparatively little importance, except possibly for one or two speciality applications.

16.2.2   PLASTICISED PVC

The wide and varied usage of plasticised PVC is largely due to the wide range of flexibility and softness which can be achieved by variation in plasticiser content, a range unmatched by any other material, with the possible exception of rubber. In some applications the desired level of these properties is often unobtainable with materials other than plasticised PVC, and one wonders how we managed before the material became available. Natural rubber and the similar synthetic rubbers are the only materials which can offer similar mechanical properties over practically the whole range, but plasticised PVC usually scores on one or more of the following counts:

(1) Ageing/weathering.
(2) Clarity.
(3) Electrical resistivity.
(4) 'Handle' and 'drape'.
(5) Ease of processing.

Most of the speciality synthetic rubbers, such as silicones and polyurethanes, are usually inferior to plasticised PVC except with respect to a few specific properties, and are invariably more expensive. A notable potential exception is the range of ethylene/vinyl acetate (E/VA) copolymers now available. Unlike polyethylene itself, these copolymers have a clarity approaching that of plasticised PVC, and flexibilities, depending on the relative proportions of ethylene and vinyl acetate, overlapping a considerable part of the harder and stiffer end of the plasticised PVC range, without the incorporation of external plasticisers. They clearly hold an advantage where extraction of external platiciser (Section 6.3.4.6) is likely to be a problem. They are also of lower density (around 0.92–0.94) than plasticised PVC, and can be compounded with fillers, though claims to greater stability to ultra-violet exposure are rather surprising in view of the inferior stability of polyethylene and vinyl chloride/vinyl acetate copolymers compared with vinyl chloride homopolymer (Section 5.3.7).

The properties of a particular grade of plasticised PVC are determined by its formulation, and in particular the nature and concentration of plasticiser (Chapter 6), though the particular polymer (Chapter 4) and any filler (Section 7.3.2) used, may also have important effects. The properties required for a given application will, therefore, determine within limits the formulation to be employed. It must be noted that those properties which are dependent on plasticiser and filler contents are interdependent. Thus changes in formulation which increase softness, flexibility, and elongation at break will generally decrease tensile strength. It is essential to consider the various mechanical properties in conjunction with each other, otherwise, if each one is considered in isolation, it is possible to arrive at a combination of properties which is practically unattainable, a situation not unknown in specifications compiled by a user without discussion with suppliers. This interdependence of properties is partly illustrated by the plots of Figure 6.6, and, by way of further example, some of the data of Figures 6.3–6.6 are replotted in Figures 16.1–16.9.

These show how variations in tensile strength, modulus at 100 per cent elongation, elongation at break, British Standard Softness, and low temperature flexibility are dependent on each other when plasticiser content is varied in an otherwise constant formulation. It must be remembered that the values quoted were determined on specific compositions and that the other ingredients besides the plasticiser contribute to some if not all of the indicated properties. Moreover the accuracy and reproducibility of the test methods used to measure the properties are generally poor. Furthermore, any consideration of the mechanical properties of plasticised PVC may be confused by the possibility (Section 6.3.4.3) that stiffening may occur with the passage of time, both short and long term. Indeed it may be that some of the inaccuracy and irreproducibility of the test methods just mentioned is due to failure to appreciate and allow for this phenomenon. It may be necessary to take this into account in particular applications, but there is too little information available to justify more than a warning that this effect may be present, and may be sufficiently significant to require formulation modifications to compensate for it.

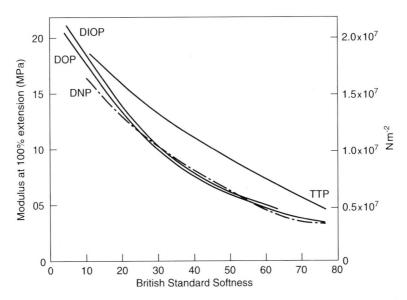

Figure 16.1. Variation of modulus at 100 per cent extension with British Standard Softness for various plasticisers

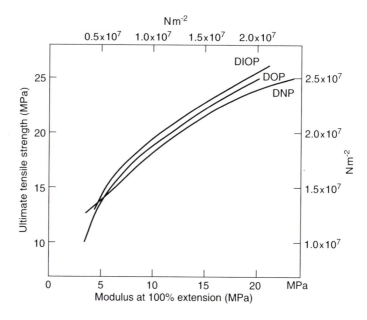

Figure 16.2. Variation of Ultimate Tensile Strength with modulus at 100 per cent extension for
some phthalate plasticisers

Bearing these considerations in mind it will be understood that Figures 16.1–16.9 can
only be taken as guides to some general trends, and any observations drawn from them must
be applied with caution. In general the trends of the individual curves are as expected, the
only possible doubts in this respect being where the slopes and shapes of curves for similar
plasticisers differ markedly. An example of this is in the plots of elongation at break against
modulus at 100 per cent extension (Figure 16.4), where DNP appears to be appreciably different in behaviour to DOP and DIOP. This behaviour is mirrored in the plot of elongation
at break against BS Softness (Figure 16.5), and possibly the plots of low temperature flexibility against elongation at break and British Standard Softness (Figures 16.8 and 16.9). In
all these cases the differences are too great to be accounted for by experimental error, and
it is very doubtful if stiffening with time would have yielded such smooth trends. The curves
of Figure 16.1 suggest that there may be a reasonable correlation between modulus and
British Standard Softness for the phthalate series of plasticisers. It should be noted from
Figures 16.3 and 16.5, that di-n-heptyl phthalate appears to be appreciably different from
the branched-chain phthalates, and further confirmatory data are required before a correlation could be suggested with confidence. In any case it is clear from Figure 16.1 that any
correlation could not be extended to other types of plasticiser, such as TTP. If the data of
Figures 16.2–16.5 can be trusted, there is a clear indication that a better combination of
mechanical properties can be achieved with DIOP than with DOP. The comparisons of these
two plasticisers in Figures 16.6–16.9 are confusing and the significance of differences
shown is not clear. The superior retention of flexibility of DOA compared with phthalates
is illustrated in Figures 16.7–16.9, though the indicated behaviour of the higher phthalates,
DIDP and DTDP, is anomalous. At relatively low concentrations, up to about 65 phr, their
retention of flexibility at low temperatures for given properties at room temperature appears

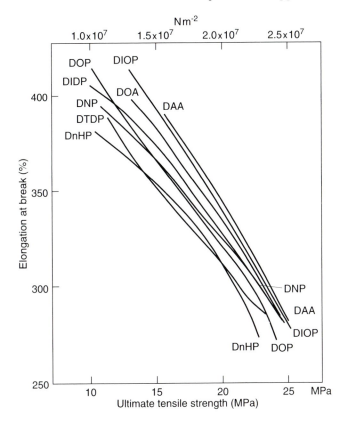

Figure 16.3. Variation of elongation at break with Ultimate Tensile Strength for various plasticisers

to be superior to the more common phthalates, but at higher concentrations their perform-ance levels off quite dramatically.

Some typical formulations for specific applications have been set out previously (Chapter 8), and some further applications are discussed later in this chapter (Section 16.3).

16.2.3   UNPLASTICISED PVC

16.2.3.1   General
Some properties of unplasticised PVC have been discussed earlier (Sections 4.7, 4.8, and 6.3), but it is convenient to consider some aspects of these in more detail here in relation to actual and potential applications.

The mechanical properties of certain specific types of unplasticised PVC have been eval-uated, described, and discussed in fair detail,[4-7] but while the data accurately describe the general order of magnitude and trends of these properties, attempts to apply actual figures to other types of unplasticised PVC are liable to be in error owing to (1) the important effects of variations in the polymer, particularly its average molecular weight, (2) the other ingredients of the composition, and (3) processing treatment.[80-84] A further difficulty in generalising about properties arises from the possibility, already referred to, that elastic properties may depend on the thermal and mechanical treatment the material under test has

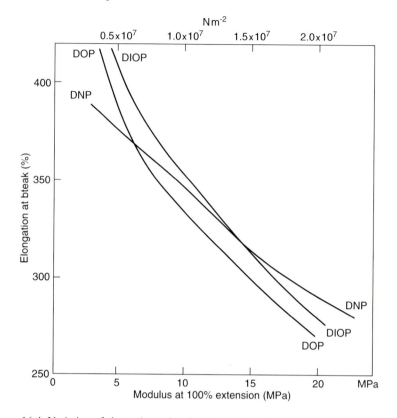

Figure 16.4. Variation of elongation at break with modulus for some phthalate plasticisers

received.[4, 8, 9] Thus static and dynamic moduli of unplasticised PVC were found to increase with repeated stressing above 60°C, though this was attributed to volatilisation of plasticising ingredients, rather than to changes in structure.

Bearing the above possible limitations in mind it may be useful to review some of the properties that might be relevant in the practical application of unplasticised PVC.

16.2.3.2 Mechanical properties

Unplasticised PVC has been likened to nylon in behaviour rather than amorphous polymers like polystyrene or polymethyl methacrylate.[7] Tensile strength is usually quoted as being around 55 MN/m² at 20°C, both for homopolymer and the copolymers normally used in unplasticised PVC. The corresponding recommended design stress level[6] is 6.9 MN/m². These values are probably based on homopolymer having an ISO Viscosity Number of about 85, although values as low as 50 MN/m² are common for compositions based on such polymers. For toughened compositions, modified by rubbers or fillers, tensile strengths are likely to be as low as 36 MN/m². Variation with temperature is typical for an amorphous rigid thermoplastic, tensile strength rising in almost linear fashion to about 124 MN/m² at −60°C and falling to about 41.5 MN/m² at 40°C. Values fall progressively more rapidly as the temperature rises above that figure, reaching about 21 MN/m² at 60°C and 14 MN/m² at 66°C. The values are, as usual, dependent on rate of straining, and the quoted values were obtained at a rate of 1 per cent per second.[6] Somewhat surprisingly, tensile yield stress appears to be

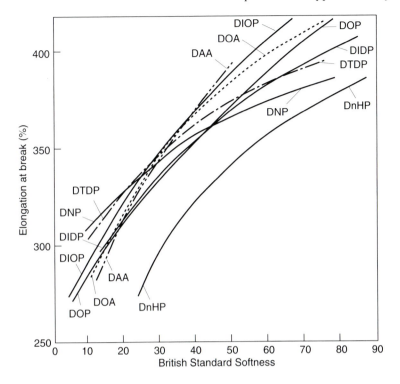

Figure 16.5. Variation of elongation at break with British Standard Softness for various plasticisers

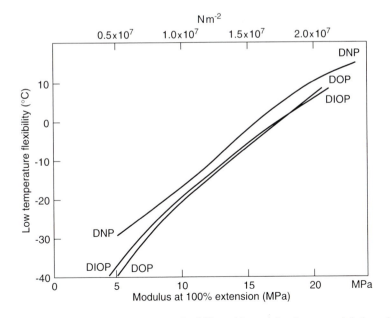

Figure 16.6. Variation of low temperature flexibility with modulus for some phthalate platicisers

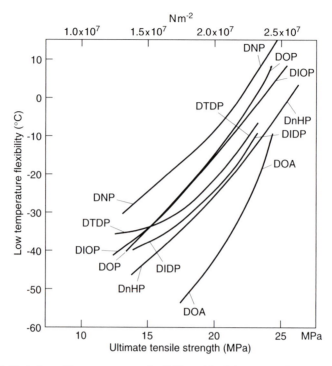

Figure 16.7. Variation of low temperature flexibility with Ultimate Tensile Strength for various platicisers

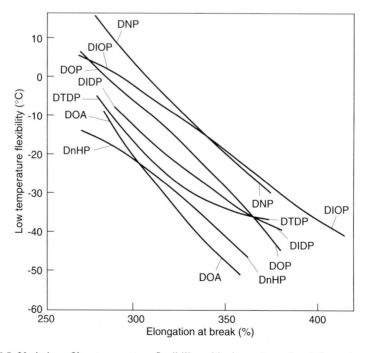

Figure 16.8. Variation of low temperature flexibility with elongation at break for various platicisers

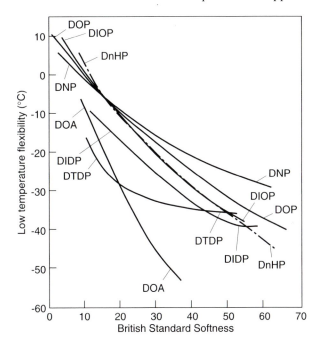

Figure 16.9. Variation of low temperature flexibility with British Standard Softness for various plasticisers

largely independent of molecular weight over a range corresponding to ISO Viscosity Numbers of 50–185, although flexural strength and energy to break increase progressively over the range of ISO Viscosity Numbers 50–125, but apparently remain constant above the latter figure.[7] It is possible that this latter observation was due to the fact that it is difficult to process unplasticised PVC based on homopolymer having ISO Viscosity Numbers above 105, so that specimens of higher molecular weight may not have achieved their full potential in mechanical performance.

Again, surprisingly, 'short-term' Young's modulus for unplasticised PVC, based on homopolymer of ISO Viscosity Number about 85, has been quoted,[6] at 3240 MN/m², as higher than that of polymethyl methacrylate (2839 MN/m²), and lower than that of unplasticised vinyl chloride/vinyl acetate copolymer (3520 MN/m²). Because of different creep behaviour, however, recommended design moduli for materials based on homopolymer and copolymer are both around 1030 MN/m², i.e. lower than that for polymethyl methacrylate (1380 MN/m²). A number of creep curves for unplasticised PVC have been published,[6] indicating that changes in apparent Young's moduli are similar to those for polymethyl methacrylate, polycarbonate, and toughened polystyrene for periods of about one month, but that after that period creep increases relatively faster. It must be noted, however, that these results were obtained on the behaviour of one specific material based on homopolymer of ISO Viscosity Number of about 85, and extrapolation to other unplasticised PVC compositions could be unwarranted. Creep behaviour is important in many unplasticised PVC pipe applications and creep curves for typical materials, together with related data on burst stress/time behaviour at different temperatures have been published.[85]

As indicated above, the behaviour of unplasticised PVC has been likened to that of nylon, and it has been pointed out[7] that the material is basically tough, the tough–brittle transition in flexure and tension occurring more than 100°C below the glass transition temperature, which is around 75°C. Plots of energy to break in flexure against temperature[7] exhibit sharp increases just below −50°C. The location of the tough-brittle transition region is dependent on the straining rate, the presence or absence of notches, and the composition. Thus in falling weight biaxial flexural impact tests a transition was revealed at about 5°C for 'normal' unplasticised PVC, and around −20°C for rubber-modified 'high impact' PVC, whereas in notched Charpy impact tests the corresponding transition temperatures were around 40°C and 0°C, respectively.[5] Corresponding tests on sharply notched specimens indicate transitions in the region of 10°C and −40°C, respectively.[7] Additives other than impact-improving rubbers can also markedly affect mechanical behaviour, and tough–brittle transitions in particular. Increase in impact strength by the addition of about 10 phr of a mineral filler of fine particle size has already been alluded to (Section 7.3.1), and this can yield impact strengths up to about 18.3 m at 23°C as measured by the Horsley method[5]. Similar proportions of titanium dioxide up to a maximum of 30 per cent increase Charpy impact strength and energy to break of notched specimens, and even small amounts of calcium stearate (up to 1.5 per cent) have been stated to increase energy to break. In both cases the effect has been attributed mainly to increase in energy of crack propagation, rather than to an increase in toughness due to reduction in yield stress.[7]

It should be noted that processing at high temperatures may induce brittle behaviour before signs of decomposition appear. This may be because phase boundaries, which provide regions for energy absorption and dispersal, have been melted out. The 'brittleness' may be removed by subsequent processing at lower temperatures, indicating that structural changes are responsible.

16.2.3.3   Thermal properties

For applicational design purposes one is mainly concerned with decrease in modulus and eventually actual softening with increase in temperature, onset of brittleness with decrease in temperature, and thermal expansion and contraction. While softening and brittle temperatures of one sort or another are widely quoted, they are usually of little value except as very rough guides. Some relationships between temperature and tensile strength and tough-brittle transition were discussed in the previous section; where it was seen that the latter is in the region of 100°C below the glass transition, which is about 75°C.

The softening temperature as measured by the method of BS 2782,[10] ranges from 74 to 79°C, dependent on the average molecular weight of the polymer, and additives used. Vicat softening points range from 72 to 77°C, while 1/10 Vicat softening points range from 82 to 90°C.

Thermal expansion and contraction, at about 7×10^{-5} per °C is sufficiently high to require consideration wherever products of fairly long length are liable to experience fluctuations in temperature such as those normally encountered even in the British climate. Examples are rainwater goods, where expansion of a long run of guttering would result in buckling, and PVC window installations, where each window has to 'float' in its aperture.

16.2.3.4   Electrical properties

For the majority of applications for which unplasticised PVC is likely to be considered its electrical properties are at least adequate, and some typical values are quoted here for completeness:

Volume resistivity	10^{15} Ω cm at 23°C
Power factor	0.02 at 1 kHz
Permittivity	3.0 at 1 kHz
Breakdown voltage	500 V/0.001 in (19.685 kV/mm)

16.2.3.5 Response to other chemicals

Here one is concerned with permeability of gases, vapours, and liquids, for applications such as packaging and materials transport; absorption to a degree that might affect the properties of the PVC product; migration of components of the PVC composition into materials with which it is in contact; and physical or chemical breakdown by strongly antagonistic media.

Permeability is very dependent on additives, and plasticisers generally greatly increase permeability to other substances (Table 16.1). Quoted water vapour permeability values are only slightly greater than that of polyethylene (483 g/μm/24h/m²), but permeability to gases is much lower, from 25 per cent down to 4 per cent of the values for polyethylene. Indeed even plasticised PVC is usually less permeable to gases than polyethylene. Low permeability is obviously important in 'modified atmosphere' packaging of foodstuffs, where shelf-life is increased by replacing the air in a package by an inert gas such as carbon dioxide or nitrogen before the package is sealed.

Table 16.1 PERMEABILITY OF UNPLASTICISED PVC[11]

Water vapour at 38°C	25 g/m³/0.001 in/24 h (0 to 90% RH) (632 g/μm/24 h/m²)
Oxygen at 20°C	0.1 cm³/m²/cm Hg/mm/24 h (0.075 cm³/m² kN/m² mm 24 h)
Air at 20°C	1.0 cm³/m²/cm Hg/mm/24 h (0.75 cm³/m² kN/m² mm 24 h)
Carbon dioxide at 20°C	0.5 cm³/m²/cm Hg/mm/24 h (0.375 cm³/m² kN/m² mm 24 h)
Hydrogen at 20°C	1.5 cm³/m²/cm Hg/mm/24 h (1.13 cm³/m² kN/m² mm 24 h)

The reaction of PVC to other solid and liquid chemicals has been reported in some detail and the potential user is referred to the extensive literature on the subject.[11-13] In general, unplasticised PVC may be regarded as being fairly resistant to attack by other chemicals.

Water absorption is low, but does produce some slight selling. Resistance to acids and alkalies is good, except for strong oxidising acids, although in general normal oxidising agents have little effect. While wet chlorine attacks PVC only slightly at elevated temperatures, bromine and fluorine will attack it at room temperature.

Most oils, fats, alcohols, and petrol have little or no effect on unplasticised PVC based on homopolymer, but the material is penetrated, swollen, and softened by aromatic hydrocarbons, and chloro-substituted aldehydes, ketones, nitro compounds, esters (such as plasticisers), and cyclic ethers. Solvent softening followed by volatilisation can be used for shaping. The only substances of any value as actual solvents are cyclohexanone, tetrahydrofuran, ethylene dichloride, and nitrobenzene, and even with these solvents solutions of around 15 per cent concentration are too viscous for most handling processes. Copolymers with vinyl acetate are much more soluble, in a wider range of solvents and solvent mixtures (Section 15.2).

Migration from a PVC product into an adjacent medium is of great importance in many applications, even when the amount of migrating substance is very small. Obvious examples are pipe for conveyance of drinking water, food packaging, and blood transfusion and other medical applications. The migrating substance could include any of the components of the PVC composition, including oligomer and degradation products. The range of possible materials involved makes it difficult to be specific, and each new case has to be examined individually for both exploratory and control purposes.

16.2.4 RESPONSE TO RADIATION

Properly formulated and processed unplasticised PVC is generally sufficiently stable to ultra-violet radiation to satisfy the requirements of exposure outdoors, at least in moderate climates. While transparent compositions are more prone to degradation than opaque ones stabilised with basic lead compounds, even the former can now be formulated for satisfactory performance outdoors, as witness the considerable quantities of transparent corrugated sheeting used for lightweight roofing. Flexible products introduce the added complication of the behaviour of platiciser, but the main problems here have been those due to loss of plastisicer by volatilisation or exudation. Extenders such as chlorinated paraffins are particularly prone to exude under ultra-violet radiation.

Atomic radiation generally cross-links vinyl chloride homopolymers but degrades copolymers.[14] The properties of unplasticised PVC change in ways that might be expected on cross-linking. It is perhaps noteworthy that impact strength can be doubled by a dose of 300 Mrad.[15] Radiation at a dosage of 2.5 to 4 Mrad is used as a sterilising medium for medical and surgical goods, and packaging. This treatment places severe demands on plasticised PVC compositions, which need to be unaffected chemically or in colour. Attempts to cross-link PVC by irradiation resulted in severe dehydrochlorination and chain scission due to high energies used (>20 Mrad).[40–42] Subsequently plasticised PVC has been successfully cross-linked by irradiation dosages around 4 Mrad of compositions containing suitable monomers, particularly polyfunctional acrylates and methacrylates.[16, 42] Cross-linking plasticised PVC in this way raises its maximum service temperature to an extent dependent on irradiation dose and concentration of cross-linking monomer, differences in tensile properties being more marked at 130°C than at room temperature.[43]

16.2.5 RESPONSE TO BIOLOGICAL AGENCIES

The interaction between PVC and biological agencies is of importance both from the point of view of breakdown of the PVC, and from the point of view of the possibility of its acting as a base for colonisation by organisms which may have harmful effects. The subject of microbial growth on PVC has received some publicity,[17] and a fair amount of work has been done on the subject,[17–25] most of it concerned with plasticisers, and mainly examining fungal rather than bacterial growth. Because of the large number and varied behaviour of different micro-organisms statements to the effect that any particular material is not attacked by them are rarely if ever justified. In so far as it is possible to generalise it may be taken that vinyl chloride polymer itself is not attacked. There is evidence to suggest, however, that residual emulsifying or suspension agents on commercial polymers can support growth of some micro-organisms under favourable conditions – favourable, that is, to the organisms!

Plasticisers vary considerably in their reactions, but it appears generally that aliphatic esters, such as adipates, azelates, ricinoleates, and sebacates are prone to microbial attack, whereas phosphates and phthalates are usually resistant. The resistant plasticisers, however, do not inhibit growth on aliphatic esters when the two types of plasticiser are mixed.[25] Lubricants such as waxes and non-toxic stearates constitute a very likely potential source of food for micro-organisms, whereas stabilisers based on lead and other metallic compounds, and possibly some of the organotin stabilisers, tend to inhibit growth.

It should be noted that observed growth of microbial colonies on surfaces of plastics articles can sometimes be attributed to contamination by some material which constitutes a suitable food for micro-organisms, and this can occur even when the material of the article is antagonistic to the organisms. Where the contamination is liable to be washed away, as for example in an unplasticised PVC water pipe, a high growth rate may be observed initially, but the organisms may disappear as the contaminant is washed away.[21]

The problems of contamination products in contact with PVC bearing microbial colonies is obvious. Effects on PVC itself are not always so obvious but microbiological attack on plasticiser can lead to stiffening,[20, 23] and it has been suggested that stiffening and cracking of PVC baby pants may be due to the same kind of effect rather than to extraction of plasticiser by repeated washing. Some types of staining of PVC have also been attributed to coloured compounds produced by microbiological agents. Thus pink staining or 'pinking' of vinyl upholstery, wall coverings, and doors has been shown to be due to a red dye produced by a species of *Streptomyces*, probably *Streptomyces rubrireticuli*.[26] A number of fungicidal and bactericidal additives, such as 1-fluoro-3-bromo-4,6-dinitro-benzene and 3,5,3′,4′-tetrachlorosalicylanilide[27] have been proposed, but clearly great care must be exercised in the use of such materials.

Also of occasional concern is attack of PVC products by termites and rodents, particularly liable to occur in tropical situations. Most observations of these attacks have been on plastics-insulated electric cables,[28] where the hazardous consequences are all too obvious. There is slight evidence to suggest that PVC may be more liable to attack than polyethylene. Repellants and poisons do not appear to be effective, the former often failing to repel and the latter killing the culprits after the damage is done. For these reasons and because of toxicity problems that arise, pesticides are generally not included in PVC compositions.

16.3   APPLICATIONS

16.3.1   GENERAL

It is not easy to find accurate information of the amounts of PVC used in different applications, but perhaps a rough indication is all that is required for present purposes. The main difficulties of acquiring the information arise from the very wide range of applications for which PVC is used and the wide range of different compositions involved. Very often figures are simply quoted for 'PVC', with no indication of the proportions of plasticiser or other additives, and hence no accurate indication of the amounts of polymer. The first three columns of Table 16.2 provide an approximate indication of the amounts used in a variety of applications in the UK during 1969.[29] There was steady if erratic growth during the following two decades, total UK consumption reaching 570 000 t/annum by 1992.[86] Up-to-date

Table 16.2 UK USAGE OF PVC

Product	1969		1992	
	1000 ts	%	1000 ts	%
Unplasticised PVC extrusions	59.5	20.4	199.5	35*
Cables	44	15.1	51.3	9
Flexible calendered sheet	45	15.4	45.6	8
Flooring	33	11.3	34.2	6
Coated fabric and paper	24.5	8.4	22.8	4
Flexible extrusions	15	5.1	22.8	4
Unplasticised sheet & film	12.5	4.3	62.7	11
Paste mouldings & solutions	11.5	3.9	—	—
Footwear	11.5	3.9	11.4	2
Records	10	3.4	5.7	1
Unplasticised PVC mouldings	8.5	2.9	34.2	6*
Belting	5	1.7	—	—
Unplasticised PVC bottles	4	1.4	51.3	9
Miscellaneous	8	2.7	28.5	5
Totals	292		570	

All figures approximate. * Estimated split.

accurate figures are difficult to obtain, but the figures in the last two columns probably give a fair indication of UK usage in 1992.[86, 87]

As might be expected the distribution of PVC uses between the various markets in North America is similar to that in the UK, but the proportion of unplasticised PVC is even higher, in spite of the fact that the major market of window profiles does not appear to have developed to the same extent as in Europe. The major US applications for UPVC are water, sewage, conduit, and drain pipes.[39]

16.3.2   ELECTRICAL AND ELECTRONIC APPLICATIONS

16.3.2.1   *General*
Although PVC can be rated as only a moderate insulator when compared with polymers like polyethylene and PTFE, because it has lower resistivities and higher power factor, insulation and sheathing of cables of various types has accounted for a major proportion of PVC consumption for many years. While a large and increasing amount continues to be taken up in these applications, their relative importance in the overall PVC situation has diminished as a number of other uses, such as unplasticised pipe and sheet in civil engineering, and a variety of forms in packaging, have expanded rapidly.

16.3.2.2   *Cables and wires*
The use of plasticised PVC for cable insulation and sheathing arises directly out of the need to find a replacement for rubber during the 1939–45 war, but the 'substitute' has shown itself to be so superior in various ways[30] that it has now largely displaced its predecessor. It is fairly easy to process in a wide range of flexibilities, is very easy to handle during installation, and, if properly formulated, is non-flammable and does not harden and crack on ageing. Many different formulations are employed,[31] a few typical ones being illustrated in Chapter 8 (Sections 8.9.1.5–8.9.1.8). As indicated in Chapters 6 and 7, volume resistivity is markedly dependent on the nature and concentration of plasticiser and filler used, gener-

ally falling with increase in the former and decrease in the latter, other things being equal. Pigments also have a significant effect, usually decreasing resistivity quite markedly. However, small concentrations of impurities are so effective in reducing resistivity as to render generalisations largely invalid.[32, 33] As an indication, volume resistivities measured at 23°C range from about 2×10^{12} to 6×10^{14} Ω cm in unpigmented insulation grades of PVC, but lower in corresponding pigmented grades, rising to a maximum of about 3×10^{14} Ω cm in high quality 'hard' insulation. Electrical resistivity is very dependent on temperature,[30] and a linear relationship between log volume resistivity and reciprocal absolute temperature is commonly observed, though departure from linearity due to polarisation is often noted at temperatures from about 40°C. Typically values for volume resistivity may be reduced by about one third over a rise in temperature from 20°C to 23°C. At 30°C the value may be as low as about one fifth of the value at 20°C, and by 50°C the value may have been reduced by a factor of around fifty. Breakdown voltage is generally in the region of 16–18 kV/mm at 50 Hz, and appears to be largely independent of plasticiser content. Power factor is relatively high, round about 0.1 (tan δ) at 1 kHz, while permittivity varies between 4 and 7 at the same frequency, though both these properties are very dependent on frequency and temperature.[31] Though lower than that of polyethylene, the resistivity of a properly produced insulation grade of PVC is adequate for most purposes, but the high power factor constitutes a limitation on the range of types of cable for which PVC can be used.[34, 35] In the UK PVC insulation has tended to be limited in power cables to a maximum of about 11 kV,[88] but considerably higher ratings are in use in the rest of Europe.[30] In spite of its limitations PVC is adequate for a wide range of low voltage power and other wiring requirements, and is used in many different types of cable. Most types are covered by standard specifications, which latter have been well summarised elsewhere.[30] Common forms of PVC cable include (1) the circular, in which a single conducting core is covered by a single layer of PVC, or a number of conducting cores are insulated, twisted together, and protected with a circular sheath of PVC, (2) twisted, in which individually insulated conducting cores are merely twisted together, and (3) flat, in which conducting cores are laid in parallel side by side in a single insulant or in separate insulating layers held together by an outer sheath. Power cables are usually variants on the first of these, mostly with metal armouring interposed within the outer sheathing or between two separate layers of sheathing. Communications, switchboard, and motor vehicle wiring are also usually of this general type. Wiring for lighting, and domestic power are usually of the third type, and certain types of instrument wiring also come in this category.

Other specific types of wiring insulated with PVC which deserve special mention are those for electric blankets and underfloor heating, both of which are of the first type mentioned above, spiralled retractable telephone leads, and wiring for coal mines.[37] Insulation for electric blanket heaters needs to be formulated to meet long running periods at slightly elevated temperatures, and polyesters and the higher alkyl phthalates are generally used as plasticisers in this application. Some care is necessary in formulating insulation for underfloor heating wiring to ensure that compatibility is maintained under running conditions, and extenders, if used, should be evaluated carefully before adoption. The elasticity of retractable telephone leads is mainly due to the spiral form of the conductor, but extra resilience is achieved by including adipate, azelate, sebacate, or similar ester plasticisers in the formulations. Colour is a problem in this application since the leads are usually expected to match accurately the instruments to which they are attached.

16.3.2.3   *Plugs and fittings*
Perfectly satisfactory domestic electric power plugs and similar fittings have been produced, but urea-formaldehyde moulding materials continue to be the most commonly employed in the UK. However, electronic and electrical goods are increasingly being supplied with a PVC plug moulded on to the cable end.

16.3.2.4   *Conduit, terminal boxes, etc.*
Extruded unplasticised PVC is now widely used as conduit for electric wiring, both in the rigid continuous tubular form, and in the flexible spirally wound form. Junction boxes and similar fittings are injection moulded from unplasticised PVC. In these applications the main attraction of PVC, compared with other insulating materials, is its freedom from corrosion and fire resistance, though ease of handling and installation are also important.

16.3.2.5   *Battery separators*
Porous PVC sheets are commonly used for the separators of accumulator batteries, and are produced by a variety of methods. Several thousands of tons of polymer are formed into this type of battery separator by direct sintering, as described in Section 15.1.2. Other types of sheet separator are produced by other methods, such as extraction of particulate fillers of appropriate size which have been compounded in before formation of the sheet by calendering or extrusion.[38]

Perforated calendered unplasticised sheeting is used to produce tubular separators.

16.3.2.6   *Business and related equipment*
Injection moulded unplasticised PVC competes with other rigid plastics in the production of cases for calculators, personal computers, monitors, printers, fax machines, photocopiers, and keyboards and disc drives etc. And there is also the ubiquitous credit card! The fire resistance of unplasticised PVC gives it an advantage in instrument cases over competitive materials in which objectionable brominated fire-retardants have to be incorporated.

16.3.3   COMMUNICATIONS

16.3.3.1   *Telecommunications*
The main use of PVC in telecommunications is as insulation and sheathing for wiring, as described in Section 16.3.2.2. Unplasticised PVC appears to be a satisfactory material for telephone instruments, but for what are almost historical reasons ABS is the main material used for these at present in the UK.

16.3.3.2   *Recorded sound*
Unplasticised PVC is used to produce both gramophone records and recording tapes. The former are produced, almost entirely by compression moulding, from low molecular weight copolymers (Section 13.2.2). In the former PVC seems to offer the best available combination of processing behaviour, playing performance, and cost. The copolymers are used mainly to achieve moulding cycle times comparable with those required for the earlier records based on shellac. Wear behaviour is superior to that of other common polymers, although the brilliance of sound reproduction may not be quite so good as with polymethyl

methacrylate. As shown in Table 16.2 consumption of PVC in records has dropped appreciably, presumably due to the competition of compact discs.

Recording tapes need, amongst other things, to have high strength and modulus in the longitudinal direction. They are therefore produced by calendering and stretching unplasticised PVC of relatively high molecular weight, usually by the 'Luvitherm' process (Section 12.6). In strength and modulus PVC tapes are somewhat inferior to polyethylene terephthalate, which now dominate the market.

16.3.4   BUILDING AND CIVIL ENGINEERING

16.3.4.1   General

The number and variety of applications of PVC under this heading are many and varied, and account to a large extent for the great expansion in PVC usage in the UK and throughout the world during the past couple of decades. The major products are extruded guttering and downpipe, pipe for water, waste, and land drainage, cladding, window and door frames, cills, and fascia boards, but other less publicised applications account for considerable amounts of polymer. Some of these products, especially cills and fascias, have a solid outer skin over a cellular core, so that they handle and machine like wood, except that the PVC products have a much more uniform structure than wood, and do not require painting or preserving. These are produced by co-extrusion of the two forms of PVC.

Some excellent trade publications on the use of plastics in building have appeared, and much literature relating to specific applications is available.

16.3.4.2   Pipework and similar products

Although unplasticised PVC was available for many years in pipe and sheet form, its use in the UK was limited for a long time to certain chemical plant,[12] and the first real breakthrough came when extruded guttering and down-pipe, and moulded fittings, became competitive with cast iron. The technical advantages of PVC over the latter are fairly obvious, and include freedom from corrosion and therefore ease of handling, transporting, and supporting. It was at one time thought that the relatively high coefficient of thermal expansion which PVC shares with other unfilled thermoplastics would lead to a tendency for distortion to occur, under the influence of even the relatively small temperature fluctuations encountered in Britain, but this has either proved to be unfounded or has been easily avoided by intelligent design. Improvements in injection moulding of unplasticised PVC have permitted relatively easy production of associated bends and other fittings, and its stiffness and superior weathering performance have now firmly established the material in this application.

Originally most guttering was made by slitting extruded circular section tube, but nowadays a wide variety of more or less complicated specially designed shapes are available, attempting to make best use of the inherent properties of PVC rather than copying existing designs in iron and asbestos. Some designs, for example, are extruded with a fascia board section which can be screwed direct to the ends of the rafters, thus eliminating the need for the usual wooden board. A limiting factor in the early days of the development was the existence of an already extensive range of designs of guttering and fittings in iron, with which PVC had to fit if an impact on the building industry was to be made. This stage has passed, and unplasticised PVC now accounts for a very high proportion indeed of the guttering, fittings, and down pipe being installed. A wide range of sizes is required in industrial

PVC piping, and with each size there is a wide range of injection-moulded fittings such as angle bends, T-joints, and so on, so that capital outlay in moulds for a PVC piping project can be enormous.

Other applications for unplasticised PVC pipe and fittings[41-43] are in conveyance of water, waste, and gas, and for land drainage, where the usual advantages over iron and steel apply. Here greater stiffness scores over polyethylene at sizes over 50 mm or so in diameter. For water conveyance the main problem has been that of avoiding any hazard of toxicity, which has severely limited the choice of stabiliser systems. This point is covered by the appropriate British Standard Specification,[44] which limits the amount of lead which may be used to concentrations equivalent to about 2 per cent of tribasic lead sulphate. With modern equipment and techniques, however, this is no longer a matter of great difficulty. Waste, soil, and land drainage pipes involve no such problem, and there is real concern only where there is a risk of very hot liquids or high concentrations of active solvents for PVC being passed into the pipes. Acceptance of unplasticised PVC for gas pipe was held up because some town gases contain relatively large amounts of aromatics which might soften the pipe, and HD polyethylene is now preferred. This material also competes with PVC for water and sewage pipes.

Some large diameter pipe for soil and waste systems is made from unplasticised PVC reinforced with filament-wound glass-reinforced polyester.

16.3.4.3   *Sheeting and similar products.*
Flat laminated unplasticised PVC sheet has been used for external cladding of buildings for decorative purposes, but because of the obvious limitations of the flat shape this use has not become very extensive. Much more development has occurred in the use of thermoformed interlocking sheets for the same application, but here sections made by extrusion processes seem more common. Lining of interiors, particularly of tunnels and similar constructions with flat sheets, curved in one direction where necessary, is quite common.

Flat or thermoformed clear unplasticised sheeting has been used to a limited extent for rooflights, but extruded corrugated sheeting is well-established and widely used for rooflights in roofs constructed from corrugated iron and asbestos, and for complete roofs, for example over balconies, terraces, penthouses, farm sheds, outbuildings, etc. Light transmission is about 80 per cent, giving similar lighting to glass.[39] Corrugations may be longitudinal or transverse to the length of sheeting, depending on the method of production, the latter somewhat unconventional form being particularly useful in some roofing situations. Although normally regarded as a 'rigid' material, unplasticised PVC in sheet form has a limited flexibility, and to some extent can be bent to conform to supporting frames which deviate from the flat. Various laminates of greater rigidity than unmodified sheet are available. Of these, laminates of galvanised mild steel woven wire mesh between layers of unplasticised PVC are mainly of interest in meeting fire resistance requirements. The woven wire is crimped so that a limited amount of thermoforming can still be performed. Laminates of unplasticised PVC sheeting with isophthalic polyester resins can also be made, thus combining the chemical and weathering resistance of PVC with the rigidity and heat resistance of glass-reinforced polyesters. Fire retardant versions of polyester are available for this type of laminate. A number of other PVC laminates are used for curtain walling and ceiling panels. These include laminates of unplasticised PVC with plywood, hardboard, aluminium, reinforced polyester, and expanded polystyrene, and some of these are being used for exterior cladding. Laminates of plasticised PVC, applied to steel an aluminium by paste-spreading or bonding of calendered sheeting, are used in curtain walling, infill panels, and outside cladding.

Translucent ceiling panels are made from plain or thermoformed unplasticised PVC sheet, or from plasticised sheeting suspended in frames.[39]

Another use of unplasticised sheeting in the building industry is as shuttering for concrete casting. It is possible to make the PVC adhere to the concrete, thus providing an *in situ* decorative finish. 'Hot melt' PVC compounds have also been used as moulds for precasting concrete landing slabs.[45]

Pressed thick clear plasticised PVC finds an interesting use in the form of flexible 'see-through' doors in industrial buildings, through which vehicles such as fork-list trucks can drive without previous opening. Similar doors are also made of extruded strips of plasticised PVC. At the other extreme laminates of fabric and plasticised PVC are used to construct folding doors and partitions which are both functional and decorative in houses, etc.

Another use for plasticised PVC sheeting in building is as a waterproof, flame-retardant roof lining material. In civil engineering somewhat similar sheet material is used as lining for reservoirs of various kinds, but here there is strong competition from other materials, such as polyethylene and butyl rubber.

16.3.4.4 Extruded sections

The use of interlocking extruded sections of unplasticised PVC was referred to in the previous section, and a wide variety of different forms is now available, some imitating wooden boarding, such as shiplap, the more adventurous attempting to take advantage of the special possibilities which PVC offers. Of rather similar nature are the forms of planking for fixing of wall copings, for wall protection and for outdoor fencing. A typical form of this is a U-shaped section with a flat base which constitutes the 'face', the two arms carrying small flanges which fit over corresponding supporting brackets. A similar principle is used for some forms of extruded coving with integral conduits for services. Other forms of coving or skirting are extruded to shapes similar to conventional forms made from wood. These are usually cellular or, more commonly cellular internally with a solid skin, produced by co-extrusion.

After a slow start some years ago, window frames, cills, and ancillary components are now available in unplasticised PVC in a wide range of designs,[40] and roller doors and shutters made from interlocking sections extruded from the same material have been available for some years.

An application for extruded PVC which has been established for several years is handrail section, which usually slips over a suitable metal or wooden supporting rail. This product is usually extruded from a fairly hard grade of plasticised PVC, without filler or with a relatively small proportion of synthetic calcium carbonate incorporated so as to obtain a high surface gloss.

In civil engineering construction of concrete structures where water retention is involved, water-stops (or bars) made by extrusion of plasticised PVC are used on a large scale. These are usually of dumb-bell or like section, one half of the section being embedded in one of two abutting concrete structures and the second half being embedded in the other, thus forming a seal between the two. Typical situations are in tanks, dams, reservoirs, conduits and underground rooms, passages, and tunnels.[39, 47, 48] Extrusion of water-stops can be difficult since their sections are relatively large. This means that the die land length required to ensure adequate back pressure to ensure proper gelation and homogenisation can be high, requiring a large die.

16.3.4.5   Window profiles

In the UK penetration of the window frame market was slow compared to some other European countries, but this has now been a major application for several years. Early frames were often unsightly at the joints, but development of the market was accompanied by improvements in mitring and welding techniques, and modern frames are of high precision and finish. PVC window frame sections are usually hollow and stiffened by internal steel or aluminium tubing, and the profiles are commonly of quite complex design. The precision with which they are produced is indicative of advances in machine and die design, formulation, and control systems during the past decade or so. Fabrication techniques, too, have progressed so that corners are cut accurately and joints are made neatly and effectively.

16.3.5   FURNISHINGS, LIGHTING AND ASSOCIATED FITTINGS

16.3.5.1   General

Some of the applications discussed in the previous section might be considered furnishings. Here we are mainly concerned with floor, wall, and window coverings.

16.3.5.2   Floor coverings

Vinyl/asbestos floor tiles were established for many years[39, 50], but asbestos is no longer acceptable because of its associated health hazard. Their main advantages over previous forms of tile were the possibilities of brighter colours, good wear properties, heat and mechanical resistance, and ease of cleaning. The asbestos has been replaced by other minerals such as talc, but the market for this type of tile has declined.

Flexible vinyl flooring, whether in tile or continuous form is quieter than vinyl/asbestos, and has better wear resistance, though it tends to be indented slightly more readily. Production of continuous flooring with printed patterns, and of fabric-backed flooring, has been referred to previously (Section 12.8.2). A variety of laminates is now available including one or two foamed layers. In those that have two foamed layers both may be PVC or one may be PVC and the other polyurethane (Sections 14.4.7.2 & 14.4.8).

Other uses of PVC in floor coverings include carpet backings using plastisols and rugs woven from extruded plasticised PVC.

16.3.5.3   Wall coverings

Uses of PVC in wall coverings fall into four distinct classes, namely unplasticised sheeting as wall panels, lightweight leathercloth, and wallpapers with a coating of clear copolymer or plasticised homopolymer. There are also various forms of embossed cellular wall-coverings providing attractive appearance with warm feel.

16.3.5.4   Curtains and fittings

While few people would accept PVC curtains for general use, calendered plasticised PVC sheeting and PVC coated fabrics do make very functional materials for curtains in damp situations, e.g. bathrooms, showers, and kitchens. Modern methods of printing permit the production of PVC curtain materials in quite complicated multi-coloured patterns, and the curtains can be very attractive as well as functional.

Many types of curtain rail are made by extrusion of unplasticised PVC and many of the associated fittings are moulded from the same material. Here PVC scores over brass, etc., by virtue of its freedom from corrosion, the ease with which it can be fabricated into quite complicated forms, its quietness, and its bright colour.

Venetian blinds are made from extruded sections of unplasticised PVC. The advantages over metals are freedom from corrosion, self-colouring, and quietness in use. A related application is the use of louvred calendered unplasticised PVC in external window blinds.[52]

16.3.5.5 Lighting fittings
The use of PVC sheeting as ceiling panels for illumination has already been referred to (Section 16.3.4.3).[39] Opaque white unplasticised PVC sheeting is also used as reflectors for fluorescent tubes, and translucent material is used for diffusers and globes where the temperatures reached are not too high.

16.3.5.6 Furniture
PVC leathercloth is used extensively for upholstery, and this application must account for a goodly proportion of paste resin manufacture. Copolymer coatings and other forms of laminate are beginning to find application in furniture construction, and shrink-fit extruded tubing from PVC is used to a limited extent. A much-publicised but controversial use for PVC in furniture is in inflated chairs and the like fabricated from plasticised PVC sheeting.[53]

There is now a widespread market for laminates of printed PVC veneers with chipboard, hardboard, plywood, and metal, in the production of furniture, shelving, and decorative panels. Typical formulations include processing aid, impact modifier, and about 25 phr of plasticiser. The veneers are produced by calendering, usually printed by rotogravure and frequently embossed.[89] Connectors between units of 'self-construct' furniture are made from extruded sections.

16.3.6 CHEMICAL ENGINEERING

Its chemical and corrosion resistance, coupled with its relatively low cost, have made unplasticised PVC eminently suitable for many applications in chemical plant,[12] though the material is sometimes barred where certain specific chemicals which attack it are likely to be encountered.[11–13] Extruded pipe and tubing, moulded fittings, pumps, and valves, made from unplasticised PVC are used for conveyance of water, chemicals, and effluent. Ducting, hoods, and ventilators for exhaust gases and fumes are usually fabricated from sheeting.[39] Troughs, tanks, and tank linings are fabricated from sheeting which can be welded together at the seams to provide a liquid-tight construction. Self-supporting tanks are made from the PVC/polyester laminates referred to earlier (Section 16.4.3). An interesting use for unplasticised PVC foil is as a protective and decorative wrapping for thermal insulation materials round piping, thus providing it with a smooth surface finish.

Uses for plasticised PVC in chemical engineering include filter cloths woven from extruded filaments, and conveyor belting made of PVC/fabric laminates. Flexible and rigid PVC piping is used in food manufacture.

16.3.7   CLOTHING

The earliest domestic use of PVC was in the form of calendered plasticised sheeting fabricated into plastic raincoats. Although many of these were unsatisfactory, mainly as a result of a proneness to tearing, and excessive stiffening in cold weather, due to incorrect formulation or inadequate processing, there is no doubt that this is a good application for the material if properly carried out. Some years ago there was a minor explosion of the use of PVC in clothing due to the introduction of attractive and functional fabrics (Section 14.4.8) and a fashion craze for vinyl clothing. Although the craze has died down there is still a market in 'fashion' clothing. In addition PVC is widely used for foul weather and industrial protective clothing such a fishermen's 'oilskins' and aprons.

A major use for PVC is in footwear, though consumption in the UK in this area has not risen over the past two decades. Various forms of PVC-coated fabric are used for uppers and top-pieces for shoes, bootees, and sandals, while unit soles,[54] direct moulded soles,[55] moulded soles and upper units, and 'Wellington' boots are in widespread use. The use of PVC in these applications has been dependent not only on the properties and price of the material, but also on the development of machines particularly suited to economic production of a variety of types of footwear (Section 13.3.2). For soles and heels the wear behaviour is obviously important. The hardness/softness of the composition is of prime importance,[50] and the BS Softness should generally be between 60 and 90, the optimum usually being around 70–75. Given proper formulation and processing, durability can be at least as good as and often appreciably better than that of previously established materials such as leather and rubber,[57] and PVC 'Wellingtons' can be recycled.

16.3.8   PACKAGING

16.3.8.1   General

One of the most widespread and oldest uses for PVC, namely as supported and unsupported flexible sheeting or leathercloth for handbags, wallets, shopping bags, suitcases, sports bags, etc., where the material scores on the grounds of attractiveness, wear behaviour, and low cost, is strictly speaking a packaging application; but here we are more concerned with conventional packaging of commodities for transport, storage, and display. Even under this definition there is another considerable application for PVC which one would not usually think of as packaging, namely covers for books and wallets for photographic transparencies. These are usually fabricated by HF sealing of calendered plasticised sheeting.

PVC, both unplasticised and plasticised, is used for packaging in many forms of a wide variety of different commodities. Some, such as for example metal tools, are completely innocuous in contact with PVC, whereas others, such as oils, may extract plasticiser and thus require special formulation (Section 6.3.4.6); others may migrate though the polymeric phase, while still others may be destined for ingestion and therefore raise questions of toxicity. Migration and extraction have received a good deal of study and a good range of data exists, particularly in the trade literature, but new materials to be packaged require special tests to be carried out because the behaviour can be very specific to particular materials.

The specific permeability behaviour of PVC is attractive for a number of packaging applications, and the slow initial development in this field can be attributed to an unnatural fear of the difficulties of processing PVC arising from its relative instability.[58] This problem has been dispelled, and PVC is now used in a wide range of packaging.

16.3.8.2   *Bottles, bags, sacks, and similar containers*

The early development of PVC bottles was much slower than was predicted in spite of some obvious advantages of the material (Section 11.3.6), but once teething problems and prejudice were overcome expansion was quite rapid; and in spite of criticisms in recent years this market still constitutes a major one for PVC (Table 16.2). Formulation and equipment for blow-moulding PVC bottles have been discussed earlier and a fair amount of information is available in the literature.[68–75, 90–93] Bottles are also blow-moulded with bi-axial orientation, which improves drop resistance, bursting strength, modulus, permeation, transparency, and brilliance.[93] Bottles can also be produced from PVC foil by blow-forming in two halves and then sealing the two halves together.[76] The main advantages of unplasticised PVC bottles have been listed[74] as:

(1) Excellent chemical resistance and resistance to stress cracking especially by oils and greases.
(2) Low permeability to air, flavours, essential oils, and freedom from panelling.
(3) Rigidity, permitting the use of thin wall.
(4) High clarity with certain grades.

These properties make the material particularly suitable for motor oils, cosmetics, toiletries, household disinfectants, spirit-based polishes, car polishes, mineral waters, fruit squashes and essences, wine, vinegar, etc.

Little interest has been shown in plasticised PVC bottles. Though they could undoubtedly be made without much difficulty, presumably they would offer no advantages over and would be more expensive than bottles of LD polyethylene. Tubular containers of the 'squeeze' variety are made from plasticised PVC by extrusion of tube followed by end-sealing and sealing in a screw neck.[77] They are used for paint concentrates, wax polishes, and creams. The compositions used usually contain sufficient plasticiser to make the tubes more flexible than could be obtained with LD polyethylene. Great care has to be exercised in selecting plasticisers to ensure that they will not be extracted too rapidly by the specific materials being packaged.

Sacks and smaller bags are essentially similar in principle to the tubes just mentioned, although some have been made by sealing calendered sheeting rather than by the conventional extrusion-blown film technique. A typical formulation for sacks is given earlier (Section 8.9.1.3). A plasticiser content equivalent to about 45 parts of DIOP and a filler content of about 25 parts appear to yield optimum properties for the application. Sacks made from such compositions are superior to those made from LD polyethylene in that they are softer, they stack rather better because the softness and the filler content reduce slipping, and they have a lower tendency to snagging. In addition the HF sealing of PVC is less susceptible to dust and more easily adaptable for the production of valved sacks. The main disadvantages are a tendency to brittleness at low temperatures and the relatively low softening temperature. The former can easily be overcome by replacing some of the phthalate plasticiser with an aliphatic ester plasticiser, but the latter fault seems insurmountable and precludes PVC sacks from packaging materials which are delivered hot, e.g. some fertilisers. In spite of the advantages of PVC sacks for some products they have made little headway in Europe except in Italy.

Clear plasticised PVC bags are used for some packaging, such as of gramophone records, but these applications are mainly filled by LD polyethylene and polyethylene terephthalate in the UK.

16.3.8.3   *Films and foils*

One of the factors which for many years held up the development of PVC films in packaging was the fact that film less than about 0.05 mm thick was not generally available, although it had been technically feasible from the early 1960s to produce film down to 0.0025 mm by blow extrusion. Usage of PVC film and foils expanded even in the UK during the mid 1970s, the main forms being unplasticised film and foil produced by the 'Luvitherm' process (Section 12.6), and unplasticised and plasticised film produced by extrusion blowing (Section 11.3.4). 'Stretch wrap' film and 'cling-film' are usually extrusion blown from plasticised PVC, slit, and then re-reeled in flat film form. It is used for wrapping foodstuffs such as meat and fruit. For this kind of product appropriate non-toxic plasticisers and stabilisers have obviously to be selected, and anti-fogging agents are often included to prevent the formation of droplets of water on the inside of the film. 'Shrink wrap' films are also extrusion blown from plasticised PVC and then slit, though the plasticiser content is generally somewhat lower than for stretch wrapping. Commercial films will shrink by about 60 per cent on heating, depending on the degree of orientation introduced during production. Shrink wrapping is mainly used for over-wrapping, and window and display packaging where toxicity is not usually important, but formulations appropriate for direct food wrapping are not difficult to devise.

Thin unplasticised film is mainly used for overwrapping, where it competes with regenerated cellulose, to which it is superior in dimensional stability because of its relatively low water absorption, while being comparable in rigidity, clarity, and sparkle.

Thicker films and foils, mainly of unplasticised PVC, are now commonly formed into a wide variety of containers and trays and vacuum-formed packaging inserts for the packaging and display foodstuffs and confectionery, such as margarine, cream, yogurt, cut meats, and chocolates.[78] Blister, skin, and snap packs are also formed from unplasticised PVC foils.

16.3.8.4   *Coatings*

Lacquers based on soluble copolymers are used for lining metal cans and drums. Coatings which are strongly adherent can be produced by including small proportions of terpolymers such as those with maleic anhydride (Section 3.2.3), with or without epoxy resin.

16.3.9   MEDICAL AND SURGICAL APPLICATIONS

Plasticised PVC now has a variety of uses in medicine and surgery. Obviously special attention has to be paid to formulation where the product is going to be in contact with a human body; and even more so where substances in contact with the PVC are to be introduced into the body. An early application was to form simulants of external parts of the body in plastic surgery and artificial limbs. Its transparency and flexibility have led to plasticised PVC being used for blood bags and blood transfusion, drip, dialysis, and urinary bags and tubing; also for oxygen tents and waterproof mattress sheets. It may be noted that blood is found to keep longer in PVC bags than in glass.

16.3.10   TOYS AND SPORTS GOODS

Play-balls of all sizes rotationally cast from PVC plastisols have made a tremendous impact on the toy and sports goods markets, since they were introduced. They can readily be made in different colours, are easy to inflate, and are much cheaper than leather balls. Soft toys, particularly figures such as animals and puppets, and even some fairly rigid toys, are also rotationally cast or slush moulded from plastisols.

16.3.11   TRANSPORT

A considerable quantity of plasticised PVC is used in cars and other road transport, much of it in forms discussed earlier (Sections 16.3.2.2, 16.3.2.5, and 16.3.5.6). PVC-coated fabric is used extensively for the upholstery, interior panels, linings and trim, and folding hoods. Clear PVC sheeting is used as windows in the latter.

Cables with PVC insulation and sheathing are used for wiring the electrical systems, where the ease with which the material can be coloured in variety is of value because of the need for easy identification of the different conductors.

Armrests, crash pads, and other padded fittings are rotationally cast from PVC plastisols, and filled with polyurethane foams.[78] Blow moulding is also used to produce these types of fitting. Sleeves and gaiters, e.g. for gear levers, are also cast from plastisols, while protective covers for coils, plugs, and distributors are applied by dip-coating.

Unplasticised sheeting is used for internal lining of vehicles and aircraft; also internal lining and cladding of vans, e.g. for transport of meat, where the mechanical toughness, chemical resistance, low thermal conductivity, and high specific heat are valuable.[78] Unplasticised sheeting supported by glass-fibre-reinforced polyester is used for road tankers.[78] Unplasticised PVC foam has been used as the core in a sandwich structure between GRP or plywood for the construction of boats, where the structure is said to be more rigid and lighter than conventional GRP, and has better buoyancy and resistance to water penetration after damage.[79] Other marine applications include buoys and fenders made by rotational casting. Highway cones have been produced from PVC pastes in a three layer structure.[94] The layers have different formulations with a base layer which is heavily filled (i.e. up to 300 phr) to provide stability of the cone when in use. The layers are fused together in the final stage of fabrication.

16.3.12   AGRICULTURE, HORTICULTURE, AND MISCELLANEOUS RELATED APPLICATIONS

PVC and LD polyethylene compete for many applications under this heading. Thus unplasticised PVC and polyethylene are both used for piping, the latter probably being used in greater quantity for water because most of this is in small sizes where polyethylene is cheaper. Flexible films for glazing, cloches, etc., are made from polyethylene, though there can be little doubt that plasticised PVC can be technically preferable because its weathering behaviour can be superior if properly formulated. It is, indeed, used in some countries for tunnel greenhouses. The two materials also compete in such applications as reservoir, channel, and canal linings, crop drying covers, rick and silage covers, and tarpaulins.

Unplasticised PVC, both plain and corrugated, is used for more permanent glazing and roofing.

Formulation requires some attention if the best weathering performance is to be obtained. Thus, in particular, the resin used should be a homopolymer, and the plasticiser should be of the relatively non-volatile primary type, such as DIOP. Similar considerations apply in the covering of cords, ropes and wires for applications like clothes lines, and stays for boats.

Fence-posts and stays the former comprising an outer tube of unplasticised PVC filled with concrete and fitted with a moulded PVC cap have been produced primarily for farm fencing.

REFERENCES

1. *Plastics and Rubber Weekly*, No. 211, 1 (9 Feb. 1968)
2. R.B. TAYLOR and A.V. TOBOLSKY, *J. appl. Polym. Sci.*, **8**, 1563 (1964)
3. R. SABIA and F.R. EIRICH, *J. Polym. Sci.*, Pt. A., **2**, 1909 (1964)
4. W. SOMMER, *Kolloidzeitschrift*, **167**, 2, 97 (1959)
5. D.R. REID and R.A. HORSLEY, *Br. Plast.*, **32**, 4, 156 (1959)
6. A.F. MILLS, I.S. Note No. 948, I.C.I. Ltd. (1962)
7. P.I. VINCENT, *Plastics*, **28**, 306, 120 (1963)
8. F.H. MÜLLER and C. SCHMELZER, *Ergebn exakt. Naturw.*, **25**, 359 (1951)
9. H. THURN, *Kolloidzeitschrift*, **165**, 57 (1959)
10. BS 2782, Pt. 1, method 102C
11. *The Chemical Resistance of PVC*, I.S. Note No. 1031, I.C.I. Ltd. (1963)
12. J.D.D. MORGAN, *Plast. Prog., Lond.*, 39 (1951)
13. *'Darvic' and 'Flovic' Vinyl Foil and Sheet*, I.C.I. Ltd. (1962)
14. *The Effect of Atomic Radiation on Plastics Materials*, I.S. Note No. 1088, I.C.I. Ltd. (1964)
15. J. BYRNE, T.W. COSTIKYAN, C.B. HANFORD, D.L. JOHNSON and W.L. MANN, *Ind. Engng Chem. analyt Edn*, **45**, 11, 2549 (1963)
16. A.A. MILLER, *Ind. Engng Chem. ind. Edn*, **51**, 10, 1271 (1959)
17. *Plastics and Rubber Weekly*, 3 June and 1 July (1966)
18. W.H. STAHL and H. PESSEN, *Appl. Microbiol.*, **1**, 30 (1953)
19. M. BOMAR, *Chemický Prům*, **6**, 12, 506 (1956)
20. S. BERK, H. EBERT and L. TEITELL, *Ind. Engng Chem. ind. Edn*, **49**, 7, 1115 (1957)
21. A. MULLER and W. SCHWARTZ, *Kunststoffe*, **47**, 10, 583 (1957)
22. E.H. HUECK-VAN DER PLAS, *Plastica*, **13**, 1216 (1960)
23. H.J. HUECK, *Plastics*, **25**, 276, 419 (1960)
24. R. BURGESS and A.E. DARBY, *Br. Plast.*, **37**, 1, 32 (1964)
25. W. SUMMER, *Corros. Technol.*, 11, 4, 19 (1964)
26. C.C. YEAGER, *Plast. Wld*, 14 (Dec. 1962)
27. *Belgian Pat.* 601 344
28. E.I. COOKE (Telcon Ltd. publications – 1956)
29. *Br. Plast.*, **43**, 1, 65 (1970)
30. D.H. BOOTH, P.M. HOLLINGSWORTH and W.H. LYTHGOE, *I.E.E. Conf. Rep.* Series No. 6, 1 (1963)
31. V. RODMAN, *Electl Rev., Lond.*, 787 (May 1964)
32. G.A.R. MATTHEWS, 'Compounding Techniques', Chapter 5 in *Advances in PVC Compounding and Processing*, (M. Kaufman, ed.) Maclaren, London (1962)
33. L.H. WARTMAN, *S.P.E. Jl*, **20**, 3, 254 (1964)
34. W.R. FREEMAN, *Electl Mfr*, **7**, 4, 5 (1962)
35. H. EMERY, *Industrie Plast. Mod.*, **16**, 2, 98 (1964)
36. E.A. MACKENZIE, *Wire, Lond.*, **38**, 7, 959 (1963)

37. D.W. BIRD, *Electrl Distrib.*, **4**, 17, 8 (1963)
38. G. ALLIEVI, *Industria ital. Elettrotec.*, **15**, 5, 335 (1962)
39. A.W. CROAKER and R.W. WYPART, 'Handbook of PVC Formulating', (E.J. Wickson, ed.) John Wiley & Sons, Inc., New York (1963) pp. 92–3
40. A. CHAPIRO, *J. Chim. Phys.*, 895 (1956)
41. P. ALEXANDER, A. CHARLESBY and M. ROSS, *Proc. Roy. Soc. Lond.*, **223A**, 392 (1954)
42. L.P. NETHSINGHE and M. GILBERT, *Polymer*, **29**, 1935 (1988)
43. L.P. NETHSINGHE and M. GILBERT, *Polymer*, **30**, 35 (1989)
44. BS 3505
45. *Concr. Bldg Concr. Prod.*, **39**, 7, 393 (1964)
46. *Plastics and Rubber Weekly*, 23 May (1969)
47. E.O. MEASER and G.M.J. WILLIAMS, *Proc. Instn civ. Engrs*, **21**, 475 (1962)
48. R.E.F. GARDNER and D.M. WATKINS, *Proc. Instn civ. Engrs*, **21**, 761 (1962)
49. Anon., *Plastics*, **32**, 357, 850 (1967); *Rubb. Plast. Wkly*, 178, 1 and 35 (1967)
50. T.J. WIGGINS, *Trans. J. Plant. Inst.*, **33**, 108, 211 (1965)
51. Anon., *Int. Plast. Engng*, **2**, 12, 554 (1962)
52. *Plastics and Rubber Weekly*, 25 July (1969)
53. D.R. JONES, *Rubb. Plast. Age*, **47**, 2, 155 (1966)
54. R.W. GOULD, *Br. Plast.*, **35**, 11, 572 (1962)
55. G.J. BAYNES, *Shoe Leath. News*, 19 May (1965)
56. *PVC for Footwear: Processing and End-products*, T.S. Note W109, I.C.I. Ltd. (1965)
57. J.M.J. ESTEVEZ, *Plast. Polymers*, **37**, 129, 235 (1969)
58. I. PHILLIPS and G.C. MARKS, *Br. Plast.*, **34**, 6, 319 (1961)
59. A.J. LEHMAN and W.I. PATTERSON, *Mod. Packag.*, **28**, 5, 115 (1955)
60. B.L. OSER, *Fd Engng*, **26**, 4, 57 (1955)
61. A.J. LEHMAN, *Proc. Am. Assoc. Dairy Fd Drug Off.*, **20**, 4, 159 (1956)
62. Reports of the 'Toxicity' Sub-Committee of the British Plastics Federation (1958 and 1962)
63. *Plastics for Food Contact Applications, a Code of Practice for Safety in Use*, BPF/BIBRA (1969)
64. J.M. BARNES and H.B. STONER, *Br. J. ind. Med.*, **15**, 15 (1958)
65. O.R. KLIMMER and L.V. NOBEL, *Arzneimittel-Forsch.*, **10**, 44 (1960)
66. A. HILL, *Plastics*, **22**, 234, 603 (1957)
67. H. DOMINGHAUS, *Plastverarbeiter*, **14**, 12, 775 (1963)
68. Anon., *Plastics*, **29**, 316, 75 (1964)
69. H.A. WILLIAMS, *Br. Plast.*, **37**, 4, 198 (1964)
70. M. SALAME and J. PINSKY, *Mod. Packag.*, **37**, 8, 209 (1964); **37**, 9, 131 (1964)
71. W. BURSIAN, *Plastverarbeiter*, **15**, 2, 77 (1964)
72. Anon., *Mod. Plast.*, **40**, 10, 116 (1963)
73. W.B. SISSON, *Plast. Polymers*, **36**, 125, 453 (1968)
74. *Plastics in Packaging*, I.C.I. Ltd. (1965)
75. Anon., *Packaging*, **34**, 402, 87 (1963)
76. Anon., *Mod. Packag.*, **37**, 9, 100 (1964)
77. W. NEITZERT, *Plastverarbeiter*, **11**, 4, 177 (1960)
78. *Plastics in the Motor Industry*, I.C.I. Ltd. (1964)
79. *Plastics and Rubber Weekly*, 4 July (1969)
80. G.H. BURKE and G.C. PORTINGELL, *Brit. Plast.*, **36**, 5, 254 (1963)
81. D. DOWRICK, *Plastics*, **30**, 328, 63 (1965)
82. D.R. MOORE, K.V. GOTHAM and M.T. HITCH, *PRI Intl. Conf., PVC Processing*, Royal Holloway College, Egham Hill, Surrey, England (6–7 April 1978) Paper 14
83. P. BENJAMIN, *ibid*, Paper B5
84. D.E. MARSHALL, R.P. HIGGS and O.P. OBANDE, *PRI Intl. Conf. PVC Processing II*, Brighton, England (26–28 April 1983) Paper 13
85. J.B. PRESS and A. TREBUCQ, *Developments in PVC Production and Processing*, (A. Whelan and J.L. Craft, eds), Applied Science Publishers, London (1977) Chapter 9
86. Anon, *Plast. Rubb. Wkly.*, 9 January 1993 p. 5
87. M. TINGLE, *PVC – a Particularly Versatile Compound* (1993)
88. V.A.C. BURTON and T.J. CLARKE, *PRI Intl. Conf., PVC Processing*, Royal Holloway College, Egham Hill, Surrey, England (6–7 April 1978) Paper B2

89. P. Baker, *PRI Intl. Conf., PVC Processing II*, Brighton, England (26–28 April 1983) Paper 21

90. J. Pickering, *Developments in PVC Production and Processing*, (A. Whelan and J.L. Craft, eds), Applied Science Publishers, London (1977) Chapter 7

91. J. Pickering, *PRI Intl. Conf. PVC Processing*, Royal Holloway College, Egham Hill, Surrey, England (6–7 April 1978) Paper A4

92. K. Whitehead, *ibid,* Paper A3

93. M. Weiss and K. Whitehead, *PRI Intl. Conf., PVC Processing II*, Brighton, England (26–28 April 1983) Paper 25

94. L.G. Krauskopf, 'Handbook of PVC Formulating', (E.J. Wickson, ed.), John Wiley & Sons, Inc., New York, (1993) p. 215

17
Toxicity and environmental considerations

17.1 INTRODUCTION

In any discussion of toxicity and environmental hazards there is a basic problem of semantics. It is a relatively straightforward matter to label known poisons, such as hydrogen cyanide, as 'toxic' or 'environmentally hazardous', though even with these there are presumably low concentrations which would not be harmful. On the other hand it is doubtful if anything can be strictly termed 'non-toxic' or 'completely free of hazard'.

Concern about these matters has quite rightly increased during the past few decades, but has often led to ill-informed and sometimes almost 'hysterical' attacks on manufacturing procedures and products. PVC has not escaped these attacks, which have been so severe that in some European countries there have been proposals to ban the material, not only in obviously sensitive applications like packaging, but in building and construction products too. In some countries there have been 'voluntary covenants' to phase out the use of PVC in packaging. Some suppliers make the specific point that their products are 'PVC free' without giving any explanation.

The legal situation is complicated not only by the fact that different countries have different legislation, but also that within one country different aspects of toxicity and environmental matters are often the concern of different official bodies. Thus, for example, in the UK safety in the workplace is a matter for the Health and Safety Executive; packaging of food the responsibility of the Ministry of Agriculture, Fisheries and Food, while control of emissions of noxious substances is a concern of the Department of the Environment and the Health and Safety Executive. Within the European Community as a whole there are Council Directives which have to be enacted as Statutory Instruments by member governments. In the USA there are more than thirty Federal Departments with significant regulatory responsibility for interpreting and enforcing public law, including the Food and Drug Administration, the Environmental Protection Agency, the Occupational Safety and Health Administration, and the Consumer Product Safety Commission.[92] At the time of writing (1994) the situation is complicated still further by being in a state of flux with potential new legislation under consideration, so that it is impossible to be completely up-to-date.

It is not intended to enter the controversy here, but to outline the facts available at present and leave the reader to make his or her own judgement, if necessary after study of the extensive literature on the subject.[1-9] 'For balanced assessments of environmental issues concerning PVC and phthalates, as well as several other materials, the reader is referred to John Emsley's *The Consumer's Good Chemical Guide*.[93]'

307

17.2   POLYMER PRODUCTION

The 1972 monograph from which this book is derived had the following passages (pages 8 and 9):

> Physiologically, vinyl chloride is non-toxic but slightly narcotic. However, the gas can cause loss of consciousness at high concentration at which its smell is still often insufficient to offer a reliable warning. Special care is therefore necessary in operations involving entry of workers into plant (e.g. autoclaves and storage tanks) where vinyl chloride might be present.
>
> Physiological effects arising from breathing air containing relatively high concentrations of vinyl chloride are only slight.

It was therefore extremely embarrassing for the author when shortly after publication evidence began to appear showing vinyl chloride to be responsible for some rare forms of tumour, mainly angiosarcoma of the liver.[1-4] The first evidence was uncovered during an investigation into the bone deficiency acro-osteolysis which had been noticed in a small number of PVC plant operators, especially those engaged in cleaning polymerisation vessels. By 1974 a clear link between the angiosarcoma and exposure to vinyl chloride monomer had been established and drastic steps aimed at eliminating exposure were in progress. By the middle of 1974, 84 cases had been confirmed world-wide, and by 1986 the figure had risen to 119. This form of cancer, like others, has a long period of latency, averaging 20 years, so that further cases are still to be expected however stringent the preventive steps may be. The majority of cases have been among process workers in polymerisation plants, especially those involved in cleaning autoclaves, rather than workers in monomer production plants.

In addition to concern about harmful effects of vinyl chloride itself, concern has also been expressed about the possibility of emission of dioxins during monomer manufacture.[10] An extensive study of a monomer and by-product incineration plant at Rafnes in Norway over the period January to July 1993 found a total emission of dioxins of 0.031 g during production of 425 000 t of vinyl chloride. This corresponds to a total emission of 0.05 g per annum.[79] The European Council of Vinyl Manufacturers has claimed that the total emissions from vinyl chloride plants is negligible compared with the dioxin release from other sources in nature and society which are impossible or very difficult to control.[11]

Until 1973 the 'Threshold Limit Value' for vinyl chloride in a work-place was 500 ppm (1.300 mg/m^2), but this does not mean that process workers who developed tumours had been exposed to levels even as high as that, because it has been estimated that the limit in American and European plants had been reduced to 200–250 ppm some years before.[3, 7] A combination of measured values and estimates suggests that vinyl chloride concentrations in the atmospheres around polymerisation plants were as high as 500–1000 ppm up to 1955, were reduced to 300 ppm by 1970, and to 150 ppm by mid 1973. Thereafter concentrations were reduced dramatically to 2–5 ppm by the end of 1976. In Europe and the U.S.A. regulations to limit exposure in work-places were introduced and have been up-dated.[90] These are more or less complicated, allowing for fluctuations but specifying upper limits, annual averages for the atmosphere and for exposure of operatives, but require exposure to be limited to concentrations around 1–3 ppm averaged over a year.[3, 7, 80]

In order to reduce vinyl chloride concentrations to these levels and to minimise exposure of workers around plants, it has been necessary to carry out modifications to existing plants, to introduce additional procedures, and also to develop sufficiently sensitive analytical methods.

Among the first of these has been the provision of devices for cleaning polymerisation vessels without requiring entry by process workers, as mentioned previously (Section 3.3.3.3). An obvious measure was to intensify inspection and maintenance of plant so that any minor leaks can be detected and rectified before becoming a serious source of monomer emission. Any opening of vessels between batches should be done only after thorough degassing. Any unavoidable emission is now conveyed through tall stack pipes so that the vinyl chloride is rapidly destroyed by the action of air and sunlight. General ventilation of plant is obviously another matter that required attention.

A major potential source of atmospheric vinyl chloride is residual monomer left in the resin after polymerisation. This amounts to 10–20% of the original charge. The major proportion of this is recovered by gasification and liquefaction, but prior to 1974 at least 1% remained, most of which was lost during drying operations. After drying, residual monomer amounted to 10–1000 ppm in suspension polymers, porous particle grades exhibiting lower levels than dense non-porous grades. Emulsion polymers contained relatively low concentrations of monomer (≤100 ppm). Bulk polymerised resins had higher concentrations because their production involved no drying stage.

The most satisfactory and common procedure for reducing monomer content in suspension polymers consists in using steam to heat the slurry of polymer in water to between 80 and 110°C, and to act as a carrier for the monomer, which latter is separated from the water in a suitable condenser.[3] The process can be carried out batch-wise,[12–15] or continuously.[16–19] The batch process is carried out after venting in the polymerisation autoclave or in a separate stripping vessel. In the continuous process, slurry containing around 3% of monomer is fed to the top of a vertical column fitted with trays which control average and spread of residence time, while steam enters the bottom of the column, passing upwards to heat the slurry and carry monomer with it, the steam/monomer mixture being separated after leaving the top of the tower.[3] The additional equipment and procedures involved add appreciably to the capital and running costs of production, but they do permit reduction of monomer content to ≤1 ppm.[3]

Removal of residual monomer from an emulsion polymer latex by means of steam is more difficult because the latex is liable to foam, so rather lower temperatures (80–90°C) are used, though foam suppressors can be used.[20] Monomer content can be reduced to <150 ppm by these means, but loss during subsequent drying and milling operations reduces the concentration to <10 ppm.

In the absence of water, removal of residual monomer from mass polymers requires high vacuum, though addition of water or organic non-solvents for PVC has been suggested. Although more difficult than with suspension polymers, mass polymers are now supplied with low vinyl chloride contents.[3]

17.3   POLYMER IN USE

It would be generally accepted that completely pure polyvinyl chloride is 'non-toxic' to the extent that anything can be so termed, but residual monomer is obviously undesirable, particularly where the polymer is destined for food-stuffs and other applications where human contact is involved.

Reduction in monomer content in polymer as described in the previous section has naturally been paralleled by reductions in monomer content of PVC products and monomer levels extracted by materials coming into contact with the PVC. By 1985 analysis of a range of natural mineral waters, spring waters, soft drinks, and edible oils, using a method sensitive down to 2 to 3 ppb for aqueous liquids and 15 ppb for oils, detected no vinyl chloride after storage in PVC bottles for 6 months.[7, 21]

Even before action had been taken to reduce monomer levels in polymer, an investigation of drinking water in PVC pipes in the UK found vinyl chloride in water only where it had remained static in small bore pipe for 9 or 10 months, and then at a concentration of only 0.015 ppm.[2] In spite of the encouraging results, further steps were taken to reduce monomer levels in addition to those aimed at achieving this in polymer production. These included increasing maximum temperatures in high speed mixing in the production of dry blends to over 110°C, together with aeration. By these means average residual monomer contents have been reduced to below 1 ppm.[2]

In 1976 European Community legislation on materials in contact with foodstuffs began to appear, initially setting out general requirements,[81] and directives specific to plastics materials and articles have followed. In 1978 the European Commission enacted rules on vinyl chloride, in the form of a Council Directive which set an upper limit of 1 milligram of the monomer per kilogram of material or article in contact with foodstuffs, and a migration limit of 0.01 milligram per kilogram into food.[82] Two further directives have given analytical methods for determination of concentrations of vinyl chloride in materials and articles, and migration into food.[83, 84]

Before closing this section it should, perhaps, be noted that, unlike some other addition polymers, polyvinyl chloride does not depolymerise to produce monomer when it degrades.[22–25]

17.4   PVC PRODUCTS

As suggested in Section 17.1, anything can be harmful in unfavourable circumstances, and it is impracticable if not impossible to preclude the possibility of any hazard whatsoever. In practice, with the proviso regarding monomer content discussed in the previous section, polyvinyl chloride itself is accepted as 'non-toxic' for all reasonable purposes. Unfortunately the necessary ancillary materials are not so free of suspicion, and all types need to be assessed wherever there is a requirement for low toxicity.

The legal position in the UK is ill-defined, but responsibility for ensuring that a material is satisfactory for public use appears to rest with the manufacturer and supplier of the retailed product. There is a Code of Practice for safety in food contact applications, which is up-dated from time to time.[26] There are also European Council Directives which set permissible upper limits on the overall amount of material that may migrate from a plastics material in contact with foodstuffs. These limits are 60 milligrams per kilogram and 10 milligrams per square decimetre.[85] Permissible simulants that may be used to represent a range of food products for migration testing are also specified in European Council Directives.[86, 87]

In the UK additional useful information is available from the British Plastics

Federation.[88–90] If a company is concerned with exporting to a number of different countries, legislation becomes complex and varied, and the situation can become rather difficult.[27] In addition to legal requirements there is of course the moral obligation to ensure that one's products are as harmless as possible, and here the assessment of the suitability of some materials may be liable to subjective conclusions which vary from one individual to another. Fortunately a fair measure of concordance has been reached amongst most countries, largely due to the influence of the Food and Drug Administration, Department of Health, Education, and Welfare, Washington (F & DA),[27–30] the Toxicity Sub-Committee of the British Plastics Federation,[26, 27, 30] and similar bodies in Europe. Within the European Economic Community there are the directives, referred to previously, which have to be enacted nationally within set periods.

The F & DA approach requires assessment and approval by the F & DA itself. The BPF, on the other hand, proposed assessment of the intrinsic 'toxicity' of each individual substance in terms of its 'Toxicity Factor', T, which was then to be related to its 'Extractability', E, in a packaging situation to give a 'Toxicity Quotient', Q, by application of the equation:

$$Q = 1000 \; E/T$$

The dimensions of E are weight per cent or mg m^{-3}, depending on the form of the package, while T is an accepted value for the LD50 or the threshold value in mg/kg of body weight. If the toxicity quotient was less than 10, the substance was regarded as suitable for use from the toxicological point of view.

All additives to PVC must be taken into account, but it is with stabilisers that most problems arise from requirements of low toxicity. While some lubricants and plasticisers are objectionable, it is usually not too difficult to find some which are satisfactory. Unfortunately until relatively recently practically all the efficient stabilisers were toxic (Section 5.4.9). This statement includes stabilisers containing lead, barium, cadmium and tin. It is generally accepted that octyltin compounds are substantially 'non-toxic', provided that they are free from toxic tin impurities,[31–34] and there are now other organotin compounds that are acceptable. This is particularly true if the packaged material is of aqueous rather than oily or fatty nature: but acceptance of these stabilisers is not universal. Stabilisers which do receive virtually universal approval include the stearates of calcium, magnesium, and lithium, zinc octoate, and various complex combinations of these metals; octyl and iso-octyl epoxy stearate, epoxidised oils of various kinds, especially soya bean oil, diphenylthiourea, α-phenylindole, and trinonylphenyl phosphite. As stated in Chapter 5, the demand for safer additives has led to the development of much more effective stabilisers based on combinations of calcium, barium, magnesium, strontium, and zinc complexes.[34–36] Coupled with improved machine design and control systems these have made it much easier to produce acceptably 'non-toxic' products while retaining the quality previously only attainable with more objectionable stabilisers.

The only lubricants likely to be suspect are those compounds of metals, e.g. lead, barium, cadmium, and tin, which are also stabilisers. It is generally possible to avoid the use of these substances where required, without much difficulty.

While some plasticisers, e.g. organic phosphates, and extenders, e.g. some mineral extracts, are toxicologically hazardous, most of the commoner plasticisers have been regarded as relatively harmless. Speciality plasticisers such as triethyl acetylcitrate, triethyl citrate, and triethylene glycol dicaprylate are offered for demanding special applications. However, there is evidence to suggest that acetyl tributyl citrate is more 'toxic' than di-2-ethylhexyl phthalate. In the UK di-2-ethylhexyl adipate is the preferred plasticiser in 'cling' film. Of more common concern is the tainting of foodstuffs by extraction of plasticisers (Section 6.3.4.6), which is, of course, just as damning when it comes to consideration of selection or rejection of specific ancillary materials.

However, during the past decade or so concern has been expressed about the possibility of a toxic hazard arising from dialkyl phthalates in PVC products.[38–40] Some of this concern has arisen from information about loss of dibutyl phthalate from products due to the relatively high volatility of this plasticiser. Dibutyl phthalate should have been phased out by now, and has been to a large extent, not only because of its pollution of the atmosphere but also because its slow volatilisation leads to deterioration in properties of the PVC product in which it has been included.

The most common phthalate, di-2-ethylhexyl phthalate (DEHP or DOP), has been subjected to extensive and detailed studies, and the reader is referred to these if the detailed evidence is required.[41–46] The studies have been aimed at evaluating possible toxic effects by animal feeding experiments, and estimating the magnitude of emissions and likely human intake. There are still some points of disagreement, uncertainty, and debate, but emissions to the environment are probably decreasing due to improved treatment of the atmosphere around processing plants. Incidentally Boo and Shaw[47] found that even with the comparatively volatile dibutyl phthalate less than 1% was lost by 10 minutes two-roll milling at 140–150°C. This is in line with an unreported evaluation in a calendering plant where the atmosphere was acrid with plasticiser vapour, but the weight loss was also less than 1%,[48] suggesting that the nose is sensitive to very low concentrations. Nonetheless, even when undetected by smell, concentrations in the atmosphere may still be high enough to constitute a hazard. The European Council for Plasticisers and Intermediates has carried out a detailed assessment of the release, occurrence, and possible effects of plasticisers in the environment.[91]

Although adverse effects have been observed in rodents, these have been at very high dosing levels, and there now seems to be a general agreement that di-2-ethylhexyl phthalate should not be classified as a carcinogen.[41, 43, 49–51] The level below which no toxic effects are to be expected in a human is 40 mg/day per kilogram of body weight i.e. at least 1000 g/annum for an average adult. The estimated intake per person is 2 g/annum, so there appears to be a very large safety factor.[43] Because of its use in food packaging materials following its supposedly non-toxic nature, similar studies have been carried out on di-2-ethylhexyl adipate (DEHA or DOA), with similar conclusions.[41, 43]

17.5   FIRE HAZARDS

It has rightly said that no two fires are exactly alike, and this makes definitive testing of fire behaviour of materials and articles particularly difficult.

In the sense that it will not sustain a flame in air i.e. continue to burn after removal of a source of ignition, polyvinyl chloride is non-flammable. Although reducing critical oxygen index, even unrealistically high concentrations of fine particle calcium carbonate sufficient to absorb all the liberated hydrogen chloride, do not make unplasticised PVC flammable[52-54] (See Section 7.3.2.7). From small and large scale tests in Europe it has been concluded that unplasticised PVC in various forms, including external cladding for buildings, window frames, pipes and fittings does not contribute to fire propagation.[55-64]

However, unplasticised PVC will degrade in a fire of sufficient intensity, emitting hydrogen chloride, and, like all organic materials pyrolysed in air[9], carbon monoxide and smoke. Toxic effects of emissions from PVC in fires are mainly attributable to the carbon monoxide.[60] Some of the hydrogen chloride could of course be retained by incorporation of fine particle calcium carbonate, but the concentrations required for anything like complete retention are impracticably high. Even 40 phr, which is very high for unplasticised PVC, will retain only about half of the total hydrogen chloride formed by complete degradation.[52-54] Although the hydrogen chloride contributes to the toxic effect, its irritant effect even at very low concentrations acts as a warning.

Incorporation of plasticiser into a PVC composition lowers the critical oxygen index i.e. lowers the proportion of oxygen in an oxygen/nitrogen mixture required to render the PVC 'flammable'.[52] This is true even of phosphate plasticisers, but even at very high concentrations these do not render the PVC flammable in air. However, with a purely organic plasticiser, such as a phthalate, there is a concentration, the precise value of which depends on other components of the composition, at which the PVC will continue to burn in air once ignited. For di-2-ethylhexyl phthalate, for example, this concentration is about 60 phr in the absence of basic fillers.[52] This value falls if the composition includes fine particle basic fillers, notably calcium carbonate of mean particle size less than 0.1 μm. For this reason it must not be assumed that a non-flammable PVC composition will necessarily remain so if fine particle calcium carbonate is added to the formulation.

Though different in detail, plasticised PVC shares with unplasticised PVC the hazards of smoke and toxic fume generation in fires.

The flammability of a PVC composition can be reduced by inclusion of fire-retardants such as antimony trioxide and its hydrate,[65-67] zinc borate,[68] magnesium hydroxide, molybdenum trioxide, and zinc phosphate.[69] However, some fire-retardants tend to increase smoke development considerably, while others reduce it. In the latter category are hydrated aluminium oxide, magnesium hydroxide, and molybdenum trioxide.[63, 69] Specific smoke suppressants such as ferrocene may be of value.[69] Fillers such as calcium carbonate also reduce smoke evolution,[63] but for the reasons discussed above care needs to be exercised when considering incorporation of basic fillers where it is important to retain low flammability.

Reviewing all the testing that has been carried out, and taking all factors into account, it has been concluded that 'When correctly used, PVC does not represent a greater fire hazard than other natural and synthetic organic materials.'[63]

17.6   WASTE DISPOSAL

There has been considerable public concern about plastics waste, as litter and with regard to disposal. Various forms of biodegradable materials have been developed with the intention of reducing the problem of litter, but these do not appear to have been widely adopted, and, in any case, surely a better solution is to persuade people to dispose of their waste in appropriate bins.

Concern about plastics waste arises to a large extent from the ubiquitous and palpable nature of plastics articles. Of the total municipal waste in Europe it was estimated that plastics accounted for only 7 per cent by weight in 1989,[70] and it is doubtful if the situation has changed significantly since then. Of the plastics waste PVC accounted for 10 per cent, i.e. only 0.7 per cent of the total waste, so PVC provides only a very small part of a much bigger general problem, even if allowance is made for the higher bulk densities of much of the non-plastics components of waste. Across Europe an average of 30 per cent of municipal solid waste is incinerated, the remainder being disposed of by landfill, but there is considerable variation from one country to another.[70]

As acceptable sites for landfill disposal become more scarce the trend is to turn more and more to incineration.

Incineration has been studied in some detail,[71] and the influence of PVC in waste incineration has received particular attention.[71, 72] The plastics content of waste is beneficial to the combustion process because of its high calorific value and the way in which the combustion proceeds, and from this point of view it is undesirable even if practicable to separate it from the rest for separate incineration. As most plastics have high calorific values, they constitute a source of recoverable energy.[71, 72] Thus for the hydrocarbon polymers e.g. polyethylene and polystyrene values are about 46 MJ/kg, compared with domestic waste at 44 MJ/kg, higher than paper at nearly 17, and 'wood' at 16 MJ/kg.[73] Presumably the heat required to dehydrochlorinate PVC is outweighed by the heat evolved in combustion of the residue. On the other hand the plastics content does add to the pollutants in the waste. The main problem is the hydrogen chloride resulting from the chlorine content, i.e. the PVC, but the hydrogen chloride is easily removed by flue gas scrubbing, which is now mandatory.[72] The major part of any heavy metals is held in clinker and fly ash which is collected. Emission of heavy metals is reduced by efficient dust collection.

Because of its rapid absorption by soil, water, and plant surfaces, cloud, and rain, hydrogen chloride emissions tend to deposit close to source, and may be a locally significant pollutant, even if emissions are small. Large concentrations have been found in rainwater near industrial areas. Coal burning appears to be the major source of emitted hydrogen chloride in Europe, amounting to 75 per cent of the whole. The proportion of hydrogen chloride from incineration emissions is uncertain, and probably varies from place to place depending on the proportion of PVC in the waste. PVC has been estimated to contribute 0.2 per cent of the total potential acidity emitted to the atmosphere over Europe.[74]

In addition to disposal by landfill and incineration the possibility of recycling has received considerable attention, because where practicable it not only reduces the waste disposal problem but could also go some way to reduce pressure on diminishing material resources. Ideally plastics waste should be separated from other waste and segregated into individual materials. It is obvious that it is much easier to recycle waste and scrap from a factory where materials can be readily segregated according to their composition, but the difficulties and cost of separation from domestic and much industrial waste can be excessively high. Separation of plastics containers is technically relatively simple and is in use to separate polyethylene, PVC, and PET bottles collected from supermarkets. The polyethylene is separated off by flotation, and the PVC and PET separated with the aid of X-ray detection which recognises the chlorine in PVC and activates an air-jet separation system.[71, 72] By this process PVC of up to 99 per cent purity can be isolated for reprocessing.[72] Such a system requires consumers to 'bring back' the containers to a collection point, but has the advantage that the necessity for separation from other household waste is avoided.

A coding scheme for identification of different plastics materials has been introduced by the Society of the Plastics Industry (USA),[71, 75] and it has been proposed that the European Community adopts a similar system.[72] Even with such a system it is difficult to visualise a multiplicity of separate containers for different plastics materials being sited in public places as bottle and paper 'banks' are at present. In any case if the materials are necessarily required to be separate it would be unwise to assume that materials were never put in the wrong container by mistake, or even deliberately. However there are some applications where even mixed plastics can be used.[71, 72] These include benches, poles, fences, plant and compost containers, and several applications in road construction, agriculture, and transport.[72] This form of recycling requires contamination by non-plastics waste to be at a low level, processing adequate to achieve a high degree of homogeneity, and removal of volatiles.

Schemes are in operation for collection of particular types of plastics waste for reprocessing. Thus in some European countries there are arrangements between PVC manufacturers and local authorities to 'take back' scrap flooring, window frame profiles, and pipes.[72]

17.7   THE TOTAL ENVIRONMENTAL IMPACT

In addition to those discussed in the previous sections it is now recognised that other factors are of importance. These include (1) the energy required to produce an article i.e. its 'energy content' or 'energy equivalent', because of its drain on energy resources and also because of consequent carbon dioxide emissions, and (2) the drain on raw material resources. Both have to take account of the service life and benefits of the product, and some less obvious factors such as energy usage in transport.

Calculation of 'energy balances' or 'ecobalances' was stimulated by the 1973 oil crisis,[9, 75] but have now become an important part of all assessments of product impact on the environment. Several detailed analyses of ecobalances for packaging, and some for building and construction applications have been published.[9] A full analysis requires consideration of all stages in the life cycle of a product, including raw materials acquisition, materials manufacture, product manufacture, transportation, product use or consumption, and final disposition (waste disposal, recycle or reuse). At each stage energy input and waste emissions have to be estimated.[9] It is apparent that accurate figures for some of these stages are extremely difficult if not impossible to calculate, and consequently the results of ecobalances can generally be only approximate. Nevertheless they can offer a good guide in the selection of the most environmentally friendly choice to make between a number of alternatives. As well as calculated ecobalances some raw data are available for the purpose of calculation of further cases.[9, 75] Of the more common plastics materials PVC is estimated to have the lowest 'energy equivalent value' at 42.5 MJ/kg compared to polyethylenes at around 47, polypropene at 50, and polystyrenes around 57 MJ/kg.[9, 57] These values may be compared with 300 MJ/kg for aluminium, 70 for zinc, 50 for steel, 20 for glass, and 2 for timber.[77] None of these values include energy contents attributable to subsequent processing, transport, etc.

Energy components for processing plastics are relatively low, ranging from 3 to 12 MJ/kg for extrusion, 5 to 15 for blow moulding, and 6 for calendering, the values depending on the dimensions of the product and the nature of the material.[75]

The Plastics Waste Management Institute, Brussels, has commissioned an independent development of a model and criteria for a data inventory for Europe. Such ecobalances as have already been completed for packaging and window frames, have led to the conclusion that PVC has a favourable environmental position, and that replacement of PVC by possible alternatives does not provide meaningful ecological gains.[9]

As far as usage of raw material resources is concerned it may be noted that only about 4% of crude oil consumption is used to produce plastics materials,[78] which means that only approximately 2% is used to produce PVC. Also it may be remembered that vinyl chloride can be produced from coal, but the process is energy intensive. However if oil and gas resources became no longer available PVC might become the cheapest plastic to produce. In any case use of fossil fuels to produce PVC in many applications is not emitting the carbon dioxide that results from their combustion in energy production.

REFERENCES

1. P.L. Viola, A. Bigotti and A. Coputa, *Cancer Research*, **37**, 516 (1971)
2. H.H. Clayton, *Developments in PVC Production and Processing – 1*, (A. Whelan and J.L. Craft, eds), Applied Science Publishers, London (1977) Chapter 3
3. R.H. Burgess, *Manufacture and Processing of PVC*, Applied Science Publishers, London (1982) Chapter 5)
4. W.J. Primselaar, *PVC Processing II, PRI Intl. Conf.*, Brighton, England (26–28 April 1983) Paper 1
5. J. Thornton, *The Product is the Poison*, Greenpeace Report, Greenpeace, USA (1991)
6. Anon, *PVC-Toxic waste in Disguise*, Greenpeace International (1992)
7. C. Maltoni, *Vinyl Chloride Carcinogenicity*, Association of Plastics Manufacturers in Europe (December 1986)
8. C. Rappe, *Polychlorinated Dioxins and Dibenzofurans*, Association of Plastics Manufacturers in Europe (December 1986)
9. R. Buhl and H. Roder, *Ecobalances*, European Vinyls Corporation, EVC 28E (1992)
10. Anon, *Plast. Rubb. Wkly.*, 2 (1 May 1993)
11. Anon, *Plast. Rubb. Wkly.*, 1 (12 June 1993)
12. Solvay, Belgian Patent No. 793 505
13. Hoechst, German Patent No. 2 429 777
14. Huls, Belgian Patent No. 832 866
15. Tekkosha, Japanese Patent No. 1 0553 90
16. Goodrich, Belgian Patent No. 843 624
17. Hoechst, Belgian Patent No. 866 469
18. Hoechst, German Patent No. 2 521 780
19. ICI Australia, Belgian Patent No. 857 124
20. Borden, U.S. Patent No. 4 168 373
21. ECC Commission, Directive 81/432/CEE (29 April 1981)
22. P. Bradt and F.L. Mohler, *J. Res. Natn. Bur. Stand.*, **55**, 325 (1955)
23. C.F. Bersh, M.R. Harvey and B.G. Achhammer, *J. Res. Natn. Bur. Stand.*, **60**, 481 (1958)
24. R.R. Stromberg, S. Straus and B.G. Achhammer, *J. Polym. Sci.*, **35**, 355 (1959)
25. I. Ouchi, *J. Polym. Sci.*, **A**, 3, 2685 (1965)
26. Anon, 'Plastics for Food Contact Applications: a Code of Practice for Safety in Use', BPF/BIBRA (4th Edn. 1991)
27. A.J. Lehman and W.I. Patterson, *Mod. Packng.*, **28**, 5, 115 (1955)
28. B.L. Oser, *Fd. Engng.*, **26**, 4, 57 (1955)
29. A.J. Lehman, *Proc. Am. Assoc. Dairy Fd. Drug Off.*, **20**, 4, 159 (1956)

30. Anon, Reports of the Toxicity Sub-Committee of the British Plastics Federation (1958 & 1962)

31. J.M. BARNES and H.B. STONER, *Br. J. Ind. Med.*, **15**, 15 (1958)

32. A. HILL, *Plastics*, **22**, 234, 603 (1957)

33. O.R. KLIMMER and I.V. NOBEL, *Arzneimittel-Forsch.*, **10**, 44 (1960)

34. P. SMITH and L. SMITH, Chem. in Britain, **11**(6) 208 (1975)

35. P.J. DONNELLY, *PVC Processing, PRI Intl. Conf.*, Royal Holloway College, Egham Hill, Surrey, England, (6–7 April 1978) Paper 1

36. Ciba-Geigy, *Plast. & Rubb. Wkly.* (9 January 1993)

37. H. DOMINGHAUS, *Plastverarbeiter*, **14**, 12, 775 (1963)

38. Anon, *New Scientist*, 763 (16 June 1983)

39. K.W. SINDEN, *New Scientist*, 139 (14 July 1983)

40. R. GOULD, *The Guardian*, (29 August 1989)

41. J. BELEGAUD, 'Di(2-ethylhexyl) phthalate; di(2-ethylhexyl) adipate; Toxicological Aspects; Evaluation of Risks', Assn. of Plast. Mfrs. in Europe (December 1986)

42. J.T. BARR, 'Encyclopedia of PVC', (L.L. Nass, ed.), Marcel Dekker (1992)

43. D.F. CADOGAN and C.J. HOWICK, 'Ulmann's Encyclopedia of Industrial Chemistry', VCH Verlagsgesellschaft mhH, Germany, Volume A20, 454-458 (1992)

44. World Health Organization, 'Diethylhexyl Phthalate', Environmental Health Criteria 131 (1992)

45. European Council for Plasticisers and Intermediates, 'Assessment of the Release, Occurrence and Possible Effects of Plasticisers in the Environment: Phthalate Esters used in PVC', Eur. Chem. Ind. Council (1993)

46. D.F. CADOGEN, M. PAPEZ, A.C. POPPE, D.M. PUGH AND J. SCHEUBEL, 'PVC 93: The Future', *IOM Intl. Conf.*, Brighton, England (27–29 April 1993) Paper 32

47. H-K., BOO and M.T. SHAW, *J. Vinyl Tech.*, **9**(4), 168 (1987)

48. G. MATTHEWS, Unpublished work (1961)

49. IARC Monograph, **29**, 281 (1982)

50. Official Journal of the European Communities, **L222**, 49, (17/8/1990)

51. BUA Substance Report, No. 4 (1986)

52. G. MATTHEWS and G.S. PLEMPER, *Brit. Polym. J.*, **13**, 17 (1981)

53. G. MATTHEWS and G.S. PLEMPER, *Brit. Polym. J.*, **15**, 95 (1983)

54. G. MATTHEWS and G.S. PLEMPER, European Conf. on Flammability and Fire Retardants, London, (9/10 June 1983) Paper 16

55. 'Report of Special Investigations on Ad Hoc Fire Tests on PVC Services in Buildings', FRO SI No. 9116 (1968)

56. E. BARTH, 'Brandversuche an Fassadenlamellen und-elementen', Unveroffentlichter Bericht der DN AG (1974)

57. E. BARTH, 'Brandversuche an Festern aus unterscheidlichen Werkstoffen', *Plasticonstruction*, **5** (1975)

58. R.W. FISHER and B.F.W. ROGOWSKI, 'Results of Fire Propagation Tests on Building Materials', BRE Report, Garston, Watford, England (1976)

59. R. BECHTOLD, J. EHLERT and J. WESCHE, 'Brandversuche Lehrte Schriftenreihe des Bundesministers fur Raumordnung', Bauwesen und Stadtebau, **04.037** (1978)

60. D. TESTER, 'The Behaviour of PVC in Fires', *Brit. Plast. Fed. Publn.* No. 299/1 (1983)

61. R.W. FISCHER, 'Fire Tests on Building Products. Surface Spread of Flame', FPA, London (1981)

62. P.J. FARDELL, J.M. MURRELL and Z.W. ROGOWSKI, 'The Performance of UPVC and Wood Double-Glazed Windows', FRS & BPF, London (1984)

63. J. TROTZCH, 'Fire Behaviour of PVC', Association of Plastics Manufacturers in Europe (December 1986)

64. J.E. DOE, 'Smoke and Toxicity', BPF & FPA Plastics & Fire Symposium, London (1977)

65. F.K. ANTIA, C.F. CULLIS and M.M. HIRSCHLER, *Eur. Polym. J.*, **17**, 451 (1981)

66. I. SOBOLEV and E.A. WOYCHESHIN, *Fire Flamm./Fire Retard. Chem.*, **1**, 13 (1974)

67. I. SOBOLEV and E.A. WOTCHESHIN, *Coatings & Plastics Reprints*, **36**(2) 497 (1976)

68. J COWAN and R.T. MANLEY, *Brit. Polym. J.*, **8**(2) 44 (1976)

69. K.T. PAUL, *PVC Processing II, PRI Intl. Conf.*, Brighton, England (26–28 April 1983) Paper 36

70. H. CLAYTON, 'XII Seminario de Plasticos', Ass. Portuguese Ind. Plast. (7–9 November 1991) Paper 1

71. V. MATTHEWS, 'Plastics, the Environment and Recycling', Plastics Waste Management Institute, Brussels (1992)

72. Anon, 'Recycling PVC' European Vinyls Corporation (1991)

73. A. Buekens and J. Schoeters, 'Refuse Incineration and PVC', Association of Plastics Manufacturers in Europe (December 1986)
74. P.J. Lightowlers and J.N. Cape, 'Hydrochloric Acid Emissions Attributable to the Incineration of PVC in Western Europe', Association of Plastics Manufacturers in Europe (December 1986)
75. Anon, 'The Energy Content of Plastics Articles', BPF Pubn. No. 309/1 (1991)
76. Anon, 'Ecobalance for Packaging Materials', Bundesamt fur Umwelt, Wald und Landschaft (BUWAL), Bern, Pubn. No. 132 (1991)
77. M.F. Ashby and D.R.H. Jones, 'Engineering Materials', Pergamon (1981)
78. Anon, 'PVC – a Product of Salt and Oil', Solvay Chem. based on information from BPF (1992)
79. Norsk Hydro a.s., Press Release (18 August 1993)
80. European Council Directive 78/610/EEC (29 June 1978)
81. European Council Directive 76/893/EEC (Offl. J. of the European Communities, No. L340, 9 December 1976) revised by 89/109/EEC (O.J.EEC, No. L40, 11 February 1989 and No. L347, 26 November 1989) (U.K. Statutory Instruments, 1987, No. 1523; 1990, No. 2487; and 1991, No. 1476)
82. European Council Directive 78/142/EEC (Offl. J. European Communities, No. L44, 15 February 1978)
83. European Council Directive 80/766/EEC (Offl. J. European Communities, No. L213, 16 August 1980)
84. European Council Directive 81/432/EEC (Offl. J. European Communities, No. L167, 4 June 1981)
85. European Council Directive 90/128/EEC (Offl. J. European Communities, No. L75, 21 March 1990 and No. L349, 13 December 1990) amended by 92/39/EEC (Offl. J. European Communities, No. L168, 23 June 1992) and 93/9/EEC (Offl. J. European Communities, No. L90, 14 April 1993) (U.K. Statutory Instrument 1992, No. 3145)
86. European Council Directive 82/71/EEC (Offl. J. European Communities, No. L297, 18 October 1982) amended by 93/8/EEC (Offl. J. European Communities, No. L90, 14 April 1993)
87. European Council Directive 85/572/EEC (Offl. J. European Communitities, No. L372, 31 December 1985)
88. British Plastics Federation, *Practical Guide No. 1*, CS/PM/2064
89. British Plastics Federation, *EC Synoptic Document*, CS/PM/2064 (6th edn plus supplement)
90. British Plastics Federation, *A Guide to Compliance with the Plastics Regulations*.
91. European Council for Plasticisers amd Intermediates, *Assessment of the Release, Occurence, and Possible Effects of Plasticisers in the Environment*, (1996)
92. L.B. Weisfeld, *Handbook of PVC Formulating*, (E.J. Wickson, ed.) John Wiley & Sons, Inc., New York (1993) Chapter 36, 902
93. J. Emsley, *The Consumer's Good Chemical Guide*, W.H. Freeman/Spektrum (1994), Corgi Books (1996)

18
Experimental and test procedures

18.1 INTRODUCTION

The intention of this chapter is to review the various experimental procedures available for the evaluation and characterisation of the various forms of PVC and the routine test procedures used for quality control testing. The possible forms of PVC may be classified as latices, polymers, dry blends, pastes, compounds, and wrought or shaped products. Some of the procedures are applicable to only one class, whereas others may be applied to two or more of the classes. Thus in the latter category, one is at various times interested in the inherent thermal stability of polymers, and of particular compounds, and of the effects on thermal stability of stabilisers, plasticisers, and other ingredients, and in some of the tests for thermal stability virtually the same procedure may be used, the particular ingredient under examination being varied while the other components of the composition are kept constant. For convenience, however, the procedures which follow are grouped under the different forms of PVC mentioned above, cross references being used to avoid duplication.

For most of the chemical analyses described hereinafter automated equipment, spectrographic or cromatographic apparatus, are now available, but the older tests are sometimes useful when sophisticated equipment is not to hand.

18.2 LATICES (Section 3.3.2)

18.2.1 GENERAL

Examination of PVC latices is not a very common requirement and is mainly the concern of the manufacturers of emulsion polymers. The properties of primary interest are latex stability and solids content.

18.2.2 LATEX STABILITY

The stability of vinyl chloride emulsion polymer latex is usually assessed in somewhat similar manner to that used for rubber latex, that is, by determination of a 'salt number', calculated from the proportion of selected electrolyte, usually common salt, required to coagulate the latex under standard conditions of temperature and agitation.

319

18.2.3   SOLIDS CONTENT

The total solids content of a latex can be determined quite simply by evaporating most of the water from a weighed sample, and then drying to constant weight at about 50°C. The main problem is to avoid splashing and loss of polymer during the main evaporation, but with care this is easily avoided. A simple procedure is to mix the latex with a known weight of dry sand before evaporating off the water.

18.3   POLYMERS (OR RESINS) (Chapter 4)

18.3.1   MOLECULAR WEIGHT CHARACTERISTICS (Section 4.3)

Complete characterisation of molecular weight distribution of vinyl chloride polymers is somewhat troublesome, largely owing to the low solubility in convenient solvents. Nevertheless there have been many fundamental studies of molecular weight and molecular weight distribution of polyvinyl chloride, using mainly osometry, solution viscometry, and light-scattering (Section 4.3). Nowadays gel permeation chromatography is more commonly used where characterisation of molecular weight and molecular weight distributions are required. For day to day practical purposes, e.g. for production control and raw material testing, single point solution viscometry is commonly employed. As pointed out previously (Section 4.3), many methods differing in various ways are employed.[1] The most important variable is the nature of the solvent. This may be cyclohexanone, ethylene dichloride, nitrobenzene, or tetrahydrofuran. Concentration are generally from 0.2 to 1 per cent, and temperatures of determination range from 20 to 30°C. The results are expressed as relative viscosities, specific   viscosities, inherent viscosities,[2] viscosity numbers,[3] K-values,[4, 5] or 'polymerisation degrees'.[6] The K-value of a resin is calculated form its relative viscosity (η_{rel}) by means of the expression:

$$\log_{10}\eta_{rel} = [75K^2 \times 10^{-6}/(1 + 1.5 \, Kc \times 10^{-3}) + (K \times 10^{-3})]c$$

where c is the concentration in g/100 ml. Particular care is necessary when quoting or using K-values, the values of which for a particular resin are very dependent on the nature of the solvent, and to a lesser extent on the concentration and temperature. Early attempts to obtain agreement between different manufacturers about a standard international or even national method of test were singularly unsuccessful, but most manufacturers now use the ISO method or quote ISO Viscosity Numbers[3] corresponding to the values of their own test methods, as estimated from a suitable correlation table.[1] Table 18.1 shows correlations for a number of different common methods and also indicates the corresponding orders of magnitude of weight and number average molecular weights.[1, 7] The data in this table were published[1] in 1963 and 1964.

Experimental determination of solution viscosities is a relatively simple matter, although care is required if accurate and consistent results are to be obtained. Accurate weighing and a high standard of cleanliness are essential. The method used to prepare the solution can be critical. Refluxing of the solvent is common but can result in inaccuracies, and dissolution in a sealed tube is often preferable. Where the polymer contains a fair proportion of impurities, e.g. if it has been 'prestabilised', it can be difficult to decide whether or not solution is complete, and a small amount of residue has often to be disregarded. For the actual viscosity measurements a U-tube viscometer is used, typically a No. 1 Ostwald or No. A Viscometer to BS 188.

18.3.2   CHLORINE CONTENT

The chlorine content of a resin gives an indication of the content of vinyl chloride units. The theoretical value for pure homopolymer is 56.8 per cent. Lower values indicate the presence of chlorine-free comonomers, e.g. vinyl acetate, whereas higher values indicate vinylidene chloride copolymers or post-chlorinated polymers. Theoretically the situation would be complicated by terpolymers of vinyl chloride with vinylidene chloride and chlorine-free comonomers, but these have little significance at the present time.

Chlorine can be determined by degrading a sample of polymer at 800–850°C in a stream of nitrogen, absorbing the liberated hydrogen chloride in sodium hydroxide solution, and back titrating the excess alkali, or by fusion in a Parr bomb[9, 10] or in an oxygen flask.[11] Another method which is quick, convenient, and quite accurate is to degrade the polymer under a layer of calcium oxide, treat the product with silver nitrate, and back titrate the excess of silver nitrate with ammonium thiocyanate in the presence of ferrous ammonium sulphate.

Methods involving silver nitrate are quite expensive. Automatic chlorine meters are available. These are of adequate accuracy and can remove the tedium and some sources of error present with conventional chemical techniques.

18.3.3   COMONOMER CONTENT (Section 4.6)

The simplest method of determining the presence and nature of comonomers is by infra-red spectrometry, though often only approximate estimates can be made by this method. Measurements are conveniently carried out on a film about 0.002 in (50.8 μm) thick cast from solution on a microscope slide, and should be extended over the 2.5–15 μm range.[8]

The chlorine content of a resin can give an indication of its comonomer content.

Though rarely done these days, vinyl acetate can be estimated chemically,[12, 13] for example by determination of the acetic acid liberated by hydrolysis or alcoholysis.[12]

Comonomers can also be conveniently identified and quantified by NMR.

18.3.4   VOLATILES

The most likely volatiles in vinyl chloride resins are water and free comonomers. Total volatile content can be determined by conventional methods involving heating under vacuum to constant weight, typically at 60°C and not more than 60 torr pressure, for about 4 h. Water can be determined by the Dean and Stark method using dry toluene as the carrier.

Accurate determination of unpolymerised vinyl chloride is obviously now a matter of great importance, and this has required the development of very sensitive analytical methods. These include gas chromatography combined with flame ionisation detection, and infrared absorption.[102] Down to 5 ppm by weight monomer content can be determined by injecting a solution of polymer in a suitable solvent directly into a gas chromatograph. For greater sensitivity head-space gas chromatography is used, capable of detecting as little as 0.1 ppm by weight of vinyl chloride in polymer.[103] Specialised automated equipment has been available for over a decade.

Table 18.1 CORRELATIONS FOR A NUMBER OF DIFFERENT COMMON METHODS OF CHARACTERISING MOLECULAR WEIGHTS OF VINYL CHLORIDE POLYMERS

Viscosity number ISO/R174-1961(E)	Inherent viscosity ASTM D1243-58T (Method A)	Specific viscosity ASTM D1243-58T (Method B)	ηr 0.5 g/100 cm3 ethylene dichloride at 25°C	I.C.I. K-value	K-value 1% cyclohexanone	K-value 0.5 g/100 cm3 cyclohexanone at 25°C	K-value 0.4% nitrobenzene at 25°C	K-value 0.4 g/100 cm3 nitrobenzene at 25°C
50	0.42	0.155	1.216	42		45	49	41
52	0.44	0.165	1.227	43		46	50	42
54	0.47	0.175	1.237	44		47	51	43.5
57	0.49	0.185	1.247	45	47	48	52	45
59	0.52	0.195	1.258	46	49	49.3	53	46
61	0.55	0.206	1.269	47	51	50.5	54	47
64	0.57	0.217	1.280	48	52	51.5	55	48
67	0.60	0.228	1.292	49	53	52.7	56.5	49
70	0.62	0.239	1.304	50	54	53.9	57.5	50
73	0.65	0.25	1.316	51	55	55	58.5	51
77	0.67	0.264	1.329	52	57	56.1	59.5	52
80	0.70	0.275	1.342	53	58	57.2	60.5	53
83	0.73	0.285	1.355	54	59	58.3	61.5	55
87	0.75	0.3	1.369	55	60	59.5	62.5	56
90	0.78	0.31	1.383	56	61	60.6	63.5	57
94	0.80	0.32	1.397	57	62	61.9	64.5	58
98	0.83	0.33	1.412	58	63	62.9	66	59
102	0.85	0.34	1.427	59	64	64	67.5	60
105	0.88	0.36	1.443	60	65	65.2	68	61
109	0.91	0.37	1.458	61	66	66.3	69	62
113	0.92	0.38	1.474	62	67	67.4	70	63
117	0.95	0.39	1.491	63	68	68.5	71	64
121	0.98	0.40	1.508	64	69	69.7	72	65
125	1.01	0.41	1.525	65	70	70.5	73	66
130	1.03	0.43	1.543	66	70.5	71.2	74	67
134	1.06	0.44	1.562	67	71	72.2	75	68
138	1.08	0.45	1.581	68	72	73	76	69
142	1.11	0.46	1.60	69	73	74	77	69.5
145	1.13	0.47	1.62	70	74	74.8	78	70
149	1.16	0.49	1.64	71		75.5		71
153	1.18	0.50	1.661	72		76.3		72
157	1.21	0.51	1.682	73		77		73
161	1.23	0.53	1.704	74		77.7		73.5
165	1.26	0.54	1.726	75		78.5		74.5
169	1.28	0.56	1.749	76		79.1		75
173	1.30	0.57	1.772	77		79.7		76
177	1.33	0.58	1.796	78		80.3		77
181	1.35	0.6	1.821	79		80.9		78
185	1.38	0.61	1.847	80		81.5		79

Specific viscosity 1% cyclohexanone at 25°C	Specific viscosity 0.4% nitrobenzene at 25°C	Specific viscosity 0.2 g/100 cm3 nitrobenzene at 20°C	Specific viscosity 0.2 g/100 cm3 cyclohexanone at 25°C	Specific viscosity 0.5 g/100 cm3 cyclohexanone at 25°C	Specific viscosity 0.2 g/100 cm3 cyclohexanone at 30°C	Polymerisation degree JIS K6721	Weight average molecular weight	Number average molecular weight
	0.23			0.25	0.08	275	40 000	20 000
	0.24			0.26	0.09	310	46 000	22 000
	0.25			0.27	0.10	350	50 000	24 000
0.56	0.26	0.097		0.28	0.105	380	54 000	26 000
0.59	0.27	0.099		0.29	0.11	415	58 000	28 000
0.62	0.29	0.1		0.31	0.115	450	60 000	30 000
0.65	0.3	0.105		0.32	0.120	495	64 000	32 000
0.68	0.31	0.11		0.34	0.125	525	67 000	34 000
0.70	0.32	0.115		0.35	0.13	560	70 000	36 000
0.74	0.33	0.12		0.37	0.14	600	74 000	38 000
0.77	0.34	0.125		0.38	0.145	640	80 000	40 000
0.80	0.35	0.132		0.40	0.15	680	86 000	42 000
0.83	0.36	0.140		0.42	0.155	720	92 000	43 500
0.86	0.37	0.145		0.44	0.16	760	100 000	45 500
0.89	0.38	0.15		0.45	0.17	800	105 000	47 500
0.92	0.39	0.155		0.47	0.175	840	114 000	50 000
0.95	0.41	0.162		0.49	0.18	885	122 000	52 000
0.98	0.43	0.168		0.51	0.19	930	132 000	53 500
1.01	0.44	0.175		0.53	0.195	975	140 000	55 000
1.04	0.46	0.18		0.55	0.20	1025	150 000	57 000
1.07	0.48	0.185		0.57	0.205	1070	160 000	59 000
1.1	0.49	0.19		0.59	0.21	1120	172 000	60 000
1.13	0.5	0.195		0.61	0.22	1175	182 000	62 500
1.16	0.51	0.202		0.63	0.225	1230	200 000	64 000
1.19	0.52	0.21		0.65	0.23	1300	209 000	66 500
1.22	0.53	0.215		0.67	0.235	1350	222 000	68 000
1.25	0.54	0.22		0.69	0.24	1420	236 000	70 000
1.28	0.55	0.225		0.71	0.25	1490	248 000	72 000
1.31	0.56	0.23		0.73	0.255	1570	260 000	73 000
				0.75	0.26	1650	272 000	75 000
				0.77	0.27	1720	288 000	76 500
				0.79	0.275	1810	304 000	78 500
				0.81	0.28	1900	324 000	80 000
				0.83	0.29	1980	340 000	82 000
				0.85	0.295	2070	360 000	84 000
				0.87	0.30	2170	388 000	86 000
				0.89	0.31	2260	418 500	87 500
				0.90	0.315	2360	444 000	90 000
				0.93	0.32	2460	480 000	91 500

Free vinyl acetate can be determined by vacuum drying under a stream of nitrogen, e.g. at 100°C and not less than 0.01 torr, condensing the volatiles in cold traps, and estimating the vinyl acetate by hydrolysis with sodium hydroxide in methanol followed by back titration with hydrochloric acid.

Chromatographic and spectographic methods are also available for vinyl acetate analyses at low concentrations.

18.3.5 PURITY (Section 4.5)

18.3.5.1   General
Impurities in vinyl resins, other than the volatiles discussed in the preceding section, comprise residues from the polymerisation, such as suspension agents, emulsifying agents, buffer salts, initiator residues, adventitious impurities such as dust and dirt, grease from bearings and metals from the equipment, and, strictly speaking, substances which may have been added deliberately, such as stabilisers. The main tests employed are contamination or speck counts, for actual solid dust, dirt and other particulate matter, ether extracts, which estimate most organic contaminants, metal content, particularly iron, and colour.

18.3.5.2   Contamination or speck counts
These usually consist of actual counts of the number of discoloured particles which can be detected visually in a given sample of polymer. Two essentially different procedures are used. One consists of spreading a quantity of the polymer in a uniform layer and counting the number of specks visible within a standard area, e.g. a 25 cm square, while the other consists in forming a slurry of polymer in isopropanol or in an aqueous detergent solution, allowing the particles to settle to the base of the container (usually a conical glass flask), and visually assessing the contamination in the bottom layer against a set of prepared standards.

18.3.5.3   Ether extract
As the title implies this consists in submitting a sample of polymer to ether extraction, usually in a Soxhlet apparatus, and measuring the amount extracted. Low values (e.g. less than 0.8 per cent) indicate a suspension polymer, or possibly a washed coagulated emulsion polymer, whereas higher values indicate spray-dried emulsion polymers or prestabilised suspension polymers. The latter can usually be further characterised by analysis for the common stabiliser metals, e.g. lead, cadmium, calcium, barium, sodium, and tin.

18.3.5.4   Metal content
The only metals which are determined fairly regularly in vinyl chloride resins are iron and lead, the former mainly from stability considerations and the latter for toxicity requirements. An idea of the overall metal content can be obtained by ashing. The nature and concentrations of any metals present can be determined conveniently and accurately by atomic absorption spectroscopy or X-ray flourescence.

18.3.5.5   Colour
Vinyl chloride resins are generally of a subjectively white colour, although occasional batches may show a slight pink, blue, or blue-green colour when wetted by plasticiser. The precise reasons for this discoloration are not usually known, although obviously the most

likely cause is contamination of some kind. If there is any question that this problem is present, it can be assessed quite simply by adding sufficient liquid to wet a small sample on a watch-glass.

An alternative and theoretically more refined method is to compare the reflectance of a layer of polymer with a white standard using a photo-electric spectrophotometer.

18.3.6   PARTICLE CHARACTERISTICS (Section 4.4)

18.3.6.1   General
The size and size distribution, shape and structure, bulk and packing densities of a vinyl resin are generally fairly simple to determine, and are useful in indicating the likely processing behaviour of the resin. Thus, porosity will indicate that a resin is likely to form plasticised powders relatively easily, but will probably give problems in unplasticised dry blend processing. Bulk and packing densities may indicate charge sizes and feed rates possible in processing, and considered together with shape may suggest whether or not a resin is likely to flow readily.

18.3.6.2   Particle size and distribution
Suspension and spray-dried emulsion polymers can be examined by more or less straight-forward sieve analysis, the main problem being to ensure that all particles pass through the appropriate screens, a process which appears to be severely hampered by static. Various procedures have been adopted to overcome this, including previous treatment of the resin with a surface-active agent, intermittent injection of steam, and spraying with water. The precise sizes of screens used are not critical provided that they are known and can be related to any required specification. Common sizes are BS mesh sizes 12, 36, 52, 60, 72, 100, 150, 200, and 300 and US standard sizes 20, 40, 60, and 80. The BS mesh indicated correspond to apertures of 1405, 420, 295, 250, 210, 150, 105, 76, and 53 μm, respectively. Some typical sieve analyses are set out in Section 4.4. It is common practice to use a stack of sieves of the selected sizes mounted on a shaking machine in descending order of decreasing size, so that a complete analysis can be effected in one operation, but occasionally more accurate results can be ensured by screening through one sieve at a time.

More refined techniques can be used for special cases. These include electronic counters coupled to microscopes or devices which measure particle volume, X-ray sedimentation and the Malvern laser differential particle sizer.

Resins having very small particles (e.g. direct from latex, coagulated emulsion polymers, or past polymers) are practically impossible to analyse by sieving, and microscopy or even electron microscopy is necessary when it is required to characterise their particle sizes.

18.3.6.3   Bulk and packing densities
The bulk density of a polymer is the apparent density measured as far as possible as the powder settles under gravity, without any external packing influence. It is measured simply by weighing a known volume of sample or by measuring the volume of a known weight of sample in a standard measuring cylinder, for example a 50 cm^3 laboratory measuring cylinder.[14] The result is usually expressed in grammes per litre, or grammes per cm^3, and is typically in the range 0.42–0.80 g/cm^3.

Packing densities are intended to provide some idea of the ways in which different poly-

mers pack down under the action of vibration or pressure. The methods of test are somewhat arbitrary, and tests are usually conducted immediately after measuring bulk density by
tapping or repeatedly dropping the measuring cylinder in a standard way or until no further
contraction occurs. Typical values range from 0.5 to 0.83 g/cm³. It seems obvious that comparisons of bulk and packing densities of different polymers should reveal something about
their particles, and especially about the way they flow, but no study along these lines appears to have been published, possibly because of the extremely arbitrary nature of the test
methods.

18.3.6.4   Powder flow

The readiness with which a resin flows is of importance in handling, particularly in modern
transportation systems, and is critical in applications such as the manufacture of porous
sheeting by sintering (Section 15.1.2). Sometimes, as in the latter application, there is a conflict of requirements where one requires ready flow of powder into a given configuration,
but retention of that configuration once it as been attained. A number of different types of
test are in use. One uses a series of glass tubes like 'egg-timers', each having two open bulbs
or tubes about 25 mm in diameter, joined by a standard capillary. The capillaries range from
about 3 to 13 mm in diameter, and about 25 to 38 mm long. The flow of a sample is characterised by the smallest capillary through which it will flow freely without external aid.

The second method involves placing a known quantity of sample into a standard funnel
and allowing it to flow on to a flat horizontal surface from a standard height above it. Flow
behaviour is assessed by noting the time taken for complete discharge and by determining
the 'angle of repose', i.e. the angle between the side of the conical heap of powder and the
horizontal. For spreading processes it is sometimes possible to determine an arbitrary range
of angles which correspond to satisfactory flow and retention of configuration under a particular set of conditions in a particular machine.

The third method is really an attempt to reproduce the conditions obtaining in a powder
spreading device used prior to sintering. It employs a 'comb' comprising a metal strip with
grooves of various sizes cut in one edge. The 'comb' is held with the grooves downwards
and drawn across a standard quantity of polymer, the ease of formation and retention of
ridges corresponding to the grooves giving some indication of the way a polymer will behave in a spreading machine.

18.3.6.5   Shape and structure

The general configuration of polymer particles is most easily examined by microscopy. The
internal structure of particles is usually revealed by wetting with plasticiser, alcohol, or
some other liquid which will adequately wet the surface. For suspension and spray-dried
emulsion polymers a magnification of about × 80 to × 100 is suitable.

The porosity of particles is revealed clearly by optical microscopy as just described, but
more refined methods are necessary if a numerical measure of the surface area is required.
These usually involve measurement of air permeability through a bed of polymer, or gas adsorption,[15, 16] and are rather too involved for routine evaluations. Typical values for surface
areas by such methods range from about 4000 cm²/g for the original type of suspension
polymer, to between 10 000 and 15 000 cm²/g for a porous particle polymer, 90 000 for a
spray-dried emulsion polymer, and over 100 000 for a paste resin.

To reveal the finer 'sub-microscopic' details of polymer it is necessary to use electron microscopy, and transmission electron microscopy has been useful in revealing the detailed fine structure described in Chapter 4.[104, 105]

18.3.7   RESPONSE TO PLASTICISERS (Section 4.4.2)

18.3.7.1   General
Four different aspects of response of resins to plasticisers are of interest, namely the nature of premixes formed by simple admixture, plasticiser absorption by individual polymer particles, gelation of resin/plasticiser compositions, and dispersion of polymer particles throughout the plasticised PVC matrix.

18.3.7.2   Wetness of premix
When a simple mixture of resin and plasticiser is made, the resultant premix may be either wet or dry, dependent on the nature of the resin, in particular its structure and surface area, and on the nature and amount of the plasticiser. Simple though it may seem the nature of these premixes is of great importance in processing, since the wetter and stickier they are, the less readily are they displaced from one piece of equipment to another, and the greater are the time and effort required in clearing the last portions from the equipment. Wetness of premix, in so far as it is an assessment of a particular resin, is determined by mixing a sample of the resin with a standard amount of a particular plasticiser, and estimating the state of the mixture visually and by touch, relating this to an arbitrary qualitative scale between 'dry' and 'liquid'. Alternatively, the amount of plasticiser required to give a standard degree of wetness can be determined.[17]

18.3.7.3   Absorption of plasticisers
The ease with which a plasticiser is absorbed by a resin is of relevance to compounding behaviour and to the possibilities of producing plasticised powders. There is limited evidence to suggest that too ready and complete absorption can lead to longer compounding cycles than might otherwise be obtained, and consequently, in heavily filled compositions where much plasticiser is absorbed by filler, non-porous non-absorbent resin particles may be preferable. Absorption of plasticiser at room temperature is, of course, relevant to wetness of premix (Section 18.3.7.2) and some tests for plasticiser absorption are indistinguishable from those for wetness of premix. Methods of examination are on a macro- or micro-scale. Macro-scale tests are generally similar to those used for wetness of premix (Section 18.3.7.2), involving mixing standard proportions of polymer and plasticiser under standard conditions and examining the product, or determining the proportion of plasticiser required to yield a standard quality of product. While some resins are claimed to absorb plasticiser to form dry blends at room temperatures, it is generally necessary to heat them to the region of 100–110°C, and standard mixing procedures for testing plasticiser absorption may involve standard heating sequences. After a predetermined standard treatment the plasticiser absorption of a polymer can be assessed by examination of the resultant product, for example by characterising its powder flow using capillary 'egg-timers' (Section 18.3.6.4).

   Absorption of plasticiser by individual particles (or 'grains') of polymer is conveniently examined under a microscope[18, 19, 20] (Sections 6.3.3 and 18.3.6.5). A hot-stage is preferable since this permits evaluation over a range of temperatures up to and above those at which gelation occurs. At an appropriate magnification (ca 80–100 diam) penetration into pores of resin particles can readily be observed, followed by swelling as more plasticiser enters the particles. Rates of swelling can be measured over a range of temperatures, and have been found[21] to be related to temperature by a simple Arrhenius relationship of the type:

$$\log(\text{rate}) = -E/RT$$

Absorption of plasticiser can also be examined in a qualitative way by observing the onset of a 'clear-point' when a sample of resin is stirred in plasticiser (e.g. 1 g in 25 cm³), for example in a boiling tube[22, 23] but interpretation of the results is obscure, although they may be of value qualitatively.

18.3.7.4 *Pastes*
With paste resins one is concerned with the ease with which they break down and disperse in plasticiser, and with the viscosity/shear rate behaviour of the pastes produced.

The usual test for fitness employs a Hagman gauge, which is essentially a metal block with a calibrated groove or grooves with sloping bottom so that they increase in depth from zero at one end to 0.05 or 0.1 mm at the other. A small sample of paste is placed at the deep end and spread towards the shallow end by means of a straight-edged scraper. Undispersed particles can be detected by viewing under suitable illumination or by scratching marks in the paste. Results are recorded in terms of the depth at which concentration of undispersed particles reaches a prescribed limit.

Measurement of PVC paste viscosities is not a particularly simple matter in that a thorough comparison of paste viscosity behaviour of different resins and plasticisers can be involved and tedious. It is really desirable to prepare a number of pastes with a number of different plasticisers, over a range of concentrations, over a range of shear rates, and at a number of different time intervals after preparation. For routine purposes a single composition prepared in a standard way is examined at a single selected shear rate, though initial viscosities and viscosities after different periods of storage are often measured.

Several different types of viscometer are employed depending on the refinement required and on the range of conditions over which viscosity is to be examined. Nowadays coaxial cylinder viscometers are most commonly used in industry. Other types that have been employed include falling plunger, rotating cylinder and bob, rotating disc, and cone and plate types.

18.3.8 GELATION AND DISPERSION

As indicated earlier (Section 9.3.2) the term 'gelation' in the present context is used to embrace the changes involved in conversion of separate particles of polymer or compound granules, or paste, to a more of less continuous polymeric matrix, and thus envisaged it clearly involves softening, deformation, and adhesion of polymer particles. It is an indeterminate as well as an ill-defined process, and methods of assessing it produce results which are not generally amenable to mathematical treatment.

'Dispersion', on the other hand, refers to the way in which polymer or, indeed, other particles disperse and lose their identities in the polymeric matrix. Methods of test often yield numerical results but these are generally somewhat arbitrary. In both cases behaviour in unplasticised and plasticised compositions is clearly different.

18.3.8.2 Gelation

When a particle of polymer in plasticiser is examined under the microscope at a suffic-iently high temperature, it is seen to swell and eventually lose its original shape and outline, often quite dramatically. It has been suggested that this latter behaviour is akin to the gela-tion process, and this method of examination might therefore be appropriate for assessing gelation. It should be noted that this method of test involves zero shear, and might therefore be of value in separating the effects of plasticiser attack from those of shear. The same is true of methods in which pastes are placed on a metal bar along which there is a tempera-ture gradient, the temperature at the point where gelation occurs being taken as the gelation or pre-gelation temperature. A number of variants of the technique have been described, some using insulated and some uninsulated bars, the main difference between the two being in the temperature gradient along the bar.[24–28] The point of gelation is usually taken as being where the strip of gelled paste breaks when stripped off the heated lock.

Measurement of viscosity as a polymer approaches and passes through the gelation stage constitutes the basis of most methods, though not always obvious. This principle is particu-larly amenable to the examination of paste resins, but some viscometers are not suitable for compositions based on other types of resin because the particles can interfere to a greater or lesser degree. A number of viscometric methods have been described.[29–34] In all viscomet-ric methods the onset of gelation is indicated by a peak in the viscosity/time curve, and the rather mediocre correlation observed between different methods arises from lack of agree-ment as to when the end-point, if any, occurs, and from differences in geometry of the ap-paratus and shear rates.

Other methods attempt to reproduce practical compounding conditions as nearly as poss-ible, but still rely essentially on changes in viscosity during the gelation process (Section 10.4.2.4). Thus the Brabender Plastograph and Plasticorder bear a close resemblance to an internal mixer, and a variety of kneaders of different designs are available. A miniature ex-trusion head can also be used. Essentially the machine follows the course of gelation, or whatever other process may be occurring by measuring the torque on the kneading rotors required to maintain them at constant speed of rotation. The mixing chamber is fitted with a jacket whose temperature can be controlled by circulating oil, either at constant level,[35–39] or at a constant rate of increase.[123, 124] Gelation imposes an increased load on the rotors so resulting in an increase in torque. A rapid rise in torque can also arise as a result of degra-dation, but the torque does not usually fall again, and in any case this rise normally occurs well after gelation. However, depending on conditions and material, additional peaks may be observed. Faulkner,[124] studying dry blends of polymer with 3 phr of dibasic lead stearate as the only additive, and using a steady temperature rise of the heating oil of 4°C/min, found three peaks. By carrying out several runs for different periods corresponding to different points on the torque–temperature curve, it was possible to assess the state of material at dif-ferent stages and thus follow the progress of changes throughout the mixing cycle. From the results it was suggested that the peaks were associated with the successive breakdown of particle structures, apparently corresponding to the 'grains', 'primary particles', and 'micro-domains' of Allsopp and Geil's scheme (Section 4.4).

Gelation or fusion can also be assessed by extrusion rheometry using a zero or low length/diameter ratio die at a temperature below the processing temperature of the material under examination, to produce fusion curves of extrusion pressure against processing tem-perature.[123, 125] An initial decrease in pressure with increasing processing temperature is attributed to 'grain' breakdown, the subsequent increase being due to gradual formation of a continuous molecular network giving increased elasticity. However the shapes of the fusion curves depend on formulation, molecular weight of the polymer, and heat and shear history of the specimens. For this reason it is necessary to produce a 'standard' curve for each composition and each piece of processing equipment.[126, 127]

Similar in principle is the use of a small laboratory internal mixer,[40, 41] preferably fitted with a dynamometer, and in both cases the measurement of most interest is the time taken to reach peak torque.

A simple, somewhat arbitrary, method of following gelation is to determine times required to form continuous crêpes or hides when compounded under standard conditions on a two-roll mill. This method does have the advantage that the progress of gelation can be followed visually.

A technique widely used, especially for unplasticised PVC, is to process material through an extruder until steady state conditions are reached, to stop the process at that point, and rapidly remove the screw with material still in it for examination.[42, 43, 106–109]

In the early stages of the process much can be seen by the unaided eye, but as gelation proceeds more searching techniques are necessary. These include the various optical microscopy techniques using white and ultra-violet light, and differential interference contrast illumination, and electron microscopy.

On a more gross scale degree of gelation may be assessed by measurement of mechanical properties and, in the case of unplasticised PVC, by solvent immersion tests.[107, 110, 112] Since a major objective in processing is to achieve optimum mechanical properties the use of these is obvious and requires no further explanation. Solvent tests involve immersion of a test piece from the unplasticised PVC product in a compatible solvent, usually methylene chloride or acetone, for two hours at 23°C, and examining the specimen for delamination. Methylene chloride is more searching than acetone,[107] but neither readily distinguishes different levels of gelation when this is at an advanced stage.[110, 113]

Differential thermal analysis and differential scanning calorimetry can also be used to asses gelation[110, 111] but these techniques are not readily available for routine testing of products in a manufacturing process. Solvent absorption has also been used to assess gelation levels.[111]

18.3.8.3 Dispersion

Even if the particles of a particular batch of polymer were identical to each other in size, shape, and molecular weight, it is most unlikely that the processes of softening, agglomeration, gelation, and dispersion throughout the polymeric matrix would be he same for all particles during compounding. One would imagine that a statistically random distribution would be obtained if it was possible to determine the time for complete dispersion for each particle. One might also imagine that towards the end of the process, by which time almost all particles have become substantially homogenised into a more or less uniform melt, conditions might be less favourable to dispersion of the remaining particles. It is, for example, observed in compounding experiments in the plastograph, or in a laboratory internal mixer, that there is a marked reduction in melt viscosity immediately after gelation, so that in constant speed machines there is a consequent reduction in shear within the melt, and the shear factor in gelation and dispersion is reduced for any remaining particles. These remaining particles have been likened to fruit pips floating in jam. When one adds to this the probability, nay certainty, that the particles of a batch of polymer will not be identical in size, shape, or molecular weight the problems involved in achieving complete dispersion of all particles seem insurmountable. A few particularly hard particles, or particles of high molecular weight, are likely to aggravate the situation. As mentioned earlier (Section 10.4.2.4), complete dispersion of all particles is not important in thick opaque sections, but in thin sheet or in clear compositions undispersed particles show up as 'fish-eyes' or 'nibs', and these can be objectionable even in thick opaque sections if the dispersion is particularly poor. Assessment of the dispersion behaviour (dispersibility) of a polymer involves preparing under standard conditions a specimen of compounded material in an appropriate form, and counting the number of undispersed particles in a given area of specimen or comparing the frequency of undispersed samples with sets of arbitrary standards. In establishing tests for dispersion, account needs to be taken of the processing to which the polymer is to be submitted.

Most tests for dispersion behaviour involve the preparation of specimens by extrusion, milling, or calendering under controlled standard conditions. In those methods involving extrusion the feedstock may be compounded granules, plasticised powders, or dry blends, prepared to standard formulations by a standard procedure. The extrusion unit must not be too efficient in respect of compounding or the differences in results between markedly different polymers may be insufficient to differentiate adequately. An extruder of length/diameter ratio not more than 15, and of low compression ratio (i.e. not greater than 2:1), will generally be suitable. The dimensions of the die are fairly critical, and once set must be strictly adhered to. For convenience of assessment of specimens the die is preferably designed to produce a narrow strip not more than about 1.6 mm thick. for critical applications a blown film die may be used. Results are expressed as the number of undispersed particles in a selected length of extrudate, or ranked against a set of arbitrary standards.

In mill dispersion tests standard formulations are milled under standard conditions for a fixed period of time, and removed as sheet at a controlled thickness between about 0.4 and 0.45 mm. The specimens are then examined under standard back lighting and counts taken of the number of undispersed particles present in a given area. Provided they are carefully selected, pigments may be used so that the undispersed particles show up more clearly. It is convenient to use an inspection box to carry out the counting. This can consist of an opaque box containing a suitable lamp and at the top an aperture of the size required to take a specimen of the specified area. Great care is necessary in maintaining constant operating conditions at all stages, if reproducible and meaningful results are to be obtained.

Dispersion test using laboratory calendering are essentially similar to mill tests.

18.3.9   MELT FLOW (Section 9.3.3)

Although the characteristics of a resin have a profound effect on melt flow, this property is so much a characteristic of a whole composition and the way in which it is prepared that the subject is best discussed later (Section 18.13.2).

18.3.10   SOLUBILITY

In the ordinary way solubility is only of any concern for polymers designed for solution or lacquer applications. The high viscosities of solutions having relatively high solids contents make the determination of solubility in the classical sense virtually meaningless, and one is really more concerned about viscosities obtained at various concentrations, and about the flow behaviour of the solutions, e.g. whether or not 'stringing' or 'blobbing' are likely to occur. In practice there is little more to be done than to attempt to make a solution of the required concentration, to see whether any insoluble residue is left, and to examine the solution.

18.3.11   CLARITY

In an otherwise essentially clear composition different polymers are liable to yield products of different clarity. This behaviour is usually examined comparatively by compounding the polymers under examination in a standard composition, and pressing sheet specimens about 3 mm thick. These are then compared side by side. Consideration should be given to the fact that different polymers, particularly those differing in molecular weight, may require different treatments if the maximum clarity is to be achieved in each case, and so a fixed procedure for preparing specimens may not be appropriate.

Users tend to be very critical about clarity of vinyl copolymer solutions, in spite of the fact that it is of no obvious significance in the vast majority of applications. The behaviour of a particular resin can be evaluated by preparing a standard solution and measuring its clarity or haze in a photometric instrument. Haze usually arises from contamination, undissolved polymer particles, and incompletely dissolved gels.

18.3.12   HEAT STABILITY (Chapter 5)

18.3.12.1   *General*

As indicated earlier (Sections 4.8 and 5.1.1) there is considerable variation in heat stability from one polymer to another, and this can be important in many processes and applications. Sometimes the intrinsic stability of a polymer is examined directly, but since there are variations in response to different ancillary materials, it is more common to examine them in a range of standard compositions.

18.3.12.2   *Polymer oven test*

This is a very straightforward test, the only problem arising from the need to ensure that all specimens receive the same treatment. The test consists simply in heating each polymer under examination on a watch glass (e.g. 75 mm diameter) in an oven at the required temperature, usually between 140°C and 200°C. If there is any difficulty about maintaining uniform temperature within the oven, the specimens may be mounted on a supporting drum which rotates within the oven. Colour development in the various samples of polymer is compared from time to time. Results can be expressed as a ranking or on the basis of the times taken to reach a selected degree of degradation as indicated by the attainment of a standard discoloration.

18.3.12.3   *Congo Red test*

This test involves essentially measurement of the time required for a sample to liberate, under standard heating conditions, sufficient hydrogen chloride to change the colour of a

Congo Red test paper located above the specimen.[44] A typical way of carrying out this test is to place about 2 g of polymer in a one-inch test tube and to suspend a moistened Congo Red test paper on a piece of wire at the top, possibly with a light plug of cotton wool to keep out contamination but loose enough to permit free passage of air. The tube is then suspended in an oil-bath at the required temperature, usually with a number of others under test at the same time. Times for colour change to appear are noted, and are often quoted as the 'heat-stabilities' of the polymers.

18.3.12.4 *Measurement of hydrogen chloride evolution*

For fundamental studies of degradation of PVC it is most common practice to measure the rate of evolution of hydrogen chloride under controlled conditions as discussed previously (Section 5.1.2), but for everyday test purposes colour changes are generally used. However, it is not difficult to devise a routine heat stability test based on measurement of hydrogen chloride evolution. A suitable procedure involves heating a 5 g sample of polymer in a 100 cm^3 conical flask fitted with a tap funnel and a lute dipping into 50 cm^3 distilled water in a 250 cm^3 flask. Heating is conveniently carried out by immersing the flask containing the sample in an oil-bath maintained at the required temperature, e.g. 160–180°C. After the required period of time, e.g. one hour, the flask is cooled, 75 cm^3 water are added through the tap funnel, and the funnel and lute are washed into the 250 cm^3 flask. The contents of the 100 cm^3 flask are filtered into the larger flask follow by washings from residual polymer, which latter may need breaking up thoroughly. The total chloride ion is new determined on the contents of the 250 cm^3 flash, using Volhard's method, i.e. by addition of excess N/10 silver nitrate followed by back titration against potassium thiocyanate with ferric alum indictor. Because it uses silver nitrate this is a rather expensive procedure. In any case automatic meters for measurement of chloride ion concentrations are now available.

18.3.12.5 *Mill tests*

The majority of routine tests for heat stability of vinyl resins involve compounding into a standard formulation, heating under the required conditions, and noting the development of colour, usually by comparison with a set of standards or against other resins of known performance. A typical formulation might comprise 45–50 phr of a suitable plasticiser of good colour and stability, such as DOP or DIOP, with 1 phr of a stabiliser, such as cadmium stearate, calcium stearate, lead stearate, or basic lead carbonate. In the latter case a small amount of lubricant, e.g. 0.25 phr of stearic acid, would also be required. The composition under test is weighed out carefully and then compounded on a two-roll mill in the usual way, the temperature of the rolls and the nip being controlled carefully to some predetermined values. Small pieces of compounded material are cut off the crêpes at regular intervals of time (say every 5 or 10 min) and the colours of the specimens related to the compounding time. It is common practice to express the results in terms of the time required to reach the same colour of an arbitrary standard, and it is even quite common to hear resins described as having 'heat-stabilities of so many minutes'. While this usage may be of value within a particular organisation using a particular test, the use of such descriptions generally can be very misleading and confusing. For control purposes it is usually only necessary for a resin to retain a prescribed level of freedom of colour for some prescribed period of compounding.

Mill heat stability tests are often thought to be particularly meaningful where the PVC is destined for calendering, the conditions in milling and calendering obviously being rather similar.

18.3.12.6   Press tests

Here the resin under test is again compounded into a standard formulation as described in the previous section, but once a properly compounded crêpe is formed it is removed from the mill and cut into specimens of suitable size, e.g. 25 mm × 20 mm. The specimens are then mounted in the cavities of a multi-cavity template mould placed between two polished plates, and pressed at the required test temperature, e.g. 160–200°C. After a predetermined period of time, the mould is cooled and removed from the press, and some of the specimens are extracted. The mould with the remaining specimens is returned to the press and heated for a further period of time, and so on. The time intervals depend on the stability of the compositions, and might be, for example, 25 min where the stabiliser is cadmium or calcium stearate, or 30 min with basic lead carbonate. Likewise, the overall period of heating will vary with the stability, and may be as little as 1.5 h and as long as 3 h. Results of press tests are sometimes supposed to relate to processing in closed machinery, e.g. internal mixers and extruders, but there is, of course, virtually no shear working in these tests. A practical complication which frequently arises in press heat stability tests is non-uniformity of discoloration in some or all of the specimens, making comparisons difficult. This non-uniformity can be due to inadequate dispersion of additives, e.g. the stabiliser, when discoloration is spotted, or to catalysis of degradation where the specimen contacts some metal surfaces or residues from previous tests, when discoloration tends to be deeper around the edges of the specimens.

18.3.12.7   Compound oven tests

As with press tests, the resins under examination are first compounded in standard formulations as discussed in Section 18.3.12.5. Specimens are usually then pressed out as in the preceding section using a short moulding cycle. This is merely to get the specimens into a convenient form; they are then heated in an oven at the required test temperature and removed at periodic intervals as described previously. The main problem is to ensure that all specimens receive the same heat-treatment, and various arrangements have been suggested to ensure this. Even where a reasonably uniform distribution is obtainable there are often sufficient fluctuations to produce anomalous results if sufficient care is not taken. For this reason if the only means of supporting the specimens is on a glass or polished metal plate, it is highly desirable to distribute the different specimens under test in a random way so that any adventitious variation can be taken into account. One very suitable device to ensure uniform treatment consist of a wheel on the periphery of which the specimens can be mounted. The wheel is mounted in the oven and rotated on a horizontal axis by a drive motor outside.

18.3.12.8   Extrusion tests

During routine extrusion tests of the processing behaviour of a resin in dry blend or compounded form, account is usually taken of any discoloration developed. The procedure, however, is not sufficiently sensitive for normal heat stability testing and will usually only pick out gross cases, such as those where stabiliser has been omitted from or incorrectly weighed into the composition. However, routine extrusion tests may be used to check colour against standard samples. In transparent, white, or paste, coloured compositions, departure from standard may be attributable to thermal degradation.

18.3.12.9   *Torque rheometers*
If a composition is kept mixing in a Brabender Plastograph or Plasticorder (Section 18.3.8.2) or similar apparatus after it has been compounded, it will usually eventually exhibit a torque peak as gross degradation ensues, and the length of time taken to reach the peak can be taken as a measure of the stability of the composition. Comparing a range of compositions based on different polymers, this technique should permit comparisons of the different polymers to be made.[37, 45 ,46] Actually the stage of degradation corresponding to breakdown in this type of evaluation is well beyond what is usually of concern in good industrial practice.

18.3.13   ELECTRICAL PROPERTIES

As was pointed out earlier (Section 4.7.4.1) volume and surface resistivities of PVC are very dependent on the nature of the polymer, but the effects of polymers on electrical properties are usually evaluated by preparing appropriate plasticised PVC specimens and testing them. A few decades ago there was an American proposal to classify the suitability of resins for electrical insulation according to the conductivity of aqueous extracts of the resins,[47] but this approach is not universally valid because some polymers with excellent electrical insulation properties yield relatively high aqueous extracts.

18.4   STABLISERS (Chapter 5)

The main tests used to evaluate stabilisers are described in the preceding Section 18.3.12, the only difference in procedure being that all the ingredients of compositions are held constant with the exception of the stabiliser. Comparisons of stabilisers are commonly done at equal concentrations (1 or 2 phr), but it is more meaningful to compare them at equal compound volume cost. Sometimes a stabiliser may have a upper concentration limit dictated in incompatibility and a tendency to plate-out, and thus too must be taken into account.

Effects on properties other than stability must also be considered. The main ones of concern are effects on clarity of otherwise clear compositions (Section 18.3.11) and on viscous behaviour of pastes (Section 18.3.7.4).

18.5   PLASTICISERS (Chapter 6)

18.5.1   GENERAL

As with resins and stabilisers most evaluations of plasticisers involve preparing a range of compositions and testing the products. With a given polymer different plasticisers differ in rates of absorption and gelation, but apart from the fact that the composition variable is now plasticiser rather than polymer, the methods of evaluation are the same as described previously (Section 18.3.7).

18.5.2   Efficiency (Section 6.3.4.2)

The concept of efficiency of plasticisers has already been discussed (Section 6.3.4.2). From the physical or mechanical viewpoint it is necessary to decide what properties are to be taken into account. The main function of a plsticiser being to impart flexibility, it seems reasonable that comparisons of efficiency should be based on measurements of flexibility, but since this is a property difficult to characterise and measure, some other property or properties must be selected. Ideally, too, the variation in the values of the selected properties should be examined over the whole range of useful concentrations. Ultimately the results need to be translated into considerations of volume cost of the finished PVC material. In practice full evaluation along these lines can be very time-consuming and expensive, and compromise is necessary. Fortunately at the present time the commercial situation is such that detailed comparison of two or more plasticisers is rarely called for. It is rarely necessary to put a numerical value to efficiency; rather it is necessary to know the proportions of different possible plasticisers which will produce the desired result. Nevertheless a number of numerical expressions for plasticiser efficiency have been proposed.[48] These are usually based on the concentrations required to yield some specific value of a particular chosen property, such as BS Softness, moduli, tensile strength, brittle temperature, and lowering of transition temperatures.

   Clearly any examination of efficiencies of plasticisers requires physical tests of some sort to be carried out on compounded PVC. These tests are discussed in later sections.

18.5.3   Viscosity

Three different aspects of viscosity are of interest with a plasticiser, namely the viscosity of the plasticiser itself, the viscosity behaviour of pastes made from it, and the melt viscosity of compositions containing it.

   The first of these is of interest in handling (e.g. weighting, transporting, and mixing) and also in that it is probably important in relation to capillary attraction in to the pores of porous particle polymers.

   Although plasticisers tend to be rather viscous liquids by usual chemical standards, measurement of their viscosities can be done by normal classical equipment, and no special description is called for here.

   Measurement of paste viscosities has already been discussed (Section 18.3.7.4), the only different here being that the plasticiser becomes the variable rather than the resin.

   While obviously the introduction of plasticiser into PVC reduces the melt viscosity of the composition under given temperature and shear conditions, little is known about the variations between plasticisers when compared at concentrations designed to give equal physical properties, or indeed when compared at equal concentrations. Evaluation of plasticisers in these respects would be based on melt rheological examination of compositions containing them, and is discussed elsewhere (Section 18.13.2).

18.5.4   Properties at extremes of temperature (Sections 6.3.4.4 and 6.3.4.5)

In general the introduction of plasticiser into PVC increases flexibility and softness at any particular temperature, whether it be atmospheric, relatively high, or relatively low, and also reduces the softening temperature however defined. Changes of flexibility with temperature below normal ambient vary from one plasticiser to another and are undoubtedly dependent on the nature of the plasticiser. Variations above room temperature have been much less well characterised. Study of the effects of different plasticisers involves essentially the same techniques as are used to study the properties of PVC compositions at relatively low and high temperatures, and these are discussed later (Sections 18.14.9 and 18.14.10).

However, volatile loss of PVC compositions at elevated temperatures is almost entirely a characteristic of the plasticisers, and is mainly dependent on their volatility and tendency to degrade to volatile products. Only the use of relatively volatile plasticisers like dibutyl phthalate creates any problem under normal room or atmospheric temperature conditions, but applications at elevated temperatures are more demanding, and a number of applications, e.g. high temperature electric cable, require the loss by volatilisation to be kept within close limits.

The tests proposed involve measurements of changes in mechanical properties or loss in weight after ageing under controlled standard conditions. Specimens may be heated in a circulating air oven, or a tube oven, alone, in contact with activated carbon, or surrounded by activated carbon, but separated from it by a wire cage.[49-54] Typical ageing conditions[58, 49] range from 24 h at 70°C to 7 days at 121°C. For specification purposes specimens are required to retain not less than a specified minimum of certain mechanical properties, such as tensile strength and elongation at break, or to lose no more than a certain percentage in weight.

18.5.5   COLOUR AND CLARITY

In compositions which are otherwise essentially colourless or only lightly coloured, and/or transparent, the plasticiser can play an important part in determining whether or not the maximum possible quality can be attained. This will depend on the colour and clarity of the plasticiser itself, its stability, and its compatibility with the polymer.

Plasticisers of poor colour or clarity can be rejected by direct observation, but more refined techniques are required for high quality products. The most common technique for evaluation is to compound plasticisers under examination into a composition of known potentially high clarity and freedom from colour, and to compare similar specimens under appropriate illustration. There seems to be no reason why spectrophotometers should not be used. The same tests will usually automatically look after any contribution of instability of the plasticisers.

Poor clarity is sometimes due to insufficient compatibility but this too is detected in the same way, the level of incompatibility involved often being much lower than would be readily detected by methods of test for incompatibility itself (Section 18.5.6).

Failure to obtain maximum clarity can also be due to inadequate compounding, and this is more likely to occur the less rapid the attack of the polymer by the plasticiser (Section 18.3.7).

18.5.6   COMPATIBILITY (Section 6.3.4.1)

As discussed earlier (Section 6.3.4.1), primary plasticisers are by definition compatible with polymer over the whole range of useful compositions, secondary plasticisers in the absence of primary plasticiser are compatible over limited ranges of concentrations only, while extenders are incompatible except in conjunction with appropriate proportions of primary plasticisers. It follows that a full characterisation of any plasticiser or extender can involve examination of a wide range of compositions and can prove both time-consuming and costly.

In many cases incompatibility results in an easily visible surface film of oily matter after storage for quite short periods of time, so that in all but extremely critical cases a number of specimens of different compositions covering the range of concentrations of interest can be compounded, pressed to sheets, stored under appropriate conditions, and examined from time to time.

A simple and more rapid method is the so-called 'loop' test. A typical specimen for this is a compounded strip about 25 mm long and 1.5 mm thick. This is folded over about halfway to form a loop about 6 mm in diameter, and the flat portions are clamped lightly together. Drops of plasticiser will usually appear at the loop if compatibility is not complete.

A perhaps more refined but more troublesome technique is to store the specimens in contact with sheets of polyethylene, polystyrene, or ABS, and to measure the changes, if any, in weight and electrical resistivity of the latter after varying periods of time.

Incompatibility can be induced by external conditions, e.g. exposure to sunlight, and this needs to be investigated if special conditions are likely to be encountered by the plasticised PVC.

18.5.7   EXTRACTION (Section 6.3.4.6)

As pointed out earlier (Section 6.3.4.6) a number of actual and potential applications for PVC involve contact with liquids which may extract ingredients of the PVC composition, particularly plasticiser. Such extraction is often very specific to particular plasticisers and particular extractants, and a proposed use of a plasticised PVC often requires special testing, although a good deal of information about the behaviour of specific plasticisers has been published.[48, 55, 56]

Extraction tests are carried out by immersing specimens in appropriate extractants under selected conditions of temperature and time. The specimens can be removed, dried, and weighed to determine loss in weight. Unfortunately some extractants permeate into the PVC leading to a gain in weight or an incorrect value of plasticiser extracted, and weight changes can be misleading. It is perhaps preferable to analyse the extractant for plasticiser, which can be done conveniently in by infra-red spectrometry.[8, 57] It is also more realistic to test the PVC in the form in which it is to be used if this is at all possible.

In the European Community there are Council Directives listing food simulants, test times, and temperatures, to be used for migration testing of plastics materials in contact with foodstuffs.[128, 129] Another lists categories of food wtih the appropriate simulants to be used.[129, 130]

18.5.8   EFFECTS ON ELECTRICAL PROPERTIES (Section 6.3.4.9)

The chemical nature, purity, and concentration of a plasticiser have profound effects on the volume resistivity, and to a lesser extent other electrical properties of PVC. These effects are examined by compounding a plasticiser under examination into a standard composition and measuring the required electrical properties in the usual way (Section 18.14.12).

18.5.9   EFFECTS ON STABILITY

The contribution of a plasticiser to the stability or instability of PVC is evaluated by compounding into standard compositions and testing as described earlier (Section 18.3.12).

18.6   LUBRICANTS (Section 7.2)

18.6.1   GENERAL

The functions of lubricants are to control the processing behaviour of PVC in terms of adhesion to processing equipment and melt flow. Neither of these is easy to evaluate by simple laboratory techniques, partly because the precise conditions encountered by material are rarely if ever known. Something can be learnt from examination in a torque rheometer such as the Brabender Plastograph or Plasticorder[35–39] (Section 18.3.8.2) and in melt rheometers (Section 18.13.2), but interpretation of results and correlation with production practice is usually arbitrary and on the basis of past experience of trial and error. There are so many differences in behaviour from one material to another, from one process to another, and even from one machine to another of the same type, particularly if of different size, that it is often now necessary to carry out full-scale plant trials at considerable expense. Very often the differences in behaviour only become revealed in the form of different efficiencies of production of 'good make', and these, too, only show up during protracted plant trials.

18.6.2   'PLATE OUT'

Many lubricating substances are prone to 'plate-out' (Sections 5.4.4, 11.2.5.12, 12.3.2.4). This can be examined broadly by a mill test in which a composition containing the lubricant under test and a pigment is compounded on a two-roll mill, stripped off, and followed by an unpigmented composition, when plate-out will usually manifest itself by more or less heavy discoloration of the second composition. The sensitivity of the test can be varied by varying the pigment used.

18.7   FILLERS (Section 7.3)

18.7.1   GENERAL

As with other additives effects of fillers are very often examined by compounding into standard compositions and testing the properties in the usual way.[58] This applies, for example, to effects on physical, mechanical, and electrical properties (Section 18.14), and on stability (18.3.12).

18.7.2   PARTICLE SIZE

Sieve analyses (Section 18.3.6.2) are generally of little value in characterising the particle sizes of fillers because they are usually too small, and microscopic examination and sedimentation analysis are preferable. If sieve analysis is used, washing with water or some other wet system is necessary, and even then only a coarse segregation is usually possible. Even with the other methods some means of dispersing fine particle fillers is desirable. For microscopy a dilute solution of surface-active agent in a 50:50 water/glycerol mixture is suitable. Optimum magnification will generally lie in the $\times$ 200– $\times$ 500 region. In sedimentation analysis the filler is conveniently suspended in dilute aqueous surface-active agent. For the common calcium carbonate fillers, concentrations in different fractions can be determined by titration against hydrochloric acid.

18.7.3   SURFACE AREA AND OIL ABSORPTION

Measurements of surface areas of fillers are fairly complicated and unsuited to routine testing. Where necessary a gas adsorption method can be used.[15, 16]

Oil absorption is generally reckoned to be important (Section 7.3.2.3) and standard methods of estimating it are in use.[58]

18.7.4   SURFACE COATING

If the nature and nominal concentration of surface coating on a commercial filler is known, its actual concentration can be estimated by measuring the weight loss of a sample after controlled heating under vacuum.

18.7.5   EFFECTS ON GLOSS

Properly processed, an unfilled PVC composition will generally be glossy and smooth, but introduction of filler reduces the gloss or introduces mattness according to the nature and amount used (Section 7.3.2.2). If properly dispersed it will generally be found that increased mattness will result from increased in particle size, but agglomeration leading to relatively large surface 'nibs' is more likely as particle size is reduced. Assessment of the effects of different fillers is dependent on subjective visual comparison of specimens prepared from standard compositions containing the fillers.

18.8   IMPACT MODIFIERS (Section 7.4)

Little can be done to evaluate effects of impact modifiers other than to prepare specimens with and without them and to examine the mechanical properties, particularly impact strength (Section 18.14.7). It must be noted, however, that additives may have effects other than those for which they are intended, and it is therefore necessary to examine these, e.g. melt flow behaviour.

18.9   PROCESSING AIDS (Section 7.5)

As was pointed out in connection with lubricants (Section 18.6.1, the beneficial effects of processing aids are often not apparent in small scale laboratory test, and while melt rheometry (Section 18.13.2) can yield useful information, prolonged full-scale plant trials are often necessary before a proper evaluation can be made. However the benefits of processing aids are very often well worth having, in spite of any cost and trouble of evaluation.

18.10   COLORANTS (Section 7.7)

18.10.1   COLOUR

Although single pigments are rarely used to colour PVC, the actual colour charactersitics of any prospective pigment are obviously important. Colour matching and formulation by eye have to a large extent been replaced by computerised spectrophotometers with which it is possible to determine the spectral characteristics of individual colorants and to prescribe formulations of various pigments to match a required colour. In addition to storing spectral data, other information, such as costs, can be stored, so that the computer can be programmed to determine the most effective formulation of possible alternatives on technical and/or cost grounds.

18.10.2   DISPERSION (Section 7.7.1)

Dispersibility of pigments is evaluated subjectively by preparing specimens under selected mixing, compounding, and shaping conditions and assessing the dispersion visually, preferably with the aid of a microscope.

18.10.3   STABILITY (Sections 7.7.2 and 7.7.4)

Some measure of the stability of a pigment can be obtained by heating a sample for a selected temperature/time cycle, and comparing the heated sample with some of the original untreated pigment. Tests on compounded specimens are also carried out as described previously (sections 18.3.12.5–7). For PVC suitable test cycles are 30–60 min at 170°C or 5–10 min at 200°C. As with other forms of behaviour it is often best to examine the materials in actual processing.

18.10.4   MIGRATION (Section 7.7.3)

18.10.4.1   General
It was suggested earlier (Section 7.7.3) that problems of pigment migration have arisen perhaps more frequently with PVC than with other materials, and although these problems can now generally be avoided by reasonable attention to formulation, it is still desirable to check that use of any new formulations, and new pigments in particular, is not going to lead to migration of some kind. The forms of migration exhibited by pigments include plate-out, blooming, bleeding, and chalking. The first of these is largely a function of other ingredi-

ents of PVC, particularly lubricants and lubricating stabilisers, and has been discussed previously (Section 18.6.2), but while the others are also dependent on the nature of the formulation, and also processing conditions, as a whole, the prime movers as it were are pigments.

18.10.4.2   *Blooming and chalking*
To test a pigment thoroughly it really needs to be compounded into all the compositions in which it might be used and processed under all the conditions which it might encounter, but this is obviously rarely possible because of the high cost involved. In practice only one or two typical formulations are selected initially, but a whole range of these is made up with varying concentrations of the pigment under test. Specimens are conveniently made by pressing or moulding into oblong slabs about 3 mm thick, although there is something to be said for using the stepped type of specimen common in colour swatches. The specimens are then merely stored and examined from time to time, usually wrapped in filter paper. If blooming occurs it can usually be seen as a surface layer and pigment can be scraped or wiped off with a finger-nail or a knife-blade, or even with cotton-wool.

18.10.4.3. *Bleeding*
Tests for bleeding are essentially similar to those for blooming and the two are usually carried out side by side, the only difference being that for bleeding the specimens are stored in contact with a second material, such as 'natural' unpigmented PVC of the same basic formulation, or polyethylene.

18.10.4.4   *Weathering*
Resistance to fading or discoloration by UV radiation is relatively easy to assess using standard UV sources, but outdoor weathering still presents difficulties. Accelerated weathering test equipment can provide useful indications, but, as well as the difficulty of producing realistic weather conditions in laboratory apparatus, the fact that conditions are 'accelerated' introduced an additional factor. Moreover there is, of course, no 'standard' weather, and it is still common practice to expose test specimens outdoors at a number of different locations that typically experience different weather.

18.10.5   EFFECTS ON ELECTRICAL PROPERTIES (Section 7.7.5)

For any electrical application choice of pigment can be critical and it is certainly necessary to ensure that any pigment used in an insulation grade of PVC will not affect any of the electrical properties to too great an extent. However, the testing involves compounding a pigment under test into a selected composition and testing in the usual way (Section 18.14.12).

18.11   DRY BLENDS (Section 11.2.4)

18.11.1   GENERAL

As an intermediate form of PVC composition one is concerned with the way in which a dry blend will process, and that the final product will have the required properties, much of which is similar to what is required to be known of granulated compounds and finished products, so here we are mainly concerned with aspects of behaviour peculiar to dry blends.

18.11.2   PARTICLE SIZE

The particle size and size distribution of a dry blend are mainly of importance in their effects on bulk and packing densities, and powder flow. Where it is desired to characterise the particle size of a dry blend this can usually be done quite satisfactorily by sieve analysis (Section 18.3.6.2).

18.11.3   BULK AND PACKING DENSITIES

These properties are of particular significance for a dry blend because of their bearing on feed rates to processing equipment and their relationship to the air content of the dry blend and the consequent likelihood of air entrapment within the melt during processing. They can be determined quite simply in the manner described previously (Section 18.3.6.3).

18.11.4   POWDER FLOW

Flow of a dry blend can be critical in determining the magnitude and uniformity of processing rates. It can be characterised by the methods described previously (Section 18.3.6.4). In general a dry blend will be found to feed satisfactorily to practically any processing machine if it flows through an 'egg-timer' flow tester with a capillary about 0.9 mm in diameter.

18.11.5   PROCESSING BEHAVIOUR

Three main aspects of processing are of particular significance with dry blends, especially in extrusion. These are flow rate into the machine, and hence output, liability to entrap air within the melt, and the ease with which an acceptable degree of uniformity can be obtained. Powder flow behaviour has been discussed in the previous section, but is also often tested adventitiously during examination of the other aspects of processing. Ideally full evaluation of a dry blend should be done by full-scale experiments in the actual processing for which it is intended, but this is obviously usually too troublesome and expensive. Some arbitrary guidance can be obtained by processing under selected conditions through a standard extruder and die, and comparing the behaviour and product quality with that of arbitrary standards. A suitable arrangement comprises a 30 mm single-screw extruder with an *L/D* ratio of 15:1 and a single-start screw of about 2:1 compression ratio, fitted with a strip die of about 1.5 mm gap.

Information on output rates is built up by experience, while dispersion and general quality is characterised against prepared arbitrary standards, or against extrudates prepared in the same experiment from dry blends of known quality.

18.12   PASTES (Chapter 14)

18.12.1   GENERAL

The properties of particular interest with PVC pastes are viscosity and gelation behaviour, including shelf-life. Other properties, such as heat-stability and mechanical properties, are evaluated by gelling a paste under examination so as to form specimens of appropriate dimensions, and then testing in the normal ways associated with PVC compositions generally.

18.12.2   VISCOSITY (Section 14.2.2)

Ideally the flow behaviour of a PVC paste should be evaluated over the whole range of shear rates likely to be encountered in practice, and changes in flow behaviour with time should also be investigated by repeating the testing at periodic intervals. Very often a single point determination is all that is carried out. While this is reasonably statisfactory as a routine control test where production procedure is strictly regulated, it is much less than adequate for a paste of new or unknown formulation. The measurement of paste viscosities has been discussed previously (Section 18.3.7.4).

18.12.3   GELATION (Section 14.2.3)

The progress of gelation of a PVC paste is conveniently followed by casting in a suitable mould at various temperatures in the range 150–190°C, and removing specimens from time to time for testing, for example by measuring tensile strengths and elongations to break (Section 18.14.5.1).

18.13   COMPOUNDS

18.13.1   GENERAL

As with dry blends and pastes, one is concerned with compounds both with their processing behaviour and with the properties of their end-products, and, since the tests for the latter are generally common to all forms of feedstock, we are concerned here mainly with aspects of the former, namely processing rate, melt flow behaviour, stability, and dispersion. Tests for some other properties, such as relative density, are often carried out on granules of compound, but are more properly discussed below (Section 18.14)

18.13.2   MELT FLOW BEHAVIOUR

Here one is concerned with apparent melt viscosity under processing conditions, and with any tendency to develop turbulent or other forms of irregular flow. Again the situation is such that a full-scale evaluation in production equipment is often the best criterion in establishing the suitability or otherwise of a given material, but much useful, even if often arbitrary, information can be gained from extrusion rheometry of one kind or another.

   Conventional rheometers, as used for other polymers, are often suitable, but a number of machines have been specifically introduced for PVC. In the latter category are the simple weight-loaded machine and the motor-operated so-called Macklow–Smith plastometer.[60] The Atkinson–Nancarrow rheometer tensometer attachment,[61] with some modifications to improve temperature control and recording,[62] is a useful cheap machine for PVC evaluation, but more sophisticated machines[63] are generally desirable for anything more involved than simple coarse assessments. Typical suitable die sizes for use in extrusion rheometry of PVC are in the region of 1 to 1.6 mm diameter and 3 to 6.5 mm long, shear rates ranging up to about 1200 s^{-1}. In these machines the viscosity behaviour can be examined over wide ranges of temperature and shear rates, and the results can often be correlated with production outputs. An important aspect of this work is careful examination of the extrudate samples. Onset of irregular flow in production can sometimes be correlated with behaviour in a rheometer, even though this may be of arbitrary nature. Using selected appropriate conditions the Macklow–Smith plastometer can be used to assess the gelation state of PVC granules, when rough, bumpy extrudates indicate a high level and smooth extrudates indicate a low level of gelation.

The high-speed adiabatic extruder[64] is also used for evaluation of melt flow behaviour of PVC, but this requires arbitrary correlation with specific processing requirements.

18.13.3 PROCESSING RATE

This is dependent on melt flow behaviour and on conventional thermal properties like specific heat, as well as on solid granule flow into processing equipment, and it is generally evaluated arbitrarily by extrusion through a standard machine under standard conditions.

18.13.4 DISPERSION

Like dry blends (Section 18.11.5) the dispersion behaviour of granulated compound is usually evaluated by extruding under standard conditions and comparing the extrudate with prepared arbitrary standards or with the extrudates from compounds of known quality.

18.13.5 STABILITY

The stability of a sample of compound can be assessed by pressing specimens and testing them in the usual way, but it is rarely necessary to do more than keep a routine check that batches of compound have been stabilised, and this can usually be done satisfactorily by assessing the colour of extrudates produced in standard extrusion processing experiments (Sections 18.3.1.7–18.3.12.9). UV and X-ray fluorescence can also be used to detect degradation, being particularly valuable for compositions whose dark colour might mask small changes.

18.14 PRODUCT PROPERTIES

18.14.1 GENERAL

Since the properties of a finished product in PVC are dependent not only on the composition and shape, but also on the way and conditions under which it was made, it would be preferable if all properties relevant to end use could be determined on the end-products themselves. While in a few cases this is possible, and indeed almost essential, in general one has to be satisfied with testing specimens prepared from the compositions concerned. In some cases, e.g. determinations of density or composition, this fact does not matter, but in others it is necessary to give very careful consideration to the form of the specimens, the way in which they are prepared, the way in which they are to be tested, and to interpretation of the results.

18.14.2   RELATIVE DENSITY

Relative density can be determined on pieces of consolidated PVC such as compounded granules or pieces of extrudates or mouldings by conventional methods, such as for example by displacement of water, but where regular measurements are to be made density gradient columns,[65-68] have much to commend themselves for accuracy, convenience, and speed. The densities likely to be encountered range from 1.15 to 1.5, which makes the preparation of suitable liquids a little difficult, but aqueous zinc chloride solutions can be made to cover this range conveniently.

18.14.3   SOFTNESS/HARDNESS

Softness or hardness are quoted widely as a main means of characterising PVC compounds, not so much because they are properties of intrinsic importance in themselves, but because there are no really satisfactory tests for characterising intrinsic flexibilities. In view of the possible change in softness with time (Section 6.3.4.3), even the small value of these tests is in doubt. However, they are simple and cheap to carry out, and with care yield reasonably reproducible results. Softness and hardness tests used for PVC are based on the measurement of indentation under controlled conditions. The BS Softness[69] is the indentation into a specimen 1 cm thick, in 0.01 mm units, of a plunger ending in a steel ball $\frac{3}{32}$ in (1.19 mm) in diameter, bearing a load of 535 g after a period of 30 s. Shore A Hardness Numbers[70] are also used.

Various correlations of the two systems have been published, and Table 18.2 has been compiled from some of these.

A particular point to note in doing softness or hardness tests is to ensure that the specimens are prepared and conditioned as nearly as possible in the same standard way.

As a result of structural changes values are likely to change as specimens equilibrate to standard conditions, and for this reason specimens are stored for 7 days at 23°C before testing.

Table 18.2 APPROXIMATE CORRELATION BETWEEN BS SOFTNESS AND SHORE A HARDNESS

BS Softness	0	5	10	15	20	25	30	35	40	45	50
Shore A hardness	100	99	97	93	89	86	83	79	77	74	71
BS Softness	55	60	65	70	75	80	85	90	95	100	105
Shore A hardness	68	66	63	61	59	57	55	53	51	50	48

18.14.4 INDENTATION

Indentation as a property in itself is mainly of concern for floor coverings and is specified as a test with limit requirements in the BS specifications for vinyl flooring.[71, 72] Tests for indentation are in principle similar to those for softness (Sections 18.14.3).

18.14.5 TENSILE PROPERTIES

18.14.5.1 Conventional tensile tests
Tensile strength, modulus, and elongation at break vary widely with formulation for PVC. Doubts have been cast on the value of measuring these properties, but in the absence of any more reliable and meaningful test, they are still measured and quoted widely.[73] Standard specimen dimensions and conditions of test, particularly straining rate, are specified in various relevant specifications.[55, 69, 74] The specimens may be prepared by moulding, stamping, etc., or by cutting from finished products, where the shape of the latter is appropriate. Typical specimens are dumb-bell shaped, 150 mm long, 1.25 mm thick, and 6 mm wide in the waist section. 450 mm per minute is a typical atraining rate. The modulus is taken as the load required to double the separation of two reference lines on the waist section of the specimen, and is quoted as '100 per cent modulus'.

18.14.5.2 Creep
Like other thermoplastics PVC exhibits creep under continuous loading,[75-77] and this limits its use in some potential applications. Creep is of particular significance with unplasticised PVC 'pressure' pipes, and specific tests on these have been developed, in which behaviour under applied internal pressure is observed, results usually being expressed as 'times to burst' at different circumferential or 'hoop' stresses, or vice versa, or extrapolated on a log-log basis over a period of fifty years.[78, 113, 114]

   A short-term cyclic pressure test is also used to predict long-term performance of unplasticised PVC pipe in terms of its ultimate elastic wall stress. In this test a section of pipe is submitted alternatively to internal air pressure for one minute, and relaxation for one minute, deformation being measured by means of a chain fixed at one end, attached to a dial gauge at the other, and passing round the pipe in between. The cycle is repeated ten times each at a series of increasing pressures. A deformation increase of approximately 0.012 per cent after a series of ten cycles indicates that the pipe is no longer behaving elastically, and the value of the pressure for the preceding value is quoted as the ultimate elastic wall stress. Results have correlated well with long-term behaviour of PVC pressure pipe.[115]

18.14.6 'DRAPE' AND 'HANDLE'

These two terms are applied in a subjective way to describe the manner in which calendered plasticised PVC film and sheeting hangs and folds. That there are intrinsic properties related to these subjective ideas, and that they are not correlatable with other properties such as tensile strength, is generally agreed, but no satisfactory special test has been devised. The most positive suggestion, based on the idea that elongations of the order of 1 or 2 per cent are all that are involved in normal draping, has been to measure modulus at 1 to 2 per cent elongation,[79] but accurate measurements are difficult, even with sophisticated instruments.

18.14.7 BRITTLENESS/TOUGHNESS

Plasticised PVC of the flexibilities encountered in practice is essentially such a tough material that measurements at room temperature of brittleness or toughness, e.g. as impact strengths, are pointless. Unmodified unplasticised PVC, however, can fail under impact, its behaviour having been likened to nylon rather than other amorphous polymers.[80] A fair amount of testing of moulded specimens or specimens cut from sheet has been carried out by the Izod and Charpy methods, but the results are of no more meaning for design purposes with PVC than they are for other polymers. Special impact tests have been devised for specific end-products, such as sheet, pipes, and bottles. Thus sheet can be tested by falling weight methods,[81] but these are obviously rather wasteful of material. Pipe, too, is tested by falling weight methods.[82] These include British Standard,[78] ISO[131] and the Horsley[82] impact test methods. In the latter a striker with a 12.7 mm diameter hemispherical nose is dropped from a height of 1–2 m on to the pipe specimen which is supported in a V-block. The striker is arrested so that the nose penetrates the pipe without flattening it. Impact energies up to 237 NM are used.

Pipes are also tested for 'fracture toughness' by a 'hoop' test[78] in which a circular section is cut from the pipe. This is then cut parallel to the axis to form a large closed 'C'. A notch is carefully cut on the opposite side, and resistance to failure on opening the 'C' is determined.

Instrumented impact testing equipment, which yields more information and requires fewer specimens are now widely used, but some of these need careful interpretation of the results.

Bottles are usually tested by dropping the bottles, partially filled with water, from various heights on to a hard flat horizontal surface, and determining statistically the height at which 50 per cent of the tested specimens would fail.[83] A particular problem with bottles is that liability to fracture depends on the way in which a bottle falls and the point at which impact is made, and perhaps there would be some merit in devising a test similar to one used for telephones, where a specimen can be presented in any desired aspects to the impact of a test pendulum hammer.

18.14.8 TEAR STRENGTH

When calendered plasticised PVC sheeting became available in the late 1940s, it was often of doubtful quality, particularly because of its proneness to tearing, and tear strength has been regarded as an important requirement for such sheeting ever since. In general PVC sheeting has good resistance to initiation of a tear, but, once started, propagation is relatively easy. A number of methods for measuring tear strengths of plastics films have been proposed.[54, 55, 84, 85] In 'crescent' and 'angle' tear tests the specimens are strained in a conventional tensile machine, and the force required to initiate tearing from a small area of high stress concentration is measured. More recently the Elmendorff ballistic method[54, 85] has been preferred.[55, 86]

18.14.9 LOW TEMPERATURE PROPERTIES (Section 6.3.4.4)

It was indicated earlier that if the temperature of a plasticised PVC composition is reduced, its modulus increases, i.e. it becomes progressively stiffer and more brittle. The same is true of unplasticised PVC, and there is some merit in studying mechanical properties of unplasticised PVC, particularly impact strength, at temperatures around 0°C or even lower,[77] but since it is already comparatively rigid in any appreciable section at normal room temperatures, changes in flexibility or onset of cracking on bending are not very amenable of test with the material. As was also indicated a number of different methods of test of behaviour of plasticised PVC as temperature is lowered have been devised,[54] and there is generally little correlation between them and practical commercial situations. They therefore need to be interpreted and applied with great care. The Clash and Berg low temperature flexibility test[73, 87] has often been preferred for sheeting. In this test a $2.5 \times 0.25 \times 0.04$ in ($6.35 \times 0.635 \times 0,1016$ cm) specimen is clamped rigidly at its lower end and in a rotatable clamp at its upper end. A fixed torque can be applied to the upper clamp through a pulley system, and the specimen cooled by means of a cold jacket. Angular deflections are measured under the standard torque at a variety of different temperatures, and the 'low temperature flexibility' or LTF is taken as that temperature at which the deflection equals a standard value (200°). Typical values for PVC range from just below room temperature down to −44°C and below.

Low temperature extensibility is fairly simply measured by applying a uniform stress of 20.7 MN/m^2 for 30 seconds at −5°C, the result being expressed as a percentage elongation.

Compounds are often tested by the 'cold bend' test[73] in which specimens 100 mm × 4.8 mm × 1.25 mm in size are cooled to a particular test temperature and wrapped around a 5 mm diameter mandrel at one convolution per second. Tests are repeated to determine the lowest temperature at which all specimens remain intact without cracking. Typical values for PVC range between −10 and −60°C.

In the Williams cold crack test,[54, 88] the specimen is looped on an anvil in a cold bath and struck by a spring-loaded hammer. The 'cold crack' temperature is taken as that at which 50 per cent of tested specimens fail by cracking.

A number of other low temperature flexing and cracking tests are used, and, since they rarely, if ever, give the same results on the same specimens, it is important to define the method of test employed whenever specifying some partiuclar value of low temperature performance, otherwise confusion and controversy are likely to arise.

18.14.10 PROPERTIES AT ELEVATED TEMPERATURES (Section 6.3.4.6)

Three aspects of properties at elevated temperatures are of interest, but they are not of equal interest or value. Softening temperatures are of interest from the point of view of processing, largely because of the need to limit heating in view of possible degradation, but this is of little consequence with plasticised compositions, and with unplasticised PVC a mere 'softening temperature' is not of much value in assessing service performance. Where this behaviour is concerned deformation at elevated temperatures is of interest, but even here 'softening temperatures' can be meaningless, because the service situation can be a major contributor to performance (Section 6.3.4.5). In general softening temperatures for PVC are measured by similar methods to those used for other thermoplastics,[89] e.g. the Vicat softening temperature test for all grades of PVC, and the heat distortion test for unplasticised or stiff plasticised grades.

The third aspect of behaviour at elevated temperatures is mainly concerned with loss of plasticiser by degradation or straightforward volatilisation. Chemical degradation of the polymer itself is of very little significance at normal service temperatures. Accurate reproducible assessment of plasticiser loss is difficult, and the techniques have been outlined earlier (Section 18.5.4).

18.14.11 FLAMMABILITY

PVC is generally regarded as non-flammable, but this is not strictly true, and compositions containing sufficient of a flammable plasticiser will support combustion. The main interest in testing PVC in this connection lies in determining whether or not unplasticised PVC will meet the fire resistance requirements of specifications, local authorities, etc., particularly for applications in building, where the methods of test are generally not directed at any one material specifically. There are, however, standard methods of test for degree of flammability for PVC sheeting.[50, 51, 90]

As pointed out previously, no two fires are exactly alike, which makes testing for fire behaviour particularly difficult. Tests range from simple application of a match to a small specimen to more or less sophisticated somewhat arbitrary small-scale laboratory tests and full-scale trials in specially constructed buildings. In fact the simple test first mentioned can be quite useful for assessing differences in 'ignitability' using specimens about 50–70 mm long and 20–30 mm^2 in cross-section. There are small-scale tests for assessing ignitability, mainly differing in specimen dimensions and disposition, source of ignition, and in procedural details.[116]

Despite the fact that its conditions are not closely related to those in real fires, the 'Limiting Oxygen Index' test has been widely used, and has provided useful information.[116-119] A similar method is the 'Setchkin Flash and Self-Ignition' test. Both these tests can be modified to use oxygen/nitrogen mixtures at temperatures above ambient, and conditions similar to those in actual fires can be further simulated by using gas mixtures depleted of oxygen.[119] The Setchkin method can also be used to assess burning rate.[119]

Other aspects of behaviour in fire situations include spread of flame, smoke, and toxic gas emission.

As with ignition, a number of tests for spread of flame exist, mainly differing in details of sample dimensions and disposition, and source of radiant heat. In these tests unplasticised PVC normally falls into the lowest or lower categories,[116, 120] i.e. it is amongst those materials that contribute least to spread of flame.

Smoke is important because of its obscuration and because of its harmful effect if inhaled. Various designs of 'smoke chamber' are used, in which the density of smoke produced under standard conditions is measured by means of photo-cells. Alternatively the air containing smoke produced under standard conditions can be drawn though a filter and the deposit assessed visually or weighed. Comparisons between different materials can be confusing, producing different rankings depending on the burning conditions.[116]

Gas emissions can be analysed by standard means, but the question of their toxicity requires assessment by experts in this particular subject. The major toxic gas emissions from PVC in a fire are hydrogen chloride and, like all other organic materials, natural or synthetic, carbon monoxide. This topic is discussed in Section 17.5.

Thorough investigation of fire behaviour of any product, including PVC, requires full-scale fire tests, and, in spite of their cost, they are quite commonly carried out. A typical installation for these tests comprises an isolated building constructed of fire- and heat-resistant materials instrumented with strategically placed thermocouples for temperature record-ing, photo-cells for obscuration measurements, etc.

There are also standard tests available for specific products, e.g. cables and floor cover-ings. These attempt to simulate on a small scale the conditions peculiar to the products.

18.14.12   ELECTRICAL PROPERTIES

Generally it is plasticised PVC which is of interest for electrical installation, and it is rarely necessary to measure the electrical properties of unplasticised PVC. However, the methods of test are the same, and for that matter are the same as those used for other materials. Standard tests are usually performed on sheets about 1.25 mm thick, but other shapes of specimen can be used provided the calculations take correct account of the dimensions.

Volume resistivity is measured in the conventional way.[69] Specimens should be carefully conditioned before test, for example by immersion in water for 24 h, followed by an ad-equate period in standard atmosphere. Particular attention must be paid to temperature as the results are sharply dependent on it. Measurements are preferably made at 23°C, but 20, 25, and 30°C are used in some laboratories. A potential difference of 500 V is applied across the specimen, and measurements should be taken at a standard time of 1 min after appli-cation of voltage.

Breakdown voltage is usually measured at 50 Hz for 1 min. Power factor is usually measured at a frequency of 1 kHz; but a full evaluation requires testing over a wide range of frequencies.

Electrostatic charging is no more nor less of a problem with PVC than it is with other plastics materials, nor is its assessment any more or less difficult, but a simple specific test for PVC films has been described.[91]

18.14.13   OTHER GENERAL TESTS

18.14.13.1   General
A number of other tests not so far mentioned specifically are sometimes carried out on PVC in various forms. These include water absorption and permeability. The former is carried out quite simply by measuring the weight increases of specimens immersed in water at vari-ous temperatures for various periods of time. The usual standard method involves immer-sion of discs about 1.25 mm thick and 50 mm in diameter in water at 50°C for 48 h. After immersion the specimens are re-dried to constant weight so as to allow for extraction of water-soluble matter.

The ISO method uses discs 3 mm thick and 50 mm in diameter immersed for 24 h in water at 23°C.

Plots of water absorption over a more extended period of time might be more meaning-ful, although it has been suggested that examination at 7 or 4 day intervals is all that is necessary.[92]

18.14.13.2   Permeability
For packaging and conveyancing applications one is concerned about permeability of liquids, while gas and vapour permeabilities are mainly of interest in packaging, being of some importance in conveyance only in rare cases, e.g. gas pipe and fittings.

In packaging there is interest in films and foils, both plasticised and unplasticised, and in bottles, at present entirely of unplasticised PVC. Methods of test are generally similar to those for other materials.[50, 89, 93–95] Gas permeability of film can be calculated from measurements of pressure and volume changes as gas passes through the sample from a chamber at specific temperature and pressure.[50, 96] Water vapour permeability can be calculated from weight increases of samples sealed over vessels containing dessicant.[50]

18.14.13.3   Sulphide staining

A number of applications for plasticised PVC, e.g. in sheet, leathercloth, or flooring form, involve possible contact with sulphur-containing substances which are liable to react with some stabilisers to cause discoloration or staining in all but the darkest colours, and various procedures have been proposed to test for this kind of reaction. These usually involve contacting the sample with a freshly prepared aqueous acidic sodium sulphide solution, e.g. 11 g of sodium sulphide nonahydrate in 200 cm^3 of distilled water, to which solution 6 cm^3 of concentrated hydrochloric acid have been added drop by drop with vigorous stirring to dissolve any precipitated sulphur.

18.4.14   SOME SPECIFIC TESTS FOR FILM AND SHEETING

18.14.14.1   Weight and thickness

Dimensions of flexible PVC film and sheeting are often expressed as weight per unit area or average thickness. The former is calculated simply by weighing a number of specimens of nominally known equal area. The number and size are commonly selected so that the total weight in grammes is numerically equal to or is a simple ratio of the weight in g/m^2.

Micrometer thickness is calculated by measuring the thickness with a micrometer to the nearest 0.0025 mm at five points across the width of the sheeting at two positions at least one metre apart in the longitudinal direction. Care must be taken to ensure that the micrometer foot is not over- or under-loaded.

Gravimetric thickness is calculated from the weight per unit are and the relative density.

18.4.14.2   Pin-holes (Section 12.9.6)

It can be difficult to differentiate between pin-holes and 'windows' (Section 12.9.5.5) or incompletely dispersed resin particles. Relatively large ones can be detected by viewing the film or sheeting with a bright light behind it, but smaller yet nevertheless significant pin-holes are not always satisfactorily shown up by such means. One method that has been used is to stretch the film by about 20 per cent on a suitable frame over a sheet of unglazed white paper, and to brush an aqueous dye solution (e.g. 0.1 per cent Rhodamine B) over the sheeting.[50] Pin-holes are disclosed by coloured spots on the paper.

18.14.14.3   Opacity

Superfically opaque sheeting can often be found to be translucent on closer inspection, and in fact it is quite difficult to produce a truly opaque sheeting as thin as 0.025 mm or less. In attempting to formulate an opaque composition it is important to be able to compare the opacifying effects of different pigments and fillers. One method that has been developed[97] is to prepare sheets to a standard formulation containing various amounts of the additive under examination (up to about 3 phr), and to measure photometrically the luminous reflectances of the samples over a white plane glass plate compared with measurements over a black plate.

18.14.14.4 Dimensional stability

Changes in dimensions of sheeting, including vinyl flooring, can arise as a result of release of internal stresses and absorption of water. Stress release generally results from elevation of temperature, and tests usually involve actual measurement of dimensional changes during scheduled heat treatment cycles. Calendered film and sheeting, for example, may be heated in water, e.g. 15 min at 50°C or 100°C, or in an oven at higher temperatures.[50, 51]

18.14.14.5 Ply adhesion

Laminated sheeting is largely used for its strength and wear properties, and any tendency to delamination must be avoided as far as possible. For control purposes circular test pieces may be immersed in stirred ethyl acetate at 23°C for 15 min and examined visually for ply separation.[50, 51] Actual values of ply adhesion can be obtained by the conventional method, by separating a short length of the plies in a one-inch-wide specimen, and pulling them apart in a tensometer.[51]

18.14.14.6 Blocking

Glossy flat film and sheeting is prone to blocking, for example when in reels, particularly at relatively high plasticiser contents, and one purpose of embossing is to reduce this tendency. Blocking is difficult to assess by test procedures. One method that is used is to press overlaps of two specimens between filter paper and glass plates under a pressure of about 7 kN/m^2 at 50°C for 24 h, the degree of blocking being assessed by hand.

18.14.14.7 Miscellaneous tests

A number of other tests of film and sheeting have been called for in various specifications.[50, 51] These include 'colour rubbing', print adhesion, resistance to soapy water, emboss retention, wear resistance, and fastness to light. For details the reader is referred to the methods described in the various specifications concerned.[50, 51, 98]

18.14.15 TESTS FOR SPECIFIC PRODUCTS

There are many specifications with standard tests for specific finished products such as pipes for water conveyance and drainage etc., window frames, bottles, and cables, and reference should be made to the specifications if further details are required.[78] An interesting special case is that of retractile spiral telephone cords. The insulation and sheathing for these are usually formulated with a proportion of aliphatic ester plasticiser, e.g. DBS, DAS, DIOZ, etc. (Section 6.3.4.2), but it seems likely that a good deal of the retractile behaviour is due to the wire and the general construction of the cords. However, a test method has been proposed using a helical test specimen formed from a strip 125 mm long, 6.5 mm wide, and 2 mm thick.[99] The specimens are prepared by wrapping in close turns around 6.5 mm diameter wooden dowels, heating at 130°C for 30 min, cooling to room temperature, and removing from the dowels. The test procedure involves stretching each helix 8 cm beyond its original length, while held vertically, realising it after 30 s, and measuring the time taken for recovery of 7.2 cm of the extension. Apparent moduli can also be measured on the helices, using elongations between 50 and 150 per cent.

18.15   ANALYSIS OF PVC COMPOUNDS

Some analytical procedures for polymers have been discussed previously (Section 18.3), but it is frequently desirable to determine the components of a PVC compound or finished product. This can be achieved by fairly routine chemical, chromatographic, and spectrographic techniques.[8, 9, 11–13, 121, 122] Accurate and detailed analysis can be a difficult and highly skilled job, best left to analytical experts, but some useful analysis can often be done with limited facilities, time, and expertise. Thus, with a reasonably sensitive and discriminating nose, the fact that a material is PVC at all can usually be easily confirmed by heating it in a flame and sniffing the resultant vapours. The presence of phthalate plasicisers can often be detected in the same way. To be a little more sophisticated the sample can be heated in a test tube fitted with a tube which conducts the vapours into a solution of alkali, which can then be tested for chloride ion. Alternatively the vapours can be absorbed in distilled water and the resulting pH measured. Beilstein's test for halide is also often useful. A convenient way of using this test is to heat the end of a piece of copper wire in a gas flame until any green colour disappears, dip or press the hot wire into the specimen, and then reheat. A green coloration in the flame indicates the presence of halogen, usually chlorine. Among more sophisticated techniques is to use a piece of copper gauze on which a small sample is placed, and the flame can be analysed spectroscopically.

Soxhlet extraction with ether will remove all but some polymeric plasticisers and extenders, lubricants, and stabilisers of the organotin and organic, e.g. epoxy, type. Residual polymeric plasticisers can be removed by extraction wtih methanol, and inorganic matter obtained as a residue by extraction with tetrahydrofuran, which dissolves the polymer and any organic modifiers. The polymer can be precipitated from this extract by addition of ethanol. Analysis of the ether and methanol extracts is most conveniently done by infra-red spectrometry.[8, 57, 100, 122]

18.16   STANDARD SPECIFICATIONS

18.16.1   GENERAL

A considerable number of official specifications exist with direct relevance to PVC materials and products, and there is no intention of discussing these in detail here. The situation is confused by the fact that international ISO standards often exist alongside national standards, and it is necessary to take note of which are the relevant standards in any particular country where manufacture is to be carried on or products sold. In the case of British Standards, the equivalent ISO standards, where they exist, are listed in the British Standards catalogue,[101] available in most university and main public libraries, or directly from the British Standards Institution, Customer Services, Linford Wood, Milton Keynes, UK MK14 6LE. As well as British Standards the catalogue lists Codes of Practice, International Electrotechnical Commission IEC and CEE, and other specifications. Copies of specifications can be purchased from the Institution at the afore-mentioned address, and enquiries may be made to the Information Group at the same address.

Some of the more relevant British and International standards are listed below.

18.16.2 BRITISH STANDARD SPECIFICATIONS

BS 476:	In several parts devoted to various tests for fire behaviour of materials and structures.
BS 573: 1993	Di-n-butyl phthalate.
BS 1763: 1975	Thin calendered flexible unsupported sheeting.
BS 1870: 1981	Part 3: Moulded safety footwear.
BS 1995: 1993	Di-(2-ethylhexyl) phthalate.
BS 2536: 1993	Di-(2-ethylhexyl) sebacate.
BS 2571: 1990	Flexible compounds for moulding and extrusion.
BS 2739: 1975	Thick calendered flexible unsupported sheeting.
BS 2782:	In several parts and sub-divisions of which the following are of particular interest for PVC:

Part 1: Method 122A (1988): heat deformation of flexible PVC.
Method 130A (1991): Congo Red stability test.
Method 130B (1991): pH stability test.
Method 131B (1989): heat ageing of flexible PVC sheet.
Method 140D (1987): flammability of thin PVC sheet.
Method 141 (1986): oxygen index method.
Method 150B (1988): cold flex test of flexible PVC.
Method 150C (1989): low temperature extensibility.
Method 151A (1992): cold bend test.
Method 153B (1991): stiffness in torsion.

Part 2: Methods for determination of electrical properties.

Part 3: Methods for assessing mechanical properties inc.
Method 365A (1989): softness number.
Method 365B (1992): shore hardness.

Part 4: Methods 430A-D (1983): water absorption.
Method 454C (1983): pH of aqueous extract of pvc resins.
Method 454D (1983): volatile matter in pvc resins.
Method 454E (1992): plasticiser absorption of resins.
Method 454F (1983): air-jet sieve analysis of resins.
Method 465A (1992): plasticiser loss (activated carbon).
Method 465B (1992): plasticiser loss (activated carbon).
Method 465C (1990): plasticiser migration.
Method 470E (1991): determination of ash from PVC.

Part 6: Method 621D (1980): compacted apparent bulk density.
Method 630A (1987): thickness of flexible sheet.
Method 631A (1987): thickness of flexible sheet.
Method 632A (1987): length and width of flexible sheet.
Method 641A (1991): dimensional stability of sheet.
Method 634A (1983): shrinkage of shrink-wrap film.

	Part 7:	Method 732B (1991): viscosity number of pvc resins.
	Part 8:	Method 820A (1992): water vapour transmission.
		Method 821A (1986): gas transmission.
		Method 822A (1992): water vapour transmission.
	Part 11:	Method 1101A (1981): measurement of pipe dimensions.
		Method 1102A (1986): longitudinal reversion of pipes.
		Method 1102B (1986): longitudinal reversion of pipes.
		Method 1103A (1987): stress relief of moulded fittings.
		Method 1108A (1989): true impact rate boundaries.
		Method 1110 (1989): tensile properties of guttering.
BS 2848: 1973	Flexible electrical insulating sleeving.
BS 3260: 1991	Semi-flexible floor tiles.
BS 3261: 1991	Flexible flooring.
BS 3424:	Part 22 (1983): degree of fusion of coated fabrics.
BS 3505: 1986	Water pipes.
BS 3506: 1969	Industrial UPVC pipes.
BS 3546:	Part 2 (1993): non-water vapour permeable coated fabrics.
	Part 4 (1991): water vapour permeable coated fabrics.
BS 3647: 1993	Specifications for plasticiser esters (refers to other numbers).
B£ 3746: 1990	Garden hose.
BS 3757: 1990	Rigid PVC sheeting.
BS 3869: 1965	Rigid expanded PVC for thermal insulation and building.
BS 3878: 1990	Flexible PVC hospital sheeting.
BS 3887: 1991	Pressure-sensitive closing and sealing tapes.
BS 3924: 1978	Pressure-sensitive tapes for electrical purposes.
BS 4203: 1987	Extruded rigid corrugated sheeting.
BS 4345:	Part 1 (1969): pressure pipe joints and fittings.
	Part 2 (1970): pressure pipe joints and fittings.
	Part 3 (1982): pressure pipe joints and fittings.
BS 4370:	Parts 1–14 (1973–1991): tests for rigid cellular materials.
BS 4514: 1983	Soil and ventilating pipes and fittings.
BS 4553: 1991	600/1000 V PVC insulated single phase split concentric cables with copper conductors for electricity supply.
BS 4576: 1989	Part 1: rainwater goods.
BS 4607: 1991	Part 2: rigid electrical conduit and fittings.
	Part 3: self-extinguishing pliable electrical conduits.
BS 4660: 1989	Below ground UPVC drainage and sewerage pipes and fittings.
BS 4735: 1984	Horizontal burning characteristics of cellular plastics.
BS 4808:	Parts 1–5 (1972–1974): PVC insulted (and sheathed LF telecommunication cables.
BS 4968: 1988	Di-isobutyl phthalate.
BS 4969: 1988	Di-isooctyl phthalate.
BS 4970: 1988	Di-isooctyl sebacate.
BS 5085: 1991	Parts 1 and 2: backed flexible flooring.
BS 5111: 1983	Part 1: smoke generation of cellular plastics.
BS 5255: 1989	Waste pipes and fittings.

BS 5308: 1986	Part 1: PVC insulated instrumentation cables.
BS 5462: 1984	Part 2: footwear with mid-sole protection.
BS 5481: 1989	Gravity sewers.
BS 5593: 1991	Oversheathed CONSAC cables.
BS 5790: 1979	Parts 1 and 2: PVC-coated upholstery fabric.
BS 5955: 1980	Part 6: installation of gravity drains and sewers (C.O.P.).
BS 6004: 1991	PVC-insulated cables for power and lighting.
BS 6159: 1987	General and industrial lined and unlined PVC boots.
BS 6231: 1990	PVC-insulated switch and control gear cables.
BS 6346: 1989	PVC-insulated cables for electricity supply.
BS 6469: 1992	Part 3: tests for PVC insulation and sheathing materials.
BS 6485: 1990	PVC-covered overhead power line conductors.
BS 6728: 1986	PVC hot water bottles.
BS 6746: 1990	PVC insulation and sheath for electrical cables.
	6746C: standard colours for same.
BS 7010: 1988	Dimensional tolerances.
BS 7412: 1991	Extruded UPVC hollow window frame profiles.
BS 7413: 1991	Extruded UPVC hollow window frame profiles: Type A.
BS 7414: 1991	Extruded UPVC hollow window frame profiles: Type B.

18.16.3   INTERNATIONAL ORGANISATION FOR STANDARDISATION SPECIFICATIONS

ISO 35:	Mechanical stability of latex.
ISO 62:	Determination of water absorption.
ISO 105:	Migration of textile colours into PVC coatings.
ISO 161:	Metric dimensions for plastics pipes for fluid transport.
ISO 174:	Viscosity Number of pvc resins.
ISO 182: 1990	Part 1: Congo Red heat stability test.
	Part 2: pH heat stability test.
ISO 183:	Bleeding of colourants.
ISO 264:	Socket fittings for plastics pipes under pressure.
ISO 265:	Socket fittings for waste pipes.
ISO 305: 1990	Heat stability of PVC by discoloration.
ISO 330:	Part II: plastics pipes for fluid transport (Inches).
ISO 511:	White lead.
ISO 580: 1990	Oven test for UPVC injection mouldings.
ISO 727:	Metric dimensions for UPVC socket fittings under pressure.
ISO 1068: 1975	Compacted apparent bulk density.
ISO 1163: 1991	Part 2: preparation and testing of UPVC specimens.
ISO 1265: 1979	Impurities and foreign particles in pvc resins.
ISO 1269: 1980	Volatiles in pvc resins.
ISO 1270: 1975	Ash content of pvc resins.
ISO 1628: 1988	Part 2: viscosity numbers of pvc resins.
ISO 2035: 1974	Moulded UPVC fittings for elastic sealing ring joints.
ISO 2044: 1974	Moulded UPVC socket fittings for solvent welding.
ISO 2048: 1990	Depths of engagement for UPVC sealing ring type fittings.

ISO 2507: 1992	Part 2: Vicat softening test for UPVC pipes and fittings.
ISO 2536: 1974	Metric dimensions of flanges for UPVC pipes and fittings.
ISO 2898: 1989	Preparation and testing of plasticised PVC specimens.
ISO 3114: 1977	Extractability of lead and tin from UPVC water pipes.
ISO 3127: 1980	Resistance of UPVC pipes to external blows.
ISO 3451: 1989	Part 5: ash determination of PVC.
ISO 3460: 1975	Metric dimensions of backing flange adapters for UPVC pipes.
ISO 3472: 1975	Acetone resistance of UPVC pipes.
ISO 3473: 1977	Effects of sulphuric acid on UPVC pipes.
ISO 3474: 1976	Opacity of UPVC pipes.
ISO 3603: 1977	Leakproofness of UPVC pipe fittings.
ISO 3604: 1976	Leakproofness of UPVC pipes under external pressure.
ISO 3606: 1976	Tolerances for UPVC pipes.
ISO 4191: 1989	Practice for laying UPVC water pipe.
ISO 4422: 1990	Specifications for UPVC water pipe and fittings.
ISO 4422: 1992	AMD 1: amendment to ISO 4422.
ISO 4434: 1977	Metric dimensions for adaptor fittings for UPVC pipes.
ISO 4439: 1979	Density of UPVC pipes and fittings.
ISO 4574: 1978	Hot plasticiser absorption of pvc resins.
ISO 4575: 1985	Severs rheometer apparent viscosity of PVC pastes.
ISO 4608: 1984	Room temperature plasticiser absorption of pvc resins.
ISO 4612: 1979	Preparation of PVC pastes.
ISO 4643: 1992	Moulded PVC boots for industrial use.
ISO 6110: 1992	Moulded industrial PVC boots with chemical resistance.
ISO 6112: 1992	Moulded industrial PVC boots with oil and fat resistance.
ISO 6451: 1982	Rapid fusion test for PVC coated fabrics.
ISO 6453: 1985	Flexible PVC foam sheeting.
ISO 6455: 1983	Metric dimensions for UPVC pipe laying lengths.
ISO 6990: 1983	Burst test for UPVC pipe fittings for fluid transport under pressure.
ISO 6991: 1983	Burst test for UPVC pipes.
ISO 6992: 1986	Extractability of cadmium and mercury from UPVC water pipes.
ISO 6993: 1990	High-impact UPVC pipes for underground gaseous fuel supply.
ISO 7387: 1983	Part 1: basic tests for solvent adhesives for UPVC pipes.
ISO 7473: 1978	Chemical resistance of UPVC pipes and fittings.
ISO 7617: 1988	Part 1: PVC-coated knitted upholstery fabrics.
	Part 2: PVC-coated woven upholstery fabrics.
ISO 7676: 1990	Dichloromethane test for UPVC pipes.
ISO 8095: 1990	PVC-coated fabrics for tarpaulins.
ISO 8096: 1989	Part 1: PVC-coated water-resistant fabrics.
	Corrigendum 1: technical corrigendum to above.
ISO 8771: 1985	UPVC pipes & fittings for sub-soil drainage.
ISO 9852: 1991	Dichloromethane resistance temperature of UPVC pipes.
ISO 9853: 1991	Crushing test for UPVC moulded pressure pipe fittings.

REFERENCES

1. G.A.R. MATTHEWS and R.B. PEARSON, *Plastics,* **28**, 307, 98 (1963): **29**, 317, 99 (1964)
2. ASTM D 1243–58T
3. ISO/R 174: 1961 (E)
4. H. FIKENTSCHER, *Cellulose-chem.,* **13**, 58 (1932); *Kolloidzeitschrift,* **53**, 34 (1931)
5. P. PLÁTZEK, *Plastica,* **5**, 19 (952)
6. JIS K 6721–1959
7. M. FREEMAN and P. MANNING, *J. Polym. Sci., A,* **2**, 5, 2017 (1964)
8. J. HASLAM, H.A. WILLIS and D.C.M. SQUIRRELL, *Identification and Analysis of Plastics,* 2nd edn., Iliffe, London, (1972)
9. J. HASLAM and W.W. SOPPETT, *J. Soc. chem. Ind., Lond.,* **67**, 33 (1948)
10. J. HASLAM and D.C.M. SQUIRRELL, *Analyst,* **79**, 689 (1954); **82**, 511 (1957)
11. J. HASLAM, J. HAMILTON and D.C.M. SQUIRREL, *Analyst,* **85**, 566 (1960)
12. M.R LARDERA, E. GERNIA and A. MARI, *Annali Chim.,* **46**, 194 (1956)
13. D.B. GURVICH, V.W. BALANDIAN and R.V. KOSMOSKOVA, *Soviet Plast.,* **47** (Dec. 1961)
14. ISO/R60: 1958
15. S. BRUNAUER, P.H. EMMETT and E. TELLER, *J. Am. chem Soc.,* **60**, 2, 309 (1938)
16. S.J. GREGG and K.S.W. SING, *J. phys. Colloid Chem.,* **55**, 42, 592 (1957)
17. ASTM D 1755–60T
18. C.E. ANAGNOSTOPOULOS, A.Y. CORAN and H.R. CAMRATH, *J. appl. Polym. Sci.,* **4**, 11, 181 (1960)
19. F. BARGELLINI, G. FABRIS and F. CHIOZZINI, *Poliplastic,* **11**, 70, 5 (1963)
20. F. BARGELLINI, *Materie Plast.,* **30**, 4, 378 (1964)
21. G. MATTHEWS, D.H. SMITH and J.A. VINEY, 'Rate and temperature relationships for plasticiser solvation of PVC particles', *PVC Plasticiser Group Meeting,* Univ. of Technology, Loughborough (8 April 1981)
22. G.A.R. MATTHEWS, E.C.A. HORNER and W.W. VINCENT, unpublished work (1956–1957)
23. E.T. SEVERS and G. SMITMANS, *Paint Varn. Prod.,* **47**, 12, 54 (1957)
24. J.A. GREENHOE, *Plast. Technol.,* **7**, 10, 142 (1958)
25. L.A. McKENNA, *Mod. Plast.,* **35**, 10, 142 (1958)
26. W.J. FRISSELL, *Mod. Plast.,* **38**, 9, 232 (1961)
27. A. WHEELER and B.V. CLIFTON, *Br. Plast.,* **35**, 640 (1962)
28. E.H. FOAKES, *Br. Plast.,* **39**, 2, 92 (1966)
29. J.F. EHLERS and K.R. GOLDSTEIN, *Kolloidzeitschrift,* **118**, 137 (1950)
30. A. HARTMANN and F. GLANDER, *Kolloidzeitschrift,* **137**, 2/3, 79 (1954)
31. A. HARTMANN, *Kolloidzeitschrift,* **142**, 123 (1955)
32. W.D. TODD, D. EASGROVE and W.M. SMITH, *Mod. Plast.,* **34**, 1, 159 (1956)
33. A.C. MEASEY, *Br. Plast.,* **32**, 2, 55 (1959)
34. C. CAWTHRA, G.P. PEARSON and W.R. MOORE, *Trans. Plast. Inst.,* **33**, 104, 39 (1965)
35. P. SCHMIDT, *Kunststoffe,* **41**, 1, 23 (1951); **42**, 5, 142 (1952)
36. K. THINIUS, W. REICHHERDT and B. HOSSELBARTH, *Plaste Kautsch.,* **10**, 6, 339 (1963)
37. N.W. TOUCHETTE, H.J. SEPPALA and J.R. DARBY, *Plast. Technol.,* **22**, 7, 33 (1964)
38. A.C. HECKER and S. COHEN, *S.P.E. Papers,* Quebec Section, 36 (Oct. 1964)
39. J. MATTHAN, *RAPRA Res. Rep.,* 165 (1968)
40. H.S. BERGEN and J.R. DARBY, *Ind. Engng Chem. ind. Edn,* **43**, 2404 (1951)
41. B.S. DYER, *Trans. J. Plast. Inst.,* **27**, 69, 84 (1959)
42. D.R. JONES and J.C. HAWKES, *Trans. J. Plast. Inst.,* **35**, 120, 773 (1967)
43. G.M. GALE, *RAPRA Bull.,* **21**, 5, 78 (1967)
44. ISO/R182: 1961
45. M. HLOUSEK and J. TEPLY, *Chemcký Prum.,* **16**, 12, 731 (1961); RAPRA Translation No. 1504
46. G. MENGES and J. MULLER, *Plastverarbeiter,* **17**, 7, 397 (1966)
47. J.B. DECOSTE and B.A. STIRATELLI, *S.P.E. 16th Ann. Nat. Tech. Conc.,* **56** (1960)
48. I. MELLAN, *The Behaviour of Plasticisers,* Pergamon, Oxford (1961)
49. J.B. DECOSTE, *S.P.E. Jl,* **16**, 10, 1129 (1960)
50. W.K. DALTON, *Br. Plast.,* **35**, 3 136 (1962)
51. BS 1763: 1967
52. ASTM D573–53; D1203–55

53. M. Royen, *ASTM Bull.,* No. 243, 43 (1960)

54. ISO/R176: 1961 (E)

55. *Plasticisers for PVC ,* A. Boake, Roberts & Co., Ltd. (1963)

56. *Plasticisers for PVC,* BP Chemicals (U.K.) Ltd (1968)

57. W. Meise and H. Ostromow, *Kunststoffe,* **54**, 4, 213 (1964)

58. P.I.A. Martin, *Br. Plast.,* **38**, 2, 95 (1965)

59. BS 1795

60. R. Hayes, *Chemy Ind.,* **44**, 1069 (1952)

61. E.B. Atkinson and H.A. Nancarrow, *Proc. Int. Rehol. Congr.,* Part 2, 103 (1948)

62. B.A. Taylor and T.C. Tunbridge, unpublished work for award of B.Sc., at Borough Polytechnic (1964–1965)

63. E.R. Howells and J.J. Benbow, *Trans. J. Plast. Inst.,* **30**, 88, 240 (1962)

64. J.S. Walker, *Int. Plast. Engng,* **1**, 5, 239 (1961)

65. L.H. Tung and W.C. Taylor, *J. Polym. Sci.,* **21**, 97, 144 (1956)

66. G. Oster and M. Yamamoto, *Chem. Rev.,* **63**, 257 (1963)

67. R.E. Wiley, *Plast. Technol.,* **8**, 3, 31 (1962)

68. Z. Sobiezewski and M. Wajnryb, *Polimery,* **8**, 2, 69 (1963)

69. BS 2571: 1990; BS 2782, Part 3, Method 265A (1989)

70. BS 2782, Part 3, Method 365B (1992); ISO 2898: 1989

71. BS 3260: 1969

72. BS 3261: 1960

73. P.I. Vincent, *Plastics,* **26**, 288, 121 (1961); **26**, 289, 141 (1961); **27**, 291, 115 (1962)

74. BS 2746: 1966

75. A.F. Mills, I.S. Note No. 948, I.C.I. Ltd., (1962)

76. S. Turner, *Br, Plast,* **37**, 6, 322 (1964)

77. C.M.R. Dunn, W.H. Mills and S. Turner, *Br. Plast.,* **37**, 7, 386 (1964)

78. BS 3505: 1968

79. J.H. Gisolf, *Plastica,* **15**, 10, 498 (1962)

80. P.I. Vincent, *Plastics,* **28**, 306, 120 (1963)

81. H.R. Reid and R.A Horsley, *Br. Plast.,* **32**, 4, 156 (1959)

82. Water Research Assoc., Techn. Pubn. No. 19

83. P.L. Veal, *Br. Plast.,* **38**, 5, 258 (1965)

84. ASTM D624–48

85. N. Stackhouse, *Br. Plast.,* **34**, 1, 34 (1961)

86. BS 2739

87. R.F. Clash and R.M. Berg, *Ind. Engng Chem. ind. Edn,* **34**, 10, 1218 (1942)

88. H.O. Williams, *Br. Plast.,* **31**, 3, 107 (1958)

89. BS 2782

90. ASTM D1433–58

91. H.J. Stern, L.E. Shadbolt and S. Mead, *Rubb. Plast. Wkly,* 790 (19 May 1962)

92. F. Blank, *Plaste Kautsch.,* **9**, 8, 391 (1962)

93. I. Phillips and D.V. Bartlett, *Br. Plast.,* **34**, 10, 533 (1961)

94. C.J. Major and K. Kammermeyer, *Mod. Plast.,* **39**, 11, 135 (1962)

95. M. Salame and J. Pinsky, *Mod. Packag.,* **37**, 8, 209 (1964); **37**, 9, 131 (1964)

96. ASTM D1434–58

97. 'Kronos' Laboratory Report, Titangesellschaft MBH (1968)

98. ASTM D1239–52; D673–44; D1044–56; D1242–56

99. J.B. De Coste, J.B. Howard, V.T. Wallder and H.M. Zupko, *Wire Assoc. Annl Conv.* (1958)

100. M. Cachia, D.W. Southwart and W.H.T. Davison, *J. appl. Chem.,* **8**, 5, 291 (1958)

101. British Standards Institution, *British Standards Catalogue* (1993)

102. H.M. Clayton, *Developments in PVC Production and Processing –1,* (A. Whelan and J.L. Craft, eds), Applied Science Publishers, London (1977) Chapter 3

103. R.H, Burgess, *Manufacture and Processing of PVC,* Applied Science Publishers, London (1982) Chapter 5, pp. 105–8

104. M.W. Allsopp, *ibid,* Chapter 7

105. M.W. Allsopp, *Pure Appl. Chem.,* **53**, 449 (1981)

106. G.M. Gale, *Plast. & Polymers,* **38**, 183 (1970)

107. G.M. GALE, *Developments in PVC Technology*, (J.H.L. Henson and A. Whelan, eds), Applied Sciences Publishers, London (1973) Chapter 3
108. M.W. ALLSOPP, *Manufacture and Processing of PVC*, (R.H. Burgess, ed), Applied Science Publishers, London (1983) Chapter 8
109. M.W. ALLSOPP, *PRI Intl. Conf., PVC Processing II,* Brighton, England (26–28 April 1983) Paper 4
110. M. GILBERT, D.A. HEMSLEY and A. MIADONYE, *ibid*, Paper 5
111. D.E. MARSHALL, R.P. HIGGS and O.P. OBANDE, *ibid*, Paper 13
112. D. WALTON and F.T. MURPHY, *ibid.*, Paper 7
113. P. BENJAMIN, *PRI Intl. Conf., PVC Processing*, Royal Holloway College, Egham Hill, Surrey, England (6/7 April 1978) Paper B5
114. J.B. PRESS and D.A. TREBUCQ, *Developments in PVC Production and Processing*, (A. Whelan and J.L. Craft, eds), Applied Science Publishers, London (1977) Chapter 9
115. H.F. SCWENKE, *PRI Intl. Conf., PVC Processing II*, Brighton, England (26–28 April 1983) Paper 14
116. D. TESTER, *The Behaviour of PVC in Fires*, BPF Publn. No. 299/1 (1983)
117. G. MATTHEWS and G.S. PLEMPER, *Brit. Polym. J.,* **13**, 1 (1981)
118. G. MATTHEWS and G.S. PLEMPER, *Brit. Polym. J.,* **15**, 95 (1983)
119. K.T. PAUL, *Fire and Materials,* **8**(3), 137 (1984)
120. J. TROITZSCH, *Fire Behaviour of PVC*, Assn. of Plast. Mfrs. in Europe (December 1986)
121. A. KRAUSSE and A. LANGE, *Introduction to the Chemical Analysis of Plastic*, Iliffe Books Ltd., London (1969)
122. T.R. CROMPTON, *Chemical Analysis of Additives in Plastics*, Pergamon Press, Oxford (1977)
123. A. GONZE, *Plastics,* **24**, 49 (1971)
124. P.G. FAULKNER, *J. Macromol. Sci.-Phys.,* **B11**, 251 (1975)
125. M. LAMBERTY, *Plast. Mod. Elast.,* **26**, 82 (1974)
126. M. GILBERT, D.A. HEMSLEY and A. MIADONYE, *Plast. Rubb. Proc. Appn.,* **3**, 343 (1983)
127. S.V. PATEL and M. GILBERT, *Plast. Rubb. Proc. Appn.,* **5**, 85 (1985)
128. European Council Directive 82/711/EEC (Offl. J. of the European Communities, No. L297, 18 October 1982) U.K. Statutory Instrument 1992, No. 3145): revised by 93/8/EEC (Offl. J. Eur. Comm. No. L90, 14 April 1993)
129. British Plastics Federation, *A Guide to Compliance with the Plastics Regulations*
130. European Council Directive 85/572/EEC (Offl. J. Eur. Comm. No.L372, 31 December 1985)
131. ISO 3127: 1990

Appendix

The following organisations have provided information used in the compilation of this book. Their cooperation is gratefully acknowledged.

Akcros Chemicals
Association of Plastics Manufacturers in Europe
Battenfeld Fischer UK Ltd
Brabender OHG
British Plastics Federation
British Standards Institution
Buss UK
Carter Bros (Rochdale) Ltd
Ciba-Geigy Marlenberg GmbH
Ciba-Geigy plc
Croxton & Garry Ltd
Engelmann & Buckham Machinery Ltd
Eurochlor Federation
European Vinyls Corporation
Farrel Ltd

Forbo-Nairn
Harcros Chemical Group
Health & Safety Executive
Hydro Polymers Ltd
I.C.I Chemicals & Polymers Ltd
Marley Tile
Martinswerk GmbH
Metalest (UK) Ltd
Ministry of Agriculture, Fisheries & Food
Reifenhauser GmbH & Co
Rohm & Haas (UK) Ltd
Schering Polymer Additives
Werner & Pfleiderer GmbH
Winkworth Machinery Ltd
Wickson Product Research Ltd

Author Index

Subject Index

All entries refer to PVC unless otherwise stated.